Evolutionary Computing in Advanced Manufacturing

Scrivener Publishing
3 Winter Street, Suite 3
Salem, MA 01970

Evolutionary Computing in Advanced Manufacturing

Edited by

Manoj Tiwari

Dept. of Industrial Engineering and Management,
Indian Institute of Technology,
Kharagpur, India

and

Jenny A. Harding

Wolfson School of Mechanical and
Manufacturing Engineering,
Loughborough University, UK

Scrivener

WILEY

Co-published by John Wiley & Sons, Inc. Hoboken, New Jersey, and Scrivener Publishing LLC, Salem, Massachusetts.
Published simultaneously in Canada.

Cover design by Russell Richardson.

Library of Congress Cataloging-in-Publication Data:

ISBN 978-0-470-63924-5

Printed in the United States of America

10 9 8 7 6 5 4 3 2 1

Contents

Preface

The increasing availability and use of computers in engineering has significantly changed production and manufacturing domains, since computer-controlled manufacturing systems can greatly improve the quality and pace of production. In the current era of highly competitive, global environments, industries are facing immense pressure to deliver new products more cheaply and quickly, with greater product variety and shorter life cycles. Industry therefore requires effective planning and optimal results in all stages of production, from raw material acquisition to final delivery. Traditional methods are often inappropriate and cannot deal with the planning demands of the advanced technology and requirements in modern manufacturing systems. In the recent years, evolutionary computing has gained popularity for solving manufacturing related problems.

Although many research papers and proceedings exist on evolutionary computing in production and the manufacturing realm, there are hardly any books which coherently present and explain both aspects (i.e. evolutionary computing in the context of manufacturing problems). In most research papers, production and manufacturing problems and evolutionary computation approaches are only loosely coupled which makes it difficult for readers to understand the implementation parts of the algorithms.

In this book, we have addressed the research issues related to evolutionary computing in the manufacturing domain. We have invited contributions from various learned researchers with significant expertise in the field of computational intelligence for advanced manufacturing. Each chapter explains explicitly the research related issues and ways of implementing computational intelligence techniques.

With this composition, we aim to provide readers with a good basis for understanding the development of mathematical models for production and manufacturing related issues. In addition to the mathematical models, various evolutionary algorithms such as Genetic Algorithm (GA), Particle Swarm Optimization (PSO) etc. have been discussed from their fundamentals to implementation aspects. This book will therefore help scholars, researchers and practitioners in understanding both the fundamentals and advanced aspects of computational intelligence in production and manufacturing.

The Structure of the Contributed Chapters

In this book, chapters 1 and 2 deal with the production and process planning issues and describe the basic Genetic Algorithm (GA) and Ant Colony Algorithm (ACO). Chapter 6 and chapter 11 introduce the Immune and Psycho-Clonal Algorithm in operation allocation and fault diagnosis problems respectively. Chapters 3, 4, 5, 7, 8, 9 and 13 provide variants of different algorithms. In chapter 3, the set up and process planning problem is described with a hybrid Genetic Algorithm. Chapter 4 uses cellular particle swarm optimization (CPSO) in the supply chain and product development domain. A variant of GA, Genetic algorithm with chromosome differentiation (GACD) in process plan selection is presented in chapter 5. A description of tool selection and hybrid simulated annealing (SA)-Tabu search algorithm has been explained in the chapter 7. In chapter 8, production planning has been integrated with automated guided vehicles (AGV). In this chapter the use of an enhanced memetic particle swarm optimization (EMPSO) has also been explained. Chapter 9 explains the assembly planning problem in the aircraft industry and uses the self guided ant colony algorithm, which is a variant of the ant colony algorithm (ACO).

Chapters 10, 12 and 13 introduce the most recent research issues in the production and manufacturing field. The benefits of additive manufacturing along with its applications using evolutionary computing have been explained in chapter 10. Chapter 12 deals with the product and platform performance issues in stochastic demand conditions. In the final chapter, a brief introduction to virtual enterprises (VE) has been presented, and the most difficult part in the formation of VE i.e. partner selection problem, has been discussed in the context of a hybrid particle swarm optimization(PSO) and ant colony optimization (ACO) algorithm.

Although there are many variants of evolutionary algorithms which we have not discussed here, the basic ideas behind the algorithms have been provided explicitly. We hope that readers will both enjoy and benefit significantly from this book.

Dr. J.A. Harding & Prof. M.K. Tiwari

List of Contributors

David Ben-Arieh is a Professor of Industrial Engineering at Kansas State University. His industrial experience includes working for AT&T Bell Laboratories and consulting for the aerospace industry and NASA. His research interests concentrate mainly on applications systems design and modeling and holds one patent in this area. In recent years Dr. Ben-Arieh has focused on applications in product development and innovation as well as in Health Care Systems Management, including patients flow, information systems integration, and patient quality and safety improvements.

Puneet Bhardwaj completed his degree in Industrial Engineering and Management in the year 2010 from Indian Institute of Technology, Kharagpur, India. He is currently a first year PhD candidate at the Department of Systems and Industrial Engineering at the University of Arizona, USA, and currently working as a Research Assistant as well as Teaching Assistant for manufacturing courses.

Nurcin Celik is an Assistant Professor in the Department of Industrial Engineering at the University of Miami. Her research interests are in the areas of architectural design and development of adaptive simulations for large scale and complex systems, and structural and functional analysis of social networks. She has received several awards such as the UM Provost Research Award (2011), International Association for Management of Technology (IAMOT) Research Project Award (2011), and Institute of Industrial Engineers (IIE) Outstanding Graduate Research Award (2009).

Felix Chan received his MSc and PhD in Manufacturing Engineering from the Imperial College of Science and Technology, University of London, UK. Dr Chan is an Associate Professor at the Department of Industrial and Systems Engineering, The Hong Kong Polytechnic University. His current research interests are Logistics and Supply Chain Management, Operations

Management, Distribution Coordination, Systems Modelling and Simulation, Supplier Selection. To date, he has published 10 book chapters, over 200 articles in SCI journals and 200 peer reviewed conference papers.

Anand Mohan Choubey is Operations and Strategy Analytics professional currently working at Bank of America. His industrial experience, prior to Bank of America, includes working for H & R Block Inc. on various operational improvement projects and strategy initiatives. Anand has a Master of Science degree in Industrial Engineering from Kansas State University. His professional and research interests include operations research, process improvement, risk management, operations and strategy analytics and consulting.

Jenny Harding joined Loughborough University in 1992 after working in industry for many years. Her industrial experience includes textile production and engineering, and immediately prior to joining Loughborough University, she spent 7 years working in R&D at Rank Taylor Hobson Ltd., manufacturers of metrology instruments. Her experience is mostly in the areas of mathematics and computing for manufacturing.

Nitesh Khilwani is a post-doc researcher at Nano-tera.ch, EPFL (Switzerland) doing research on semantic technologies for social networking, knowledge management and community development. He completedhis PhD from Loughborough University (UK) in 2010. His research interests are industrial engineering and management, semantic web and ontologies, text analysis and mining, programming language and database, algorithms and AI techniques.

Sri Krishna Kumar is pursuing his PhD in the Wolfson School of Mechanical and Manufacturing Engineering department of Loughborough University, UK. He received his bachelors degree in Marine Engineering from Jadavpur University, India. His research interests include knowledge management, mathematical modelling and artificial intelligence.

Vikas Kumar received PhD in Management Studies from Exeter Business School, University of Exeter, UK. Dr. Kumar is a lecturer in the Management Department at the Dublin City University Business School. Previously he has held the position of Research Assistant at Exeter Business School and The University of Hong Kong. His research interests include Supply Chain Management, Lean and Agile Systems, and Service Operations Management. To date he has published 4 book chapters and more than 30 peer reviewed articles in leading international journals and conferences.

Roberto Lu is the Vice President in Advanced Manufacturing of the TE Connectivity (formerly Tyco Electronics). He co-authored this paper while he was a Technical Fellow at The Boeing Company. He taught part-time Industrial and Systems Engineering at the University of Washington as an Affiliate Assistant Professor. His research focuses on advanced manufacturing, decision analysis, discrete event simulation, analytical process optimization, global logistics, large scale production systems integration, lean manufacturing, robotic and machine vision applications, and mass customization.

Candice Majewski is a post-doctoral research associate at Loughborough University, UK. She obtained her PhD in a polymer materials-related subject in 2007. She has spent over ten years working in the area of Additive Manufacturing with a particular focus on polymer materials and processes. She has published over twenty refereed journal and conference publications, and is a member of several committees including the ASTM task-force charged with the production of standards for use throughout the Additive Manufacturing industry.

Nishikant Mishra is lecturer in School of Management and Business, Aberystwyth University. His PhD research focused on the development of a decision support system for radiotherapy planning. He has worked in collaboration with NHS (National Health Service) and developed several software systems. He has published more than 25 articles in leading international journals and conferences. His research interests include development of decision support systems, mathematical modeling, heuristics and algorithms development for a variety of real world manufacturing, healthcare, supply chain and fault diagnosis problems.

Sai Srinivas Nageshwaraniyer is a PhD student in the Department of Systems and Industrial Engineering at The University of Arizona. His research interests are in the areas of Coal mining and transport logistics, distributed simulations of complex systems, and Meta-heuristics.

A.Y.C. Nee is Professor in the Department of Mechanical Engineering, and currently the Director of Research Administration, National University of Singapore. He received his PhD and DEng from Manchester and UMIST respectively. He is a Fellow of the International Academy for Production Engineering (CIRP) and the Society of Manufacturing Engineers (SME). He has published over 350 refereed journal papers and 10 books. He received the National Technology Award in 2002, and National Day Award in 2007. Other awards include: Kayamori Award from IEEE, Norman Dudley Award from IJPR, Joseph Whitworth Award from the IMechE, UK

Niu Sihong received a B.Eng. degree in Mechanical Engineering and Automation from Xi' an Jiaotong University, P.R. China in July 2006. She is a PhD candidate in the Digital Manufacturing Group in Mechanical Engineering Department, National University of Singapore since Jan. 2007. Her current research interests include solving partnership selection in Virtual Enterprise and single/multiple objective scheduling using Meta-heuristics, as well as multi-agent based systems.

S.K. Ong is currently lecturing in the Mechanical Engineering Department at the National University of Singapore. Her research interests are virtual and augmented reality applications in manufacturing and assistive technology and distributed digital manufacturing. She has published over 170 international refereed journals and conference papers. She has received many accolades including the 2002 Norman Dudley Award, the 2004 Outstanding Young Manufacturing Engineer Award and the 2009 Emerging Leaders Award in Academia by the US Society for Women Engineers. In 2005, she was elected an Associate Member of CIRP

Mayank Kumar Pandey is pursuing his PhD in the Mechanical Engineering department of University of Alberta, Canada. He received his M.Tech. degree in Manufacturing Engineering from National Institute of Foundry and Forge Technology, India. His research interests include reliability and maintenance of systems, condition monitoring for prognosis and diagnosis, flexible manufacturing systems and artificial intelligence.

PKS Prakash received his PhD in Industrial Engineering from University of Wisconsin-Madison, WI, US, in 2010. His other educational background includes a B. Tech. (Metallurgy and Materials Engineering) which he gained in 2005 from the National

Institute of Foundry and Forge Technology, Ranchi, India, and an M.S. (Industrial and Systems Engineering) from the University of Wisconsin-Madison, WI, US awarded in 2006. He is currently working as a Researcher at University of Warwick, UK and is supporting Warwick Analytics Limited, UK with the development of advance in-database analytics. His current research focuses on the advance process control with focus toward developing fundamentals of self-resilient systems for fault isolation and adjustment of NDF (No-defect found)

Nagesh Shukla has joined the Digital Lab under the Warwick Manufacturing Group (WMG), University of Warwick, UK as a PhD scholar in October 2007. He received the prestigious Dorothy Hodgkin Postgraduate Award, from UK Research Councils. He has worked on various collaborative projects with GE-Healthcare and University Hospitals for systems modelling & simulations. He has published various journal papers, peer reviewed conferences, book chapters and patents. Currently, he is a member of Digital Lifecycle Management Lab in WMG. His primary research interests are operational research, business process modelling, operations management, evolutionary computation, and fault diagnosis. Other interests involve the development of evolutionary algorithms, and engineering applications of evolutionary algorithms.

M. K. Tiwari is Professor at the Department of Industrial Engineering and Management, Indian Institute of Technology, Kharagpur, India. He is associate editor of the *Journal of Intelligent Manufacturing* and *International Journal of System Science*. He has published over 140 articles in leading international journals. His research interests lies in the area of Evolutionary Computing, Modeling and Simulation of Manufacturing Systems, Supply Chain Management, Planning and Scheduling of Automated Manufacturing System.

Young-Jun Son is Professor of Systems and Industrial Engineering and Director of Advanced Integration of Manufacturing Systems and Technologies Center at the University of Arizona. His research focuses on the coordination of a multi-scale, networked-federated simulation and decision model needed for design and control in manufacturing enterprise, renewable energy network, homeland security, and social network. He has received several research awards such as the SME 2004 Outstanding Young Manufacturing Engineer Award, the IIE 2005 Outstanding Young Industrial Engineer Award, the IERC Conference Best Paper Awards (2005, 2008, 2009), and Best Paper of the Year Award in 2007 from IJIE.

1

Production Planning Using Genetic Algorithm

S.K. Kumar[1] and M.K. Tiwari[2]

[1]*Wolfson School of Mechanical and Manufacturing Engineering, UK*
[2]*Indian Institute of Technology Kharagpur, India*

Abstract
Production planning and control, in manufacturing industries, generally addresses the issues of acquisition, utilization and allocation of resources to satisfy customer requirements in the most efficient and effective way. Therefore efficient management of the production function is of the utmost importance to achieve this objective. The production function includes production level, inventory level, work force level, assignment of overtime and transhipments. Mathematical models developed in this context are widely accepted and can act as decision support systems. This chapter initially focuses on developing the optimization model in different scenarios such as multi-item, multi-period, un-capacitated, capacitated, backorder, fixed and variable workforce. Improving the decision quality in those fields gives rise to complex combinatorial optimization problems, which mostly fall into the class of NP-hard problems. Finding a satisfactory solution in an acceptable time is also an important factor. Genetic Algorithms (GA) provide potent methods for solving such optimization problems and steps for GA implementation, and an example optimization problem solved by using GA, are also included in this chapter.

Keywords: Production planning, inventory, work force, genetic algorithm

1.1 Introduction

Production planning is a set of functions to undertake the efficient and effective utilization of available resources over a time horizon,

Manoj Tiwari and Jenny A. Harding (eds.) Evolutionary Computing in Advanced Manufacturing, (1–18) © Scrivener Publishing LLC

with the motivation of satisfying customer needs whilst creating profits for the organization. In manufacturing industries these functions can be defined as acquisition, utilization and allocation of resources such as raw material, man power, machines, cash flow, information etc. In today's globalized environment, with rapid technological advancement, highly customized demand, short life cycles and high competitiveness, market demand is highly volatile. Production planning therefore plays a crucial tool for any organization trying to satisfy highly customized demand and to remain competitive in the market. Poor production planning can be devastating for any organization and can lead to excessive inventory, high operational cost, poor cash flow, stock outs or lost sales and their subsequent effects. Therefore, managers must utilize their resources efficiently to avoid such consequences.

Production planning in general develops an operational plan based on the required level of manufacturing output. The plan broadly includes the raw material planning, demand management, capacity planning, work force management, inventory management, scheduling and sequencing of jobs etc., with the objective of reducing work in progress, optimizing shop floor timing, reducing inventory, managing lead time and delivery date. Production planning works across a broad domain, but this chapter will concentrate on planning of raw material, man power, inventory, production and transhipment levels. Further production planning realms will be discussed in the subsequent chapters.

The intention of this chapter is to provide an overview of developing the production planning optimization models in different scenarios and the application of genetic algorithms (GA) for solving them. There is an exhaustive range of problem contexts and solution methodologies for production planning problems, but these are not covered in detail here as this chapter concentrates more on mathematical model development in different scenarios. The next section presents various production planning models.

1.2 Production Planning Models

In general, production planning starts with a fulfilment of customer demands whilst minimizing the overall production costs. In most cases, future demand forcasting is done over a short period of time compared to the total planning horizon, in order to minimize the errors. Subsequently production planning is also divided into short time intervals to follow the forcasted demand. Demand is assumed to be deterministic, known and updated periodically.

The first mathematical model presented below has infinite manufacturing capacity i.e. the market demand is always assumed to be less than the capacity. Following are the notations used in this model:

1.2.1 Mathematical Model

A production planning problem can be formulated as follows. Suppose that I types of items are to be produced from R resources for T (1,2,3,.....T) periods. A deterministic demand D_{it} for the ith item in the tth period is assumed. a_{ir} and b_{rt} are the required input of resources 'r' for the production of the ith item and its availability in the tth period. Let C_{it} and h_{it} be the production and inventory cost. Now, the integer vectors $X_t\{x_{1t},........x_{it}\}$ and $Y_t\{y_{it},.......,y_{It}\}$ must be determined, as these represent the production and inventory quantity respectively, that will minimize the total production cost. This problem can be represented by a mathematical equation as:

1.2.1.1 *Model 1: (Assuming Infinite Capacity of the Plant)*

$$\text{Min} \sum_{t=1}^{T}\sum_{i=1}^{I}[X_{it}C_{it} + Y_{it}h_{it}] \tag{1.1}$$

Subject to the following constraints

$$Y_{i,t-1} + X_{it} - Y_{it} = D_{it}, \quad i = 1,2,..,I, \quad t = 1,2,3,...T \tag{1.2}$$

$$\sum_{i=1}^{I} a_{ir}X_{it} \leq b_{rt} \quad \forall r,t \tag{1.3}$$

$$X_{it}, Y_{it}, D_{it} \geq 0 \quad \forall i,t \tag{1.4}$$

Equation (1.1), also called the objective function, minimizes the total cost. Constraint (1.2) ensures that demand is met. In any production system, the required resources should not exceed the available resources and this has been shown in the constraint (1.3), finally constraint (1.4) is the non-negative constraint for demand, production and inventory.

Results obtained from the above model will determine the production plan of

1. Production level of each item in each time period.
2. Inventory level of each item in each time period.

It can easily be seen that model 1 assumes that the plant has an infinite production capacity or that the demand will never exceed the production capacity (mathematically both are same). However, in realistic conditions this assumption is unfounded and it is therefore imperative that capacity is included in the model.

In most of cases, e.g. Fast Moving Commercial Goods (FMCG products), unmet demand for any time period is considered as lost sales, leading to penalty costs for the lost opportunity. Nonetheless, demand for any period can be deferred to subsequent periods and can be considered as back orders with discounts or reduced prices, indirectly increasing the production cost through penalties.

In next mathematical model we will consider the lost sale condition.

1.2.1.2 Model 2: (Assuming Lost Sale)

Extending the model 1, let MC_{it} be the maximum production capacity for the ith item in the tth time period. If L_{it} and δ_{it} are the penalty cost and unmet demand for the ith item in the tth time interval, then the objective function will be:

$$\text{Min} \sum_{t=1}^{T} \sum_{i=1}^{I} \left[C_{it} X_{it} + h_{it} Y_{it} + L_{it} \delta_{it} \right] \tag{1.5}$$

Subject to

$$Y_{it-1} + X_{it} - Y_{it} - D_{it} = \delta_{it} \quad \forall i, t \tag{1.6}$$

$$X_{it} \leq MC_{it} \tag{1.7}$$

$$X_{it}, Y_{it}, D_{it}, \delta_{it} \geq 0 \quad \forall i, t \tag{1.8}$$

and constraint (1.3).

In the above model, constraints (1.6) and (1.8) are the extensions of constraint (1.2) and (1.4), considering the lost sales. Constrail. nt (1.7) restricts the production under its capacity. In such production planning, there is no control to prevent demand exceeding the supply, but the effect can be mitigated by proper inventory management i.e. production and storage of items in a period when demand is less than the capacity, provides a buffer for the future periods with excessive demand.

Sometimes in production planning, deferring the demand until a subsequent period can be economical (i.e. if the penalty cost is less than the extra production cost) when compared to the costs of fulfilling the demand in the same period. Such circumstances can also arise due to varying raw material cost, man-power cost or machine breakdowns. The next model developed considers that all unmet demand can be fulfilled in the future. Assuming the back order penalty cost P_{it}- for the ith item for the tth time interval then the objective function can be formulated as:

1.2.1.3 *Model 3: (Assuming Backorders are Allowed)*

$$\text{Min} \sum_{t=1}^{T} \sum_{i=1}^{I} [C_{it} X_{it} + h_{it} Y_{it} + P_{it} \delta_{it}] \tag{1.9}$$

Subject to

$$Y_{it-1} + X_{it} - Y_{it} - D_{it} - \delta_{it-1} = \delta_{it} \tag{1.10}$$

and constraint (1.7) and (1.8).

The constraints of model 2 and 3 remain the same except for constraint (1.10) which is the extension of constraint (1.6) with back orders in the flow balance equation.

1.2.1.4 *Model 4: (Partial Lost Sale and Back Orders)*

In the real world it is sometimes not practical to consider all lost sales as back orders. Consumers will not wait if alternatives are available in the market. This leads to a production planning scenario where a fraction of unmet demand is held as back orders. Assuming α_{it} is the fraction of unmet demand allowed as backorders for the ith item for the tth time period. The objective function will be:

$$\text{Min} \sum_{t=1}^{T} \sum_{i=1}^{I} [C_{it} X_{it} + h_{it} Y_{it} + \alpha_{it} P_{it} \delta_{it} + (1-\alpha_{it}) L_{it} \delta_{it}] \tag{1.11}$$

Subject to

$$Y_{it-1} + X_{it} - Y_{it} - \alpha_{it} \delta_{it-1} - D_{it} = \delta_{it} \tag{1.12}$$

$$\delta_{it} \in [0,1] \tag{1.13}$$

and constraints (1.7) and (1.8).

This model is a hybrid of models 2 and 3. If we take $\delta_{it} = 0$ (all lost sales) objective function (1.11) changes to (1.5) and with $\delta_{it} = 1$, it changes to objective function (1.9).

All the models built in the previous discussion considered capacity and resources as constraints whereas, man-power management is also a vital aspect for robust production planning and its execution. In the next two models man-power is also considered a constraint.

1.2.1.5 Model 5: (Fixed Man-Power)

The simplest way to address production planning with man-power constraints is to consider the man-power available to be fixed. If z_i and W_t are the required manpower input for the ith item and its availability in the tth period, then the objective function will be:

$$\text{Min} \sum_{t=1}^{T}\sum_{i=1}^{I}[C_{it}X_{it} + h_{it}Y_{it} + \alpha_{it}P_{it}\delta_{it} + (1-\alpha_{it})L_{it}\delta_{it}] \tag{1.14}$$

Subject to

$$\sum_{i=1}^{I} z_i X_{it} \leq W_t, \forall t \tag{1.15}$$

and constraints (1.7), (1.8), (1.12) and (1.13).

Constraint (1.15) ensures that the required man–power should not exceed the available man-power in each time period. Although we have added an extra constraint (15), the objective function remains the same as in model 4. The reason behind this is that manpower is considered to be fixed, i.e. a fixed cost will be incurred in each time period irrespective of the degree of utilization of manpower, therefore it becomes irrelevant for further consideration in the objective function. So the objective function will remain the same as presented in the model 4 but with one extra constraint (1.15).

1.2.1.6 Model 6: (Variable Man-Power)

In many practical cases, additional workforce may be required at some periods to cope up with unexpected or seasonal high demand. A company can usually hire temporary workers but this is likely to incur a high cost for a short period of time. In this scenario production planning needs to decide the workforce level for each period, as a variable workforce is required to cope with the unexpected or seasonal high demand.

In this model, two types of work force are considered, fixed (W_t) and variable (M_t), with H_t as the cost of hiring in the tth time period. With this consideration, the objective function will be:

$$\text{Min} \sum_{t=1}^{T} \left[\sum_{i=1}^{I} \{C_{it} X_{it} + h_{it} Y_{it} + \alpha_{it} P_{it} \delta_{it} + (1-\alpha_{it}) L_{it} \delta_{it}\} + M_t H_t \right] \tag{1.16}$$

With constraints as

$$\sum_{i=1}^{I} z_i X_{it} \le M_t + W_t \tag{1.17}$$

$$M_t \ge 0 \tag{1.18}$$

and constraints (1.7), (1.8), (1.12), (1.13) and (1.15).

1.2.1.7 Model 7: Multiple Demand Location

All the models introduced so far are for a single production unit and a single demand location. However, it is unlikely that demand is centralized. However, in many situations a production plan is needed for multiple production units and demand locations. Therefore, a production planning model with a single production unit and multiple demand locations is now presented and subsequently, this is extended for multiple production units.

While considering multiple demand locations, production planning not only decides the level of production but also the level of supply to the market. In such scenarios it is worth considering shipment or logistics cost as this varies with demand location. Let $j\{1,2,\ldots,J\}$ be the index for the demand location and μ_{it}^{j} and χ_{it}^{j} be the logistic cost per unit and level of supply of the ith item to the jth location during the tth period. The notation for other variables also changes from single dimension to J-dimensional.

Now the objective function will be:

$$\sum_{t=1}^{T} \left[\sum_{i=1}^{I} \{C_{it} X_{it} + h_{it} Y_{it} + \left\{ \sum_{j=1}^{J} \mu_{it}^{j} \chi_{it}^{j} + \alpha_{it}^{j} P_{it}^{j} \delta_{i}^{j} + (1-\alpha_{it}^{j}) L_{it}^{j} \delta_{it}^{j} \right\}\} \right] + M_t H_t \tag{1.19}$$

Subjected to

$$\chi_{it}^{j} \le D_{it}^{j} \tag{1.20}$$

$$\delta_{it}^{j} = D_{it}^{j} - \chi_{it}^{j} - \alpha_{it}^{j}\delta_{it-1}^{j} \tag{1.21}$$

$$Y_{it-1} + X_{it} - Y_{it} = \sum_{j=1}^{J} \chi_{it}^{j} \quad \forall i,t \tag{1.22}$$

$$\{X_{it}, Y_{it}, \chi_{it}^{j}, \delta_{it}^{j}, D_{it}^{j}\} \ge 0 \quad \forall i,j,t \tag{1.23}$$

$$\alpha_{it}^{j} \in [0,1] \quad \forall i,j,t \tag{1.24}$$

and constraints (1.7), (1.17) and (1.18).

Where,

D_{it}^{j} - demand at the jth location

δ_{it}^{j} - unmet demand at the jth location ,

α_{it}^{j} - fraction of unmet demand changed into backorders,

P_{it}^{j} - backorder penalty cost in the market,

L_{it}^{j} - lost sales cost in the jth market in the tth time period.

Constraint (1.20) ensures that supply to the market should not exceed the demand. Constraint (1.21) and (1.22) are the flow balance between demand location and production unit. Constraints (1.23) and (1.24) are for non-negativity and fractional behaviour.

1.2.1.8 Model 8: Multiple Plant and Multiple Demand Locations

Let k {1,2,.….,**K**} be the index for plants. Now, with the objective of minimizing the total cost, production planning will decide: (i) production (X_{it}^{k}) and inventory level (Y_{it}^{k}) for items in each period. (ii) shipment level (χ_{it}^{jk}) from production units to markets. (iii) workforce (M_{t}^{k}) level for each production unit, for every period. If μ_{it}^{jk} is the per unit shipment cost then the objective function can be written as:

$$\text{Min}\left\{\sum_{t=1}^{T}\sum_{k=1}^{K}\sum_{i=1}^{I} C_{it}^{k}X_{it}^{k}\right\} + \left\{\sum_{t=1}^{T}\sum_{i=1}^{I}\sum_{j=1}^{J}\left[\sum_{k=1}^{K}\mu_{it}^{jk}\chi_{it}^{jk}\right]\right.$$
$$\left. + \alpha_{it}^{j}P_{it}^{j}\delta_{it}^{j} + (1-\alpha_{it}^{j})L_{it}^{j}\delta_{it}^{j}\right\} + \left\{\sum_{t=1}^{T}\sum_{k=1}^{K} M_{t}^{k}H_{t}^{k}\right\} \tag{1.25}$$

Subject to the following constraints:

$$Y_{it-1}^{k} + X_{it}^{k} - Y_{it}^{k} = \sum_{j=1}^{J} \chi_{it}^{jk} \quad \forall i,k,t \tag{1.26}$$

$$\delta_{it}^{j} = D_{it}^{j} + \alpha_{it}^{j}\delta_{it-1}^{j} - \sum_{k=1}^{K} \chi_{it}^{jk} \quad \forall i,j,t \tag{1.27}$$

$$\sum_{i=1}^{I} z_i^k X_{it}^k \le M_t^k + W_t^k \quad \forall k,t \tag{1.28}$$

$$X_{it}^{k} \le MC_{it}^{k} \quad \forall i,k,t \tag{1.29}$$

$$\sum_{k=1}^{K} \chi_{it}^{jk} \le D_{it}^{j} \tag{1.30}$$

$$\{X_{it}^{k}, Y_{it}^{k}, \chi_{it}^{jk}, \delta_{it}^{j}, D_{it}^{j}\} \ge 0 \quad \forall i,j,k,t \tag{1.31}$$

and constraint (1.24).

As explained earlier, constraints (1.26) and (1.27) are the flow balancing equations between plants and demand locations. Constraint (1.28) is for man-power management. Demand and maximum production level are taken care of by the constraints (1.29) and (1.30) and finally constraint (1.31) is the non-negativity condition.

In this section production planning models have been developed based on various scenarios. More exhaustive models can be found in the literature discussing various factors such as, convex/concave cost functions, lead time, various transportation facilities, outsourcing etc., but these will not be discussed here. This is because the purpose of the model introduced in this chapter is to provide examples which focus on the steps required for model development and the way in which solutions can be obtained through Genetic algorithms (GA). In the next section GA is introduced as a powerful tool for solving production planning problems.

1.3 Genetic Algorithm

In the last two decades, as the complexity of the optimization problem has increased, evolutionary algorithms have gained popularity.

Evolutionary algorithms are based on the principle of evolution i.e. survival of fittest. A Genetic Algorithm (evolutionary algorithm) mimics the natural evolution (Darwin's Theorem) process that is (i) adaptation to the environment and (ii) discarding the weaker population. In the initialization step, the GA starts with some feasible solutions (called the initial population), represented in the form of a string (chromosome). Based on a fitness value (better the objective value, better the fitness), individuals undergo transformations (crossover and mutation). These individuals then strive for survival and selection as part of the next generation, survival is biased towards the fittest. Transformation using a crossover operation tries to store the best characteristics of the previous generation. It selects a pair of "parents" based on their fitness values to produce members of the next generation. Mutation is then used to change part of a solution and generate a new set of solutions. Mutation avoids the localized search. Crossover and mutation can be seen as exploitation and exploration of the solution space. Balancing between exploration and exploitation is achieved through a probability factor (which will be discussed in the next section). After a few generations, the program converges to a near optimal solution. Many variants of GA can be found in literature; however this chapter concentrates more on a simple application of GA rather than on examining these multiple variants. In the next section, an example application of a GA in the production planning domain is demonstrated.

1.3.1 Procedure of Genetic Algorithm (GA)

In order to illustrate the steps inside the GA, model 2 is examined with the number of items and time periods = 2. The first step of GA generates the initial feasible solutions in the form of a string (chromosome) as shown in Figure 1.1.

There are two decision variables (Production and Inventory) for two items and two periods, so the length of the string will be ($2 \times 2 \times 2 = 8$) as shown above. N such solutions are generated and these are called the initial population of solutions. In the above figure, decision variables are shown in the form of ${}^{n}X_{ij}$ and ${}^{n}Y_{ij}$ with X as production and Y as inventory units, n as solution number in the population and i,j are items and time period respectively. Care should be taken while generating initial solutions to ensure that they are feasible, as in the above example the production is kept below maximum capacity level and inventory is kept below the production level. The next step of GA calculates the fitness value.

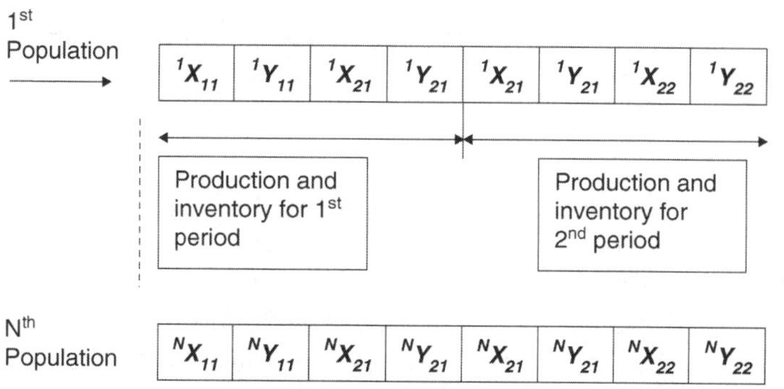

Figure 1.1 String representation in GA.

1.3.1.1 Fitness Calculation

The objective value is calculated for each generated population by using the equation (15). Hence, for each solution let it be f_i (for i = 1,...., N). The fitness calculation (sometimes it called fitness function) of any solution is based on this fitness value. It is clear that the better the fitness function, the better will be the solution. In case of minimization, a lower value of f_i should have superior fitness and this can be achieved by making fitness (F_i) = $1/f_i$, assuming $f_i \neq 0$. With maximization $F_i = f_i$. However, it is imperative to check the feasibility of the solution, i.e. all X and Y should strictly follow the constraint equations (1.2), (1.3) and (1.4) before the fitness calculation. Although, initial feasible solutions have been generated, in subsequent generations the application of crossover and mutation transformations may lead to infeasibility. Numerous methods have been suggested in the literature for dealing with the infeasibility and two basic methods are – (i) Penalty mechanism and (ii) Repair mechanism.

1.3.1.2 Penalty Mechanism

For minimization, this mechanism simply adds an extra cost in the objective function, which reduces the profit for maximization and increases the cost for minimization to reduce its fitness and probability of this solution (string or chromosome) taking part in the consequent activity (crossover). Again there are various methods by which an extra cost can be added in the objective functions. One could be (assuming minimization function): $f_i' = (1+\lambda)\, f_i$ with $\lambda > 0$ if the ith solution violates the constraints. Another method could be, $f_i' = f_i + \lambda\theta_i$, where $\lambda > 0$ and θ_i is the degree of violation. For example, in case of checking constraint (5) in case of model 2 we can write: $\theta_i = |\delta_{it} - Y_{it-1} - X_{it} + Y_{it} + D_{it}|$. The fitness value will now be $F_i = 1/f_i'$ assuming $f_i' \neq 0$.

1.3.1.3 *Repair Mechanism*

Using a repair mechanism, the infeasible solutions are made feasible and the way of achieving this is problem dependent. For example, as discussed earlier, for model 2 there are two decision variables, X and Y. X can be infeasible if it is more than the production capacity, similarly, Y can be infeasible if it is more than X (inventory cannot be more than production). Either situation can lead to infeasibility of that particular solution. In order to make a solution feasible two steps can be followed: (i) if X is more than the production capacity, change the value of X randomly to a value which is below the production capacity. (ii) if Y is more than X, change the value of Y randomly to below X.

The next step of GA includes selection, crossover and mutation.

1.3.1.4 *Selection Procedure*

As discussed earlier, next generation solutions are obtained through crossover of the members of the present generation (called parent solutions). Selection of the parents depends on the fitness value and there are various selection procedures. One of them is probability based selection, widely known as roulette wheel selection. The roulette wheel selection procedure assigns a probability factor for each solution by the equation: probability factor

$$(p_i) = F_i / \left(\sum_{i=1}^{N} F_i) \right),$$

clearly $0 \leq p_i \leq 1$. The probability factor is higher for the solutions with higher fitness. The next step involves the defining of a selection range (S_i) for each solution.

$$S_i = \left(\sum_{j=1}^{i-1} p_j + p_i \right)$$

and clearly $0 \leq S_i \leq 1$ and $S_i = S_{i-1} + p_i$. A number R is randomly generated between 0 and 1 and if $S_{i-1} < R \leq S_i$ the ith solution is selected as a parent. Then crossover takes place between a pair of parent solutions.

1.3.1.5 *Crossover*

In crossover, part of the solution is exchanged between the pair of parents to create the child solutions. Various procedures have been successfully implemented for the crossover such as, one-point crossover, two point crossover, multi-point crossover etc. In the following example for model 2 the single point crossover operation is used to illustrate the process. Let the ith and the kth solutions be selected for crossover as shown in Figure 1.2.

The crossover point is defined within the string length. In the next step the solutions are swapped after the crossover point (Figure 1.2) to generate the child solutions (Figure 1.3):

The most important factor which controls the crossover operation is the "crossover probability (P_C)". It generally defines the percentage of population which will go through the crossover i.e. if it is 0.8, then 80 percent of the population are likely to go through the crossover. To implement this in the algorithm, after selection of parents and before crossover is performed, a random number (R) is generated, such that $0 \le R \le 1$, and if $R \le P_C$, crossover is performed.

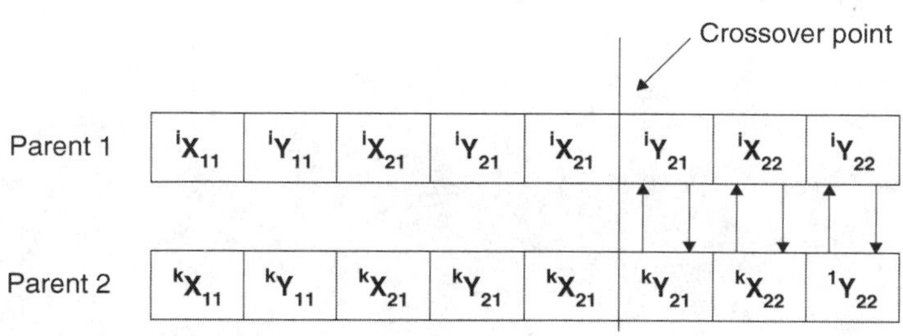

Figure 1.2 Crossover.

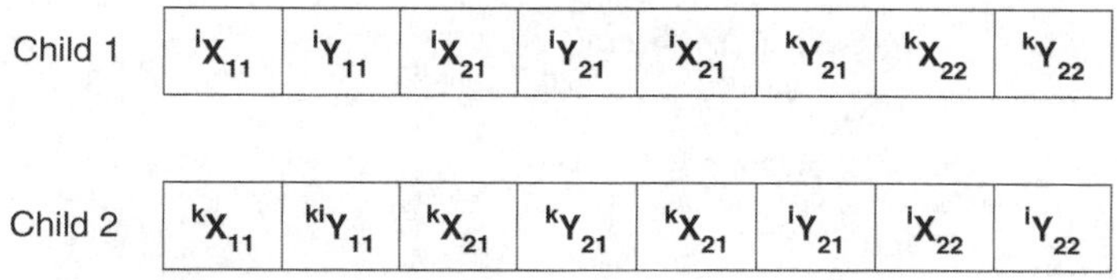

Figure 1.3 Child string after crossover.

1.3.1.6 Mutation

In the mutation operation the solution is randomly changed in order to create a new possibility of exploration of solution space. Mutation is carried out on a solution that has been shown in Figure 1.4.

Selection of the mutation point is carried out randomly and the exchanged value (in the above case A) should be feasible. Like crossover, the mutation operation is controlled by the mutation probability (P_m) and is implemented in the algorithm in a similar fashion.

Selection, crossover and mutation operations are carried out in the algorithm unless the termination criteria have been reached. Generally two conditions are adopted for termination of the algorithm: (i) when the maximum number of generations has been reached (ii) when the solution is not improving by a predefined value. The pseudocode for GA has been shown in Figure 1.5.

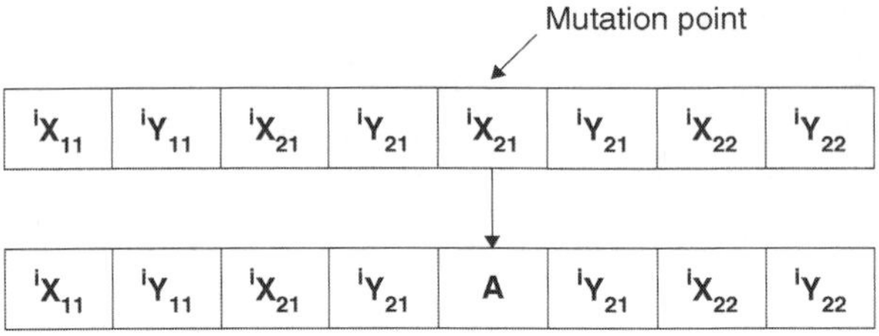

Figure 1.4 Mutation operation.

Procedure of GA

```
Begin
    t ← 0
    Initialize population
    Evaluate population
    While (termination criteria not reached) do
    t ← t + 1
    Selection of population
    Crossover
    Mutation
    Evaluate population (t)
    end
end
```

Figure 1.5 Pseudocode for GA.

1.4 Implementation of GA

To illustrate the effectiveness of GA, discussed above, a complex production planning problem is used to evaluate the computational efficiency of a GA. Assuming (Model 7) a single production plant with four items and four demand locations, with the following normal distribution for data generation is used in the example (This example uses the normal distribution in order to generate the randomized data for checking the consistency and robustness of the algorithm):

Data	Demand	Max. Prod. Cap	Prod. Cost	Inventory Cost	Trans. Cost	Back Order Coeff.	Man power Cost
Normal Dist.	N (150, 10)	N (750, 25)	N (10, 2)	N(2.5, 1)	N(5, 2)	N(0.6, 0.1)	N(7, 2.5)

1.4.1 Algorithm Parameters: (Population Size, Probability of Crossover and Mutation)

Although, there is no straightforward way to configure the GA parameters, a generic setting with population size 100, 0.8 and 0.05 as crossover and mutation probability is used as the starting point. The convergence graph obtained using single point crossover and one point mutation is shown in the Figure 1.6.

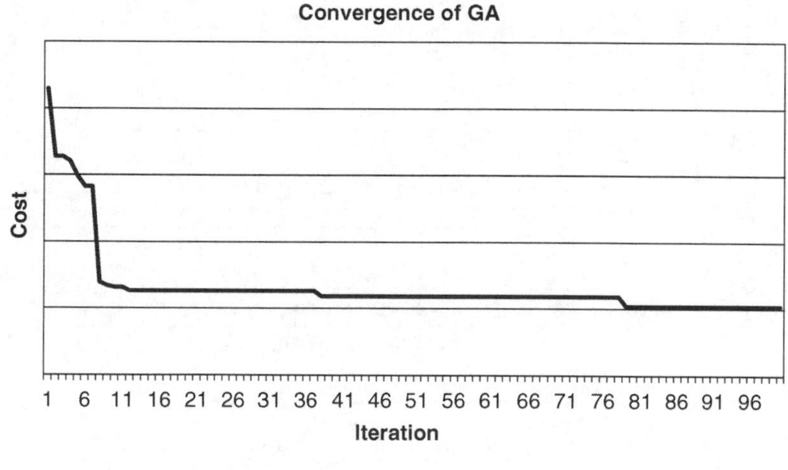

Figure 1.6 Convergence of GA.

The graph seems to be converging towards a near optimal value, exhaustive experiments are required to finalize the parameters and this is called parameter tuning.

1.4.2 Parameter Tuning

Parameter tuning requires several experiments with different configurations to obtain the best result. The effect of varying population size from 50 in steps of 50 has been studied. Results obtained are shown in the Figure 1.7. It is clear that the population size of 100 gives better and faster (less iterations are required in comparison to lower population sizes and less infeasible solutions are generated when compared to higher population sizes) results, hence, it can be inferred that a population size of 100 is sufficient to get the near optimal solution.

The next step requires the tuning of crossover (p_c) and mutation (p_m) probability. Although it is not possible to show all the experimental results, two graphs (Figures 1.8 and 1.9) where **pm** was kept as a constant (0.01 and 0.5) and varied **pc** from 0.6 to 0.8 are provided. As shown in Figures 1.8 and 1.9. The algorithm with pc 0.8 gives better results than the others. Further experiments are needed to get the best value.

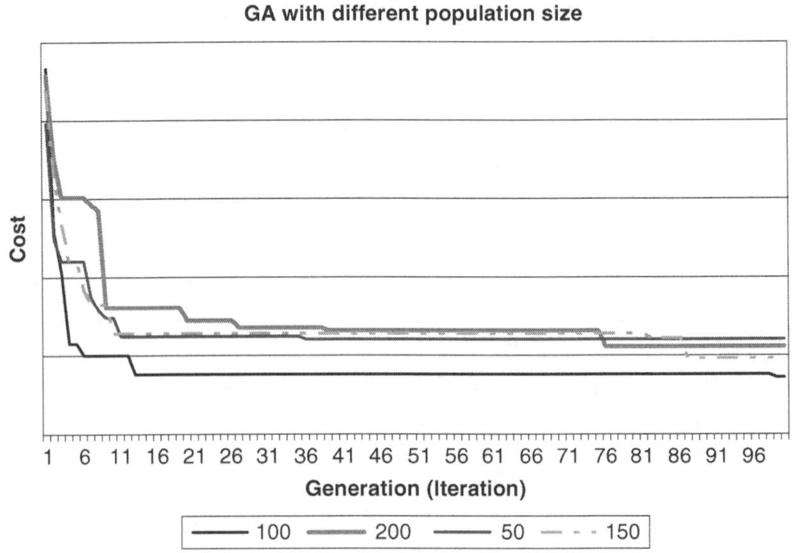

Figure 1.7 Parameter tuning with different population size.

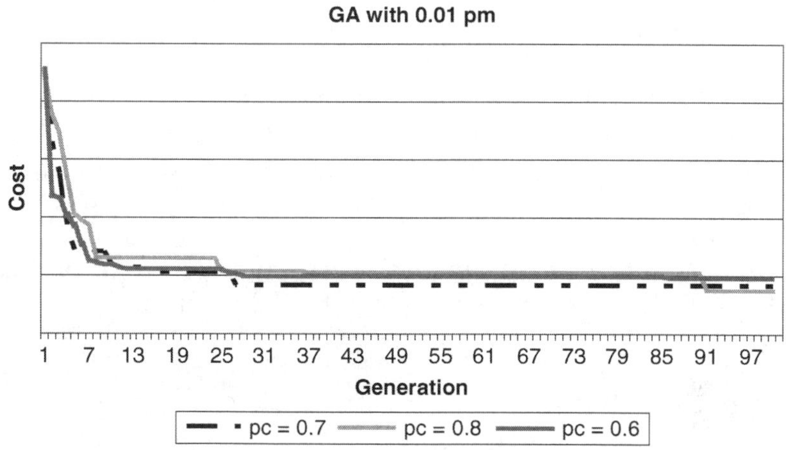

Figure 1.8 Parameter Tuning.

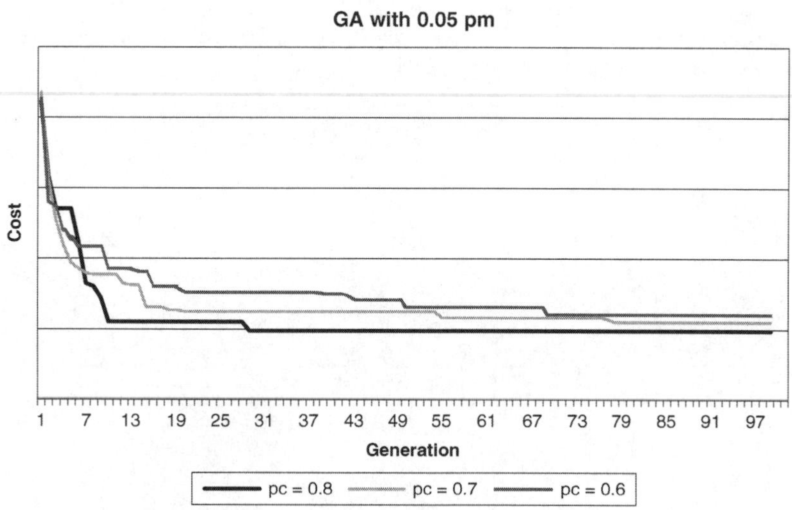

Figure 1.9 Parameter Tuning.

An important point to notice is that parameter tuning cannot be generalized and it varies from problem to problem making it a paramount step for the successful implementation of the algorithm.

1.5 Summary

In this chapter, different production planning models have been developed and the fundamentals of genetic algorithm (GA) and its successful implementation have been discussed. Although, the models developed may not cover every specific production planning consideration, nonetheless they will help readers to understand and mathematically map practical production planning problems and then successfully apply GAs to get good solutions.

Further Reading

1. Gen, M., Cheng, R., 1997. *Genetic Algorithms and Engineering Design*. John Wiley and Sons, New York, USA.
2. Goldberg, D.E., 1989. *Genetic Algorithms in Search, Optimization and Machine Learning*. Addison-Wesley.
3. Mula, J., Poler, R., García-Sabater, J.P., Lario, F.C. Models for production planning under uncertainty: A review. *International Journal of Production Economics*, 103 (2006), pages- 271–285.

2

Process Planning through Ant Colony Optimization

Puneet Bhardwaj and M.K. Tiwari[a]

Department of Industrial Engineering and Management, Indian Institute of Technology, Kharagpur 721302, India

Abstract

The broad platform of process planning involves the act of organizing a detailed plan for the production of subparts, parts and assemblies. In major cases, process planning revolves around the routes, processes, parameters, machines and tools. Efficient process planning improves two major facets of production: firstly it helps in automation in advanced manufacturing systems and secondly it increases the efficiency of the whole production system. In this chapter, we discuss alternative process plans with sequencing, their evaluation and selection, i.e. machine, machining operation, processing sequence, etc., combined with the objective of minimizing total processing, waiting and setup costs. We present a shop-floor case problem describing the shop floor scenario and demonstrating the applicability of the proposed approach. Shop-floor process planning can be considered as routing a part through the sets of fixtures, machines, tools, etc. The Ant Colony Optimization (ACO) technique is based on the foraging behaviour of the ants in which the ants communicate via a hormone called pheromone in order to find the shortest route to the food resource. We use ACO to resolve the process plan selection problem and present the analysis with conclusions.

Keywords: Process plan evaluation, sequencing, ant colony optimization

2.1 Introduction

Within the manufacturing systems, two of the most important subsystems are process planning and scheduling. Conventionally, process

---

Manoj Tiwari and Jenny A. Harding (eds.) Evolutionary Computing in Advanced Manufacturing, (19–36) © Scrivener Publishing LLC

planning deals with the selection of manufacturing processes and process parameters required to transform a part from its initial design phase to the final realized product. Process planning therefore acts as a missing link between the product design and product manufacturing.

The process planning results in the information about manufacturing processes and their requirements like machines, tools, and fixtures to generate the desired part or material. These process plans then serve as inputs for the sequence planner or scheduler. The operation sequencing is carried out to ensure the optimum use of resources with the precedence constraints within the process plans being satisfied. Thus it accomplishes the two tasks of preparing theoretical process plans, as well as their practical realization (Sugimura *et al.*, 2001). The integration of process planning and operation scheduling is still a challenge in manufacturing systems and the related community.

Conventionally, process planning and scheduling were considered as segregated tasks. First process planning was done, and then sequencing was carried out based on the process plans generated. The drawbacks resulting from addressing the two challenges separately include, but are not limited to, the time gap between the process planning and scheduling, mismatching of planned and available resources, conflict of objectives with process planning emphasizing technical requirements whilst scheduling focuses on time, and many more (Saygin and Kilic, 1999; Kumar M, Rajotia, 2003). With the advance of computing capability and expansion of planning horizons, both process planning and scheduling have come to be seen as a single problem. The state-of-the-art technology, computer integrated manufacturing systems (CIMS), aspires to integrate various phases of manufacturing, by proposing to use online information about manufacturing resources for the selection of an integrated process plan and scheduling plan from a set of alternative planning solutions.

With the decrease in product life-cycle time and increase in demand for customized products, companies are opting for smaller batch production. Customized demand also calls for volatile customer-supplier arrangements to meet the orders. These requirements for responsive environments to satisfy demand and deadlines, put pressure on the manufacturing systems to be more flexible and responsive over time. The ability to quickly convert standard product designs into process plans, and integrate new orders with the available resources and production schedules is called for. In order to address the aforementioned challenges, computer aided process planning (CAPP) systems were developed in the 1990's to fill the gap between computer aided design (CAD) and computer aided manufacturing (CAM). These systems were meant to provide the current status of the manufacturing resources, to evaluate the feasibility of the process plans based on the product design specifications.

As product life cycles are reducing in time, and the variety of products on the market is increasing, manufacturing systems have become more and more flexible. Dedicated and specialized machines have been replaced by non-dedicated and more sophisticated ones. The planned production has been changed to flexible manufacturing systems (FMS). The FMS can be found in most leading industries today, associated with the manufacturing systems, and include predefined lists of parts with required standards and specifications. Flexibility here refers to the rapid conversion of one part from the aforementioned defined set of parts to another work piece through the use of appropriate NC code.

Further, competition and increasing pressure on the productivity of manufacturing systems have called for the increased flexibility and changeability, which also require corresponding changes to be made in the manufacturing support functions as well. The new technologies are being employed to modify the current manufacturing systems into ones which is more responsive and adaptable. The future of current manufacturing systems is seen to be as the changeable and reconfigurable manufacturing systems (RMS). Enabling rapid, specific and cost-effective adjustment to the manufacturing systems considering the capacity, structure, technology and function requires reconfigurable manufacturing systems (Koren *et al.*, 1999). RMS are also expected to reduce functioning costs by minimizing the overhaul and repairs. Reconfigurability aims at achieving more competitiveness by offering enablers in terms of technology and supporting business paradigms (ElMaraghy, 2006). In 1998, a body of international experts regarded the reconfigurable manufacturing processes as the most crucial enabling technology for 2020 (Bollinger 1998). Since then, world-wide scientific efforts to design and establish reconfigurable manufacturing systems (RMS) have been growing.

The reconfigured system is obtained from the modification of the configured system. The configured system refers to the design, the selection and the collection of modules from a predefined set of construction modules, as per the specs and requirements of the customer. The reconfigured system is derived from the configured system by converting and modifying the structure, capacity, technology and functionality of the latter. The general ways to achieve reconfiguration are replacement, supplement or removal of the discrete autonomously operating components.

As mentioned earlier, reconfigurability is a key technology for 2020, hence, proving its benefits are not easy to achieve. The manufacturing community is inexperienced and still not certain about the differences between the flexible manufacturing systems (FMSs) and reconfigurable manufacturing systems (RMSs). The FMS is a relatively more rigid system with respect to the system configuration i.e., capacity and technology, and settings can last for months. RMS on the other hand is comparatively faster to adapt to unexpected and unscheduled requirements regarding structure, capacity, functionality and technology. FMSs happen to have surplus capacity and features, some of which may be left unused in many cases. FMSs can respond to the changing demand but the system configuration remains intact, whereas RMSs can undergo changes in system configuration to meet the requirements. The observed difference between the physical configuration of FMS and RMS is that FMS generally has a general purpose CNC machine and programmable automations as compared to the customized CNC machines and reconfigurable machine tools in RMS.

Reconfiguration could be implemented in soft (logical) form or hard (physical) form, and it may be at the system level or machine level. One of the important soft type reconfiguration facilitators is process planning for RMSs. Like other manufacturing systems, it is a vital function for the operation planning, sequencing and system design in RMSs. There are few publications addressing the reconfigurable process planning (RPP) problem. Agile process planning is suggested for the manufacturing of individualized products and constantly reconfiguring facilities (Zaeh *et al.*, 2005). The various process planning concepts and approaches can be classified on the basis of granularity, level of automation and scope (ElMaraghy, 2006). Some indicate the need for evolvable and reconfigurable process plans capable of responding to both minor and major modifications in evolving parts/products families, and changeable RMSs.

Some of the most relevant RPP approaches underpinning the changeable and agile manufacturing paradigms to a varying extent are:

a) Variant Process Planning Systems

These are based on the valid conjecture that similar parts have similar process plans. Thus, a process plan to manufacture the entire family is created. When a part is identified belonging to a particular family, the generic process plan for the family is retrieved and modified for that part. The variant concept of the RPP is being increasingly realized in power-train manufacturing engineering (Hetem, 2003). Generalized product models and features technology can be used for an integrated variant management concept to meet frequently changing needs (Bley and Zenner, 2005).

b) Generative Process Planning Systems

In these systems, the process plan for a new component is automatically synthesized or generated. In literature, mostly a mathematical or rather a procedural approach is suggested and solved using heuristics or other near-optimal optimization methods. A clustering method for multi-part operations planning based on the analysis of process plans has been presented (Xu, *et al.* 2004). The variant process planning systems, due to their rigid concept of part families, are not best suited for RMS, and generative process planning can better accommodate such technology. Since purely generative process planning systems have not been implemented in reality yet, a semi-generative process planning system was developed and solved using Simulated Annealing (Azab *et al.*, 2006).

c) Macro-Level or Automatic Process Planning

Macro-level process planning is tedious owing to its dependence on declarative process knowledge including geometry, cutting and machine tools, fixtures and technological requirements, and also its implicit time dependency constrained by the sequence in which the given features would be machined. The criteria addressed in most of the optimization problems ranges from the minimization of part movement between the machines to the minimization of changes in cutting parameters and tool movement. Earlier such scenarios were addressed by rule-based knowledge bases developed through experience in common machining practices (Lin *et al.*, 1998). Currently, the focus is on the application of geometric information and constraints for the development of precedence relationships. The consideration of more than one factor for developing precedence relations has also been reported in some research. The classical version of the problem has been devised into a mixed integer programming problem (Koulamas, 1993). However, most of the mathematical models formulated in literature are based on the travelling sales-man problem (TSP) (Lin & Wang, 1993).

Like flexible manufacturing systems (FMS), reconfigurable manufacturing systems (RMS) are also developed for a range of parts but with a wider sphere of influence. Hence, variant process planning techniques can easily be extended to RMS. But generative process planning techniques can address more unplanned variations, and automatic process planning is even better. Thus, a hybrid reconfigurable process planning (RPP) system which can generate process plans for parts' features beyond the scope of a part family, with an increased extent of automation can best address the current manufacturing challenges (Azab *et al.*, 2006).

The inability to obtain the optimal process plan solution through algorithms, because of the NP-hard nature of the problem, has paved the way for heuristics in order to get near-optimal solutions. The development of evolutionary and nature-inspired approaches like genetic algorithms (Dereli *et al.*, 2001; Giudice and Fargione, 2007; Guan *et al.*, 2009; Hua *et al.* 2007; Li *et al.* 2003;), particle swarm optimization (Guo *et al.*, 2009a,b), simulated annealing and tabu search (Huang and Liao, 2008; Li *et al.*, 2003, 2004) have proven to be effective for problems with near-optimal solutions like process planning. Dorigo *et al.* (1996) proposed ant colony optimization (ACO) algorithm which is another nature-inspired approach for solving NP-hard problems. The ACO algorithm has been used to solve TSP (Dorigo *et al.* 1997), vehicle routing problem (VRP) (Donati *et al.* 2008), hole-making process optimization (Ghaiebi and Solimanpur 2007), etc.

The process planning problem for a prismatic part has been modelled as a combinatorial optimization problem and solved using an integrated genetic algorithm and simulated annealing approach (Li *et al.*, 2002). The solutions generated by genetic algorithms are fed to the simulated annealing algorithm for and extended search of the optimal solution. An integrated process planning and scheduling problem similar to the one addressed in this work has been developed and optimized through a particle swarm optimization algorithm (Guo *et al.*, 2009a,b).

2.2 Ant Colony Optimization (ACO)

A process plan selection process can be seen as identifying the optimal route through a set of jigs/fixtures, tools, machines, etc. In addition to the constraints resulting from machines, tools and jigs/fixtures, the minimization of costs has been the main objective addressed so far. The graph theoretical approach makes it easy to model the aforementioned features of the process plan scheduling (PPS) problems. Thus, the PPS problems are reduced to the classical travelling salesman problems (TSP) (Chen *et al.* 1998), and then solved by heuristics and other programming approaches. The material handling costs, set-up costs, tool changing costs, costs associated with waiting, etc. are all mapped to the distances between the nodes or cities in the corresponding TSP. The precedence matrix and state matrix are used as constraints for solving the problem.

The ACO applied to the process planning problem in this chapter is an evolutionary metaheuristic. Thus like other evolutionary algorithms, it overcomes the limitation imposed by the NP-hard nature of the large PPS problems. Maniezzo *et al*, 1995, showed the results from eight evolutionary heuristics when applied to the quadratic assignment problem (QAP). The ACO is claimed to show better results when compared to the others used in the experiment.

The ACO mimics the approach used by ants for locating the shortest path from their nest to the food resources. The ants use the environment as a means of communication with each other. Such a system is called 'stigmergy', and is found in many social animal societies. Each ant leaves a chemical substance called 'pheromone' on its way while traversing the various paths leading to the food resources. The concentrated level of pheromone on a path results from all the ants traversed through that path. Thus the individual ants dynamically update their memory through the experiences of the others by perceiving the levels of pheromones on various paths. The selection of a path is determined probabilistically based on the level of the pheromone on that path. The amount of the pheromone on a path is inversely proportional to the length of the path. In this fashion, the shortest path ends up with the maximum level of the pheromone and is traversed by the maximum number of ants.

Besides sensing the pheromone level, ants also have local visibility that serve as positive feedback, i.e. encouraging ants to follow the path being traversed by most ants, resulting in all the ants following the same shortest path.

The ACO metaheuristic requires the problem to be presented in the form of nodes and edges, which graphically represent the food resources and paths being found in the ants' colonies. The optimization process is initialized by generating a certain number of random ants. Each ant represents a solution stored in the sequence of nodes in the string associated with that ant. The solutions are evaluated based on the weights of the edges joining the consecutive nodes of the sequence in the string. The initial pheromone trail density, $\tau_{i,j}^{k} = \varepsilon$, on each edge (i, j) joining node i and node j is assumed to be equal for each ant k. This non-zero pheromone trail density ensures the non-zero chances of selection of each edge for further solutions. The pheromone trail density is updated after each iteration according to the following equation:

$$\tau_{i,j} = (1-\delta)\,\tau_{i,j} + \sum_{k=1}^{n} \Delta\tau_{i,j}^{k} \tag{2.1}$$

Where δ is the pheromone evaporation rate; n is the number of ants; $\Delta\tau_{i,j}^{k} = \Theta / W_k$ if the edge (i, j) is included in the tour of ant k, 0 otherwise, $\Theta > 0$ is a constant, and W_k is the objective function value of the path traversed by ant k. The probability of a node being selected by an ant during his tour depends on the pheromone concentration and the visibility ($\vartheta_{i,j}$) for that edge. The selection probability ($\xi_{i,j}$) for an edge is calculated as follows:

$$\xi_{i,j} = \begin{cases} \dfrac{(\tau_{i,j})^{\alpha}(\vartheta_{i,j})^{\beta}}{\sum_{n \in Tabu2^k} (\tau_{i,j})^{\alpha}(\vartheta_{i,j})^{\beta}}, & if\ j \in Tabu1^k \\ 0, & otherwise \end{cases} \tag{2.2}$$

Where α and β are the parameters tuned for the relative dependence on the pheromone trail density and visibility (local heuristic information), $Tabu1^k$ and $Tabu2^k$ are the two memory vectors for storing the nodes that need to be visited and those already visited respectively during the tour of any ant k. Thus, the solution evaluation, and pheromone and probability updating is performed after the end of each iteration. The aforementioned process is repeated until the termination criterion is met. The pseudocode for the ACO procedure has been described in Figure 2.1.

2.2.1 Problem Description

The present work addresses the scenario in which the alternative process plans for each type of part already exist. The process plans are presented using the 'AND' and 'OR' links in the graphical format, and need to be decoded for possible alternative process plans. So, the evaluation of the feasibility of process plans or rearrangement is not required. The machines have been categorized into several categories, and each category has one or more machines to carry out a set of operations associated with that category. The process plan gives information about the required set of operations and their precedence relationships. The machines that are capable of executing these operations are identified using the category to which the machine belongs. Based on the current availability status of machines, the ACO will evaluate the possible sets of process plan combinations for all the parts and suggest the best combination

```
Initialize the ants (n in number), pheromone trails and other parameters;
Generate random solutions (sequence strings) for each ant;
For each ant k ∈ n: calculate W_k;
Update pheromone;
Begin iteration
        Begin for each ant k
                Initialize ant k with some random starting node;
                Choose the next node probabilistically;
                If solution completed: break loop; Else Continue;
        End
        Determine the best iteration ant (with best objective function value);
        Update the pheromone trail;
        If W_(best global ant) > W_(best iteration ant): best global ant = best iteration ant;
        Check if termination = true;
End;
```

Figure 2.1 Pseudocode for Ant Colony Optimization Metaheuristic.

of process plans along with the associated processing and set-up costs for it. The Constraints matrix and Machine state matrix are used for data required during the computing stages and are used to determine the feasibility from a machine point-of-view.

2.2.2 Case Problem

The case problem for the proposed shop-floor mathematical model consists of 4 types of machines. One or more machines of each machine type exist on the shop-floor. Each type of machine is capable of performing a defined set of operations. Some of the operations may be performed by more than one type of machines. A total of 6 machines belonging to the 4 categories of machines exist on the shop-floor. An exhaustive list of 9 different operations is assumed, and a subset of listed operations is distributed to each category of machine. The machines are numbered from m1 to m6 and operations are numbered from o1 to o9. The detailed scenario for the shop floor has been described in Table 2.1. For the sake of simplicity, only one part has been considered for the demonstration of the approach. However, a similar approach could be extended to multiple types of parts for manufacturing. Only one unit of the part is considered for production as the emphasis here is on the demonstration of the process plan evaluation approach. A sample process-sequencing graph has been constructed for the part. A total of six alternative process plans are obtained based on the information provided by the graph. The proposed mathematical model not only identifies the most suitable process plan based on the mentioned metrics but also allocates the machines required for carrying out each operation in that process plan.

Table 2.1 Shop Floor Scenario.

M/c Type	Name (Quantity; Label of m/c in Shop Floor)	Operations that can be Performed on m/c
M1	CNC Lathe (2; m1 and m2)	{o1, o2, o5, o6}
M2	Milling Machine (2; m3 and m4)	{o2, o4, o7}
M3	Forming Machine (1; m5)	{o3, o9}
M4	Shaping Machine (1; m6)	{o2, o4, o7, o8}

In the process sequencing graph as shown in Figure 2.2 below, the node abbreviations used are as follows:

'so' stands for split OR, which marks the beginning of OR alternatives.
'jo' stands for join OR, which marks the end of OR alternatives.
'sa' stands for split AND, which marks the beginning of AND alternatives.
'ja' stands for join AND, which marks the end of AND alternatives.

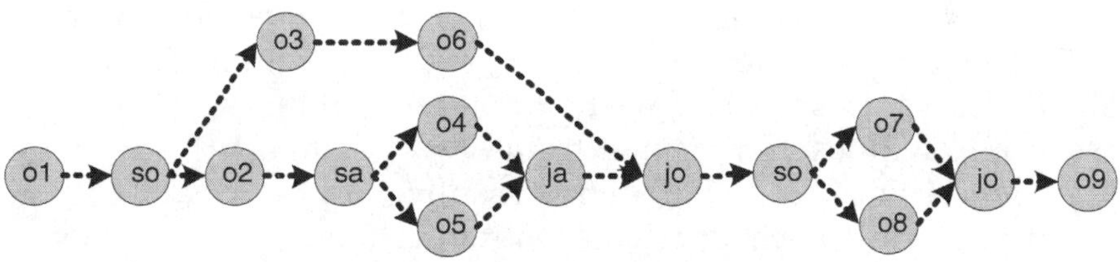

Figure 2.2 Process Sequencing Graph for a part with and, or alternatives.

For the above process sequence graph, a total of 6 alternative process plans exist, as mentioned below,

PP1: {o1, o2, o4, o5, o7, o9}
PP2: {o1, o2, o5, o4, o7, o9}
PP3: {o1, o2, o4, o5, o8, o9}
PP4: {o1, o2, o5, o4, o8, o9}
PP5: {o1, o3, o6, o7, o9}
PP6: {o1, o3, o6, o8, o9}

The probability for each 'OR' path is assumed to be equally distributed, and the same is assumed for 'AND' options. The infeasibility in operations sequence is addressed by assuming the zero probability towards that move. So, if one 'OR' path has been initiated then the probability to visit the alternative 'OR' path becomes zero. The precedence constraint matrix has been shown in Table 2.3. The binary variables Po_io_j have been introduced for defining the precedence relationships. Po_io_j is 1 if the operation o_i precedes operation o_j in the process plan sequence, and 0 otherwise.

For the simplicity of the problem, we assume that the entire manufacturing of the part is completed in 9 equal time-steps. Each time-step defines the combination of an operation and a machine state. This also assumes that the processing time for each operation is the same. The availability of the machines in each time step is shown in Table 2.2. The binary variables m_kT_i is 1 if the kth machine is available in the ith time-step, and 0 otherwise. It should be noted that the effect of the aforementioned assumptions can be cancelled out by associating the appropriate costs for the operation setups. Similarly the binary variables m_ko_i (Table 2.4) is used to define the feasibility of the machines that can be used to perform the required operation. m_ko_i is equal to 1 if the kth machine can be used as an alternative in the ith time-step for ith operation, and 0 otherwise.

Each node in the network represents the operation. Thus there are 9 nodes corresponding to the total 9 operations. The costs are associated with each transition from one node to another node. A fixed processing cost of 100 units is assumed for each feasible transition as per the precedence table shown in Table 2.3. A fixed waiting cost of 50 units is incurred if all the machines of the required machine type are unavailable based on the status of each machine as shown in Table 2.2. A setup cost of 50 units is also incurred if the current machine selected is different from the previous one. The status of each machine at different time-steps has been generated randomly using the MS Excel 2010 random number generator as shown in Table 2.2. It should be noted that the time-step T_i corresponds to the machine allocation and processing of operation o_i.

Objective function (F): Processing Cost (PC) + Waiting Cost (WC) + Setup Cost (SC)

F = PC + WC + SC

where,

$$PC = \sum_{i=1}^{9} \sum_{j=1}^{9} 100/(0.001 + Po_io_j) \tag{2.3}$$

$$WC = \sum_{i=1}^{9} \sum_{k=1}^{6} 50 * m_ko_i * (1 - m_kT_i) \tag{2.4}$$

$$SC = \sum_{i=2}^{9} \sum_{k=1}^{6} 50 * | m_ko_i * m_kT_i - m_ko_{i-1} * m_kT_{i-1} | \tag{2.5}$$

Table 2.2 Machine States in various Time-steps (1: Available, 0: busy).

Machine	Machine Statuses in various Time-steps ($m_k T_i$)								
	T_1	T_2	T_3	T_4	T_5	T_6	T_7	T_8	T_9
m1	1	1	0	1	0	1	1	1	1
m2	0	1	1	1	0	0	0	0	0
m3	0	0	0	0	0	1	0	0	1
m4	1	1	1	1	1	0	1	0	1
m5	1	1	0	1	0	0	0	1	1
m6	1	0	1	1	1	1	1	1	1

Table 2.3 Constraint Matrix (Po_io_j = 1 if row to column movement of ants is permissible, else 0).

Po_io_j	o1	o2	o3	o4	o5	o6	o7	o8	o9
o1	0	**1**	**1**	0	0	0	0	0	0
o2	0	0	0	**1**	**1**	0	0	0	0
o3	0	0	0	0	0	**1**	0	0	0
o4	0	0	0	0	**1**	0	**1**	**1**	0
o5	0	0	0	**1**	0	0	**1**	**1**	0
o6	0	0	0	0	0	0	**1**	**1**	0
o7	0	0	0	0	0	0	0	0	**1**
o8	0	0	0	0	0	0	0	0	**1**
o9	0	0	0	0	0	0	0	0	0

2.2.3 Results

As mentioned earlier, the ant colony optimization (ACO) algorithm requires that the problem be expressed in the graphical form. The distance matrix or position coordinates for the nodes are required for the optimization by the ACO algorithm. The node coordinates have been measured as the costs. The x-coordinate is equal to the sum of waiting cost and set-up cost, and y-coordinate is equal to the processing cost for that node (Table 2.5). The final tour plot for the

Table 2.4 Machine Alternatives for various Operations (1: Feasible, 0: Non-feasible).

Machine	Machine Alternatives for various Operations ($m_k o_i$)								
	o_1	o_2	o_3	o_4	o_5	o_6	o_7	o_8	o_9
m1	1	1	0	0	1	1	0	0	0
m2	1	1	0	0	1	1	0	0	0
m3	0	1	0	1	0	0	1	0	0
m4	0	1	0	1	0	0	1	0	0
m5	0	0	1	0	0	0	0	0	1
m6	0	1	0	1	0	0	1	1	0

Table 2.5 Node Coordinates.

Node	X - Coordinate	Y - Coordinate
1	701.99	51
2	701.99	201
3	801	201
4	92.98	151
5	502.98	201
6	701.99	101
7	501	201
8	801	51
9	900.01	101

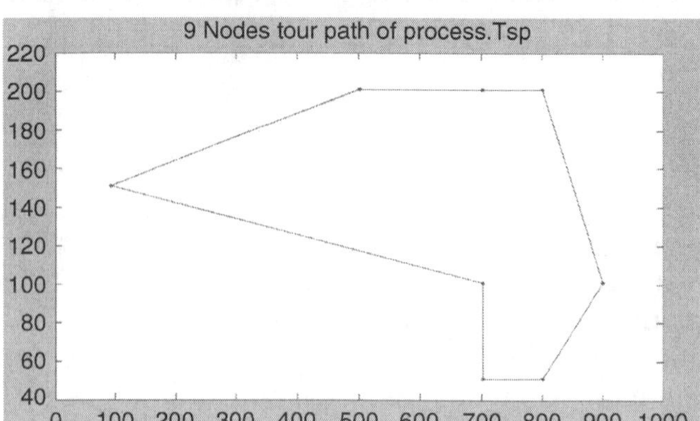

Figure 2.3 Tour Path.

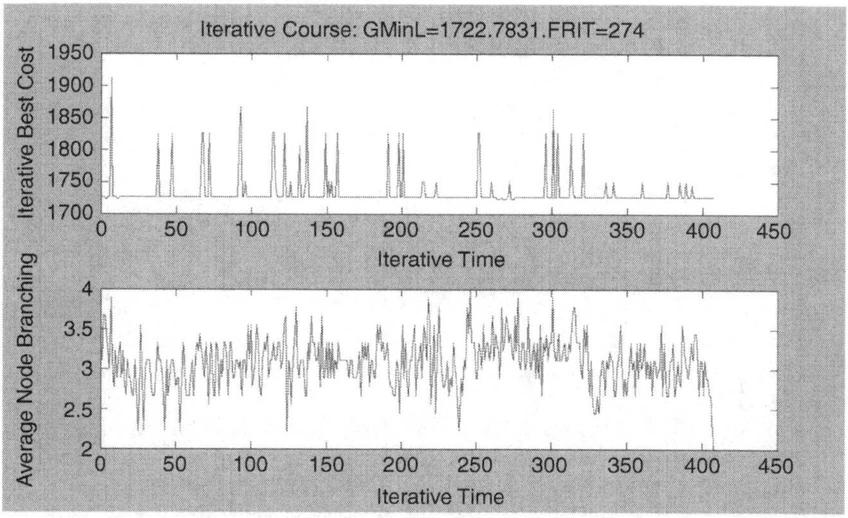

Figure 2.4 Best Cost and Average Node Branching.

solution has been shown in Figure 2.3. Figure 2.4 shows the best cost obtained and the average node branching for the various iterations.

References

Azab, A., Perusi, G., ElMaraghy, H. and Urbanic J., 2006, Semi-Generative Macro-Process Planning for Reconfigurable Manufacturing, Proceedings of CIRP International Digital Enterprise Technology (DET) conference, Setubal, Portugal.

Bley H. and Zenner, C., 2005, Feature-based Planning of Reconfigurable Manufacturing Systems by a Variant Management Approach, CIRP 3rd International Conference on RMS, Ann Arbor, MI.

Bollinger, J., Visionary Manufacturing Challenges for 2020, National Research Council Report, National Academy Press, Washington, D.C (1998).

Chen, J., Zhang, Y. F., & Nee, A. Y. C. (1998). Setup planning using Hopeld net and simulated annealing. *International Journal of Production Research*, 36(4), 981–1000.

Dereli, T., Feliz, I. H., & Baykasoglu, A. (2001). Optimizing cutting parameters in process planning of prismatic parts by using genetic algorithms. *International Journal of Production Research*, 39(15), 3303–3328.

Donati, A. V., Montemanni, R., Casagrande, N., Rizzoll, A. E., & Gambardella, L. M. (2008). Time dependent vehicle routing problem with a multi ant colony system. *European Journal of Operational Research*, 185(3), 1174–1191.

Dorigo, M., & Gambardella, L. M. (1997). Ant colony system: A cooperative learning approach to the traveling salesman problem. *IEEE Transactions on Evolutionary Computation*, 1(1), 52–56.

Dorigo, M., Maniezzo, V., & Colorni, A. (1996). The ant system: Optimization by a colony of cooperating agents. *IEEE Transactions on Systems, Man, and Cybernetics-Part B*, 26, 1–13.

ElMaraghy, H.A., 2006, "Reconfigurable Process Plans for Responsive Manufacturing Systems", Proc. CIRP Int.l Digital Enterprise Tech. (DET) Conf., Setubal, Portugal, (keynote paper).

Ghaiebi, H., & Solimanpur, M. (2007). An ant algorithm for optimization of hole-making operations. *Computers & Industrial Engineering*, 52(2), 308–319.

Giudice, F., & Fargione, G. (2007). Disassembly planning of mechanical systems for service and recovery: A genetic algorithms based approach. *Journal of Intelligent Manufacturing*, 18(3), 313–329.

Guan, X. S., Wang, Y. Q., & Tao, L. Y. (2009). Machining scheme selection of digital manufacturing based on genetic algorithm and AHP. *Journal of Intelligent Manufacturing*, 20(6), 661–669.

Guo, Y. W., Li, W. D., Mileham, A. R., & Owen, G. W. (2009a). Applications of particle swarm optimisation in integrated process planning and scheduling. *Robotics and Computer-Integrated Manufacturing*, 25(2), 280–288.

Guo, Y. W., Li, W. D., Mileham, A. R., & Owen, G. W. (2009b). Optimisation of integrated process planning and scheduling using a particle swarm optimisation approach. *International Journal of Production Research*, 47(14), 3775–3796.

Hua, G. R., Zhou, X. H., & Ruan, X. Y. (2007). GA-based synthesis approach for machining scheme selection and operation sequencing optimization for prismatic parts. *International Journal of Advanced Manufacturing Technology*, 33(5–6), 594–603.

Huang, K. L., & Liao, C. J. (2008). Ant colony optimization combined with taboo search for the job shop scheduling problem. *Computers & Operations Research*, 35(4), 1030–1046.

Koren, Y., Heisel, U., Jovane, F., Moriwaki, T., Pritschow, G., Van Brussel, H., Ulsoy, A.G., 1999, Reconfigurable Manufacturing Systems, Annals of the CIRP, 48/2: 527–540, (keynote paper).

Koulamas, C., 1993, Operation Sequencing and Machining Economics, *IJPR*, 31: 957–975.

Kumar M, Rajotia S. Integration of scheduling with computer aided process planning. *Journal of Materials Processing Technology* 2003;138:297–300.

Li, J. R., Khoo, L. P., & Tor, S. B. (2003). A Tabu-enhanced genetic algorithm approach for assembly process planning. *Journal of Intelligent Manufacturing*, 14(2), 197–208.

Li, W. D., Ong, S. K., & Nee, A. Y. C. (2002). Hybrid genetic algorithm and simulated annealing approach for the optimization of process plans for prismatic parts. *International Journal of Production Research*, 40(8), 1899–1922.

Li, W. D., Ong, S. K., & Nee, A. Y. C. (2004). Optimization of process plans using a constraint-based tabu search approach. *International Journal of Production Research*, 42(10), 1955–1985.

Lin, C.-J. and Wang, H.-P., 1993, Optimal Operation Planning and Sequencing: Minimization of Tool Changeovers, *IJPR*, 31: 311–324.

Lin, C., Lin, S.-Y., and Diganta, D., 1998, An Integrated Approach to Determining the Sequence of Machining Operations for Prismatic Parts with Interacting Features, *J. Materials Proc. Tech.*, 73: 234–250.

Maniezzo, V., Colorni, A., and Dorigo, M. Algodesk: an experimental comparison of eight evolutionary heuristics applied to the quadratic assignment problem. *European Journal of Operational Research*, 1995, 181, 188–205.

Saygin C, Kilic SE. Integrating flexible process plans with scheduling in flexible manufacturing systems. *International Journal of Advanced Manufacturing Technology* 1999;15:268–80.

Sugimura N, Hino R, Moriwaki T. Integrated process planning and scheduling in holonic manufacturing systems. In: Proceedings of IEEE international symposium on assembly and task planning soft research park, vol. 4, Japan, Fukuoka; 2001. p. 250–4.

Xu, H., Tang, R.-Z. and Cheng, Y.-D., 2004, Study Of Process Planning Techniques For Reconfigurable Machine Tool Design, *Journal of Zhejiang University*, 38: 1496–1501.

Zaeh, M. F., Rudolf, H. and Moeller, N., 2005, Agile Process Planning Considering a Continuous Factory Reconfiguration, Proc. CIRP 3rd International Conference on RMS, Ann Arbor, MI.

3

Introducing a Hybrid Genetic Algorithm for Integration of Set Up and Process Planning

S.H. Chung and F.T.S. Chan

Department of Industrial & Systems Engineering, Hong Kong Polytechnic University, Hong Kong

Abstract

Scheduling problems are easily found in real life situations such as publishing houses, hospitals and airports. Production scheduling is one of the most difficult problems that significantly affect the production planning system. Traditionally, a production manager makes decisions on production planning, especially scheduling, with his intuition, experience, and judgment. However, the difficulty of determining a good production schedule increases along with the problem scale. As a result, nowadays, various optimization techniques and information systems have been developed to improve scheduling in production planning.

In solving production scheduling problems, Genetic Algorithms (GAs) are an efficient optimization method based on the evolutionary computing paradigm that has emerged in recent years. GAs can obtain near optimal (or sometimes optimal) solutions from large solution spaces for many different engineering problems, especially in practical environments. This chapter introduces a hybrid genetic algorithm to deal with practical production scheduling problems.

The studied production models are subject to capacity constraints, precedence relationships, and alternative machining with different processing times. More importantly, we will take into consideration the processing time, transportation time between resources, and especially machine set-up time between different processes. The introduced hybrid genetic algorithm is demonstrated through some examples for minimizing the makespan of a set of given tasks.

Manoj Tiwari and Jenny A. Harding (eds.) Evolutionary Computing in Advanced Manufacturing, (37–50) © Scrivener Publishing LLC

Keywords: Hybrid genetic algorithm, process planning, set-up time, makespan, scheduling

3.1 Introduction

In today's global competitive environment, pressures are placed on manufacturing organizations to compress order fulfillment times, meet delivery commitments consistently and also maintain efficiency in operations to address cost issues. However, managers always meet many optimization problems such as poor utilization of resources and fluctuating demands. Thus, production planning is needed.

Production scheduling is one of the major issues in process planning and control of all different units and is at the heart of the performance of manufacturing organizations. Traditionally, people make process planning decision, especially scheduling, with their intuition, experience, and judgment. However, humans are not well equipped to control or optimize large and complex systems. When the size of the planning task becomes large, the difficulty of finding a good solution will obviously increase. As a result, nowadays, various optimization techniques and information systems have been developed to support process planning.

Scheduling problems are easily found in real life situations like factories, publishing houses, shipping, hospitals, airports, etc. Manufacturing scheduling is one of the most difficult tasks of all scheduling problems. Typically, most of those problems are NP Hard problems, and it is impossible to find an optimal solution without the use of an essentially enumerative algorithm.

3.2 Process Planning

With the challenge of global competition, pressures are placed on manufacturing organizations of household electrical appliances to compress order fulfillment times, meet delivery commitments consistently and also maintain efficiency in operations to address cost issues. However, managers always meet many optimization problems such as poor utilization of resources and fluctuating demands. Thus, process planning is needed.

Process planning is concerned with determining operations and allocating resources to meet fluctuating demands. In reality, resources cannot be acquired as needed and production capacity cannot infinitely change with zero cost. Companies have to make decisions on the best utilization of resources based on the forecast of demand. As Godfrey (2002) mentioned, the aim of a production plan is to achieve the goals set by the business plan to meet the fluctuating market demand requirements and minimize the expected total production costs. Scheduling is a decision-making process that plays an important role in the production planning system. (Pinedo, 2002).

Therefore, what is the meaning of scheduling? Baker (1974) defines it as "the allocation of resources over time to perform a collection of tasks". Also, Wight (1984) defines it as "establishing the timing for performing a task". Another definition has been put forward by Pinedo (2002): "Scheduling deals with the allocation of scarce resources to tasks over time. It is a decision-making process with the goal of optimizing one or more objectives."

As a result, scheduling is the process of setting priorities for work. In the process, we always consider how to allocate our limited resources in order to optimize our efficiency or meet our objectives.

3.3 Machine Set-up Time

In addition to the production modes, machine set-up time also plays an important role in scheduling. However, the facts indicate that a majority of the literature deals with problems without sequence dependent setups, some surveys and studies indicate that setups are important in a majority of practical situations, and must be accounted for in the design of algorithms for scheduling problems. In a survey of industrial schedulers, Panwalkar *et al.* (1973) report that about 70% of the schedulers stated that setup times depended on processing sequence in at least 25% of the jobs they scheduled. Kim and Bobrowski (1994) study the impact of setup times on the performance of scheduling systems using simulation; they conclude that to better model practical situations, setup times should be considered explicitly whenever they are significantly greater than the processing times (Skylab and Jeffrey, 2005).

In most research on scheduling problems, setup times are assumed to be independent of the sequence of tasks on a machine. It is also assumed that setup times are negligible or are directly added into the processing times of the tasks. However, significant setup times are incurred in some situations whenever a machine switches service from one task to another (Skylab and Jeffrey, 2005). Wilbrecht and Prescott (1969) state that this is particularly true where production equipment is being used close to its capacity levels. Wortman (1992), states that the efficient management of production capacity requires the consideration of setup times.

Many practical industrial situations require the detailed consideration of setups and the development of appropriate scheduling tools. For example, Pinedo (1995) describes a manufacturing plant making paper bags where setups are required when the type of bag changes. The duration of a setup depends on the similarity of the bags made in the preceding lot. A similar situation was observed in the plastics industry by Das *et al.* (1995) and Franca *et al.* (1996). The printing industry also has setups that are sequence dependent because different cleaning operations are required when the print colors are changed. The aluminium industry has casting operations where setups, mainly affecting the holding furnaces, are required between the castings of different alloys (Gravel, *et al.*, 2000). The textile, pharmaceutical, chemical and metallurgical industries present other practical examples where sequence-dependent setups are frequently observed.

Some notable examples are found in petroleum producing plants, car spraying facilities, textile dying plants and pharmaceutical industries and are described in Allahverdi *et al.* (1999). In the casting centre, holding furnaces contain molten metal ready to be poured. Certain draining and cleaning operations of varying durations between the castings of two successive jobs for various metal alloys may be required. These operations may be seen as the setup operations (Gagne, *et al.*, 2002).

Machine setup time is actually a significant factor for production scheduling in all flow pattern manufacturing environments. One of the most widely used assumptions in real scheduling configurations is the consideration of sequence-dependent set-up times (SDST). After finishing the processing of one job and before starting processing of the next job, some sort of set-up such as changing tools or cleaning up must be carried out. In SDST, the magnitude of time required to do the set-up depends on both the immediately preceding and current job to be performed on the same machine (Baker 1974). The application of SDST could be seen in many real-life situations such as chemical, printing, pharmaceutical, and automobile manufacturing (Zanchieh, *et al.*, 2006).

Pinedo (1995) cited machine set-up time as a significant factor for production scheduling in all flow patterns, and it may easily consume more than 20% of available machine capacity if not well handled. Scheduling problems with SDST are among the most difficult classes of scheduling problems.

The magnitudes of setup times often depend on the similarity of the process technology requirements of two consecutive operations. Typically, large setup times are associated with two consecutive operations if they differ significantly in processing requirements or utilize different process technologies (Srikar, B.N. and Ghosh, S. A., 1986).

3.3.1 Optimization Methodology: Genetic Algorithms (GA)

"Getting the orders out the fastest way possible is rarely a production manager's only objective."(Tapan, 1999) Not only makespan must be considered, he/she should also try to minimize tardiness and costs and maximize the utilization of capital equipment and human resources. To find a good solution according to these criteria, many researchers have proposed many different optimization methodologies such as pure mathematical optimization and heuristic optimization approaches.

Genetic algorithms (GA) is one of the hottest heuristic optimization methodologies, and is widely recognized as an appropriate and efficient approach in scheduling especially for practical industrial applications (Vignaux and Michalewicz 1991, Al-Hakin 2001). The idea of the Genetic Algorithms (GA) approach was first developed by John Holland in 1975 (Holland 1975). Nowadays, it is widely adopted by many researchers for solving various problems, such as cargo loading, job shop scheduling, transportation, etc. GA is a heuristic approach. Its heuristic optimization algorithms mimic the mechanism of genetic evolution in biological nature, which is based on the principles of evolution and genetics to guide the search which results in the "survival of the fittest". It is a stochastic search algorithm and optimization technique that imitates a natural selection and biological evolutionary process. (Goldberg 1989, Davis 1987).

Applications of the GA approach can be found in many publications in different areas. Aytug *et al.* (2003) provided a detailed review of the use of GAs to solve operation problems. The reviewed papers are classified according to the nature of the problem, such as production control, facility layout design, line balancing, production planning and supply chain management.

In the area of production control, Al-Hakin (2001) used GA to solve job-shop scheduling problems. His paper developed an analogous electrical system to measure the performance of the system, and a new measure for the fitness function of GA. In the area of process planning, Onwubolu and Mutingi (2001) used GA to optimize a multiple constrained resources product mixed problem, compared with the results of linear-integer programming. Yoon and Ventura (2002) used GA to solve lot-streaming flow shop scheduling problems with earliness lateness penalty. Burdett and Kozan (2000) used GA to minimize makespan in a flow shop for m jobs with operations on n machines. Onwubolu and Mutingi (1999) used GA to minimize tardiness in flow shop with permutation schedules. Janiak and Portmann (1998) used GA to minimize makespan in a flow shop with linear models of operations. Neppalli *et al.* (1996) used GA to schedule n jobs on two machines in a flow shop to minimize flow time with optimal makespan. The GA results in these papers proved that the technique of GA is reliable and robust.

Some researchers have combined GAs with other methodologies to enhance their searching ability for their problems and these are called hybrid GAs. For example, some researchers combined GA with Tabu Search (TS) to enhance the optimization results. (Glover 1989, Abdinnour-Helm 1999). Cavalieri and Gaiardelli (1998) applied a hybrid GA with the combination of GA and the dispatching rule of earliest due date, to solve multi-objective scheduling problems. Sakawa (2002) combined GA with fuzzy logic to maximize the fulfillment reliability of job's due date in a job-shop scheduling problem.

To plan a production schedule, some information must be acquired for decision making. For example, before scheduling the production, a production manager has to know the due date of orders, finished and semi-finished goods inventory, material consumption, labor availability, production capacity etc. That information will be acquired from different departments and input into the GA program for production scheduling.

In general, a simple GA approach consists of several steps as described below. The first step in GA is to design and encode a chromosome. A number of chromosomes will be generated randomly to form an initial pool of solutions. Once the pool is formed, the chromosomes generated will be evaluated by a fitness function which is defined in order to check the strength of the chromosomes. Next, some chromosomes will be selected to form a mating pool with the function of Roulette Wheel Selection. Then, the chromosomes in the mating pool will be randomly paired for the process of crossover and mutation. After that, the chromosomes will be evaluated again. Finally, GA will determine whether the generation is completed. If the predefined stopping condition is reached, the best solution found is output, which will be transformed into a suggested production schedule. Otherwise, it will go back to the Roulette Wheel Selection and continue the process.

3.4 Chromosome Representation

In GA, it is presumed that the potential solution of any problem is an individual and can be represented by a set of parameters which are usually regarded as the genes of a chromosome and each chromosome represents a potential solution of the problem. Each gene represents a number of values called alleles. In general, there are two types of alleles, binary and real parameter. Figure 3.1 is a sample of a binary chromosome which is represented by values of 0 or 1. A sample of a real parameter is shown in Figure 3.2. In real parameter chromosomes, each gene is represented as a real number.

According to the nature of the problem, the coding of the chromosome representation will vary in different problems. Real parameter representation is most commonly applied. Figure 3.3 demonstrates how chromosomes can be encoded and decoded as an example. A gene consists of three numbers. The first number represents the machine number, the second number represents the job number and the last values of alleles represent the operation of jobs. The gene "233" indicates that Operation 3 of Job 3 is allocated to Machine 2.

1	0	0	1	0	1	0	1

Figure 3.1 A sample of binary chromosome.

5	0	0	8	0	3	0	5

Figure 3.2 A sample of real parameter.

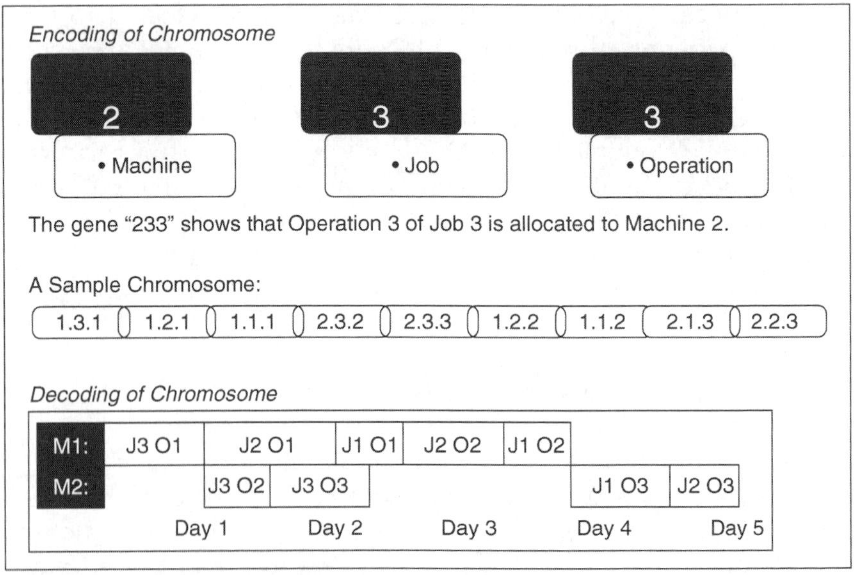

Figure 3.3 Encoding and Decoding of Chromosome.

3.5 Fitness Value Evaluation

To evaluate the relative strength of a chromosome compared to others, a fitness value is defined. It is estimated by the chromosome structure and a positive value results. The larger the value, the stronger the chromosome is. For example here, we want to minimize the makespan of all the processes, meaning that all the processes can be finished as soon as possible as shown in Equation 3.1 and Table 3.1.

$$F_i = 1 - (C_i / C_{max}) * 100 \tag{3.1}$$

where

F_i = fitness value of chromosome i

C_i = makespan of chromosome i

C_{max} = the largest total earliness and total tardiness in mating pool

The makespan is equal to the completion time (C_i) of the last operation of all the processes. It consists of its processing time plus the transportation time required to transport the component from the preceding operation's machine to the will-be performed machine. In addition, the machine set-up time required to perform the operation from the previous one is also required.

Table 3.1 Sample calculation of fitness value.

Chromosome	Total EA and TA (Day)	Fitness Value
C1	128	26
C2	47	73
C3	56	67
C4	60	65
C5	172	0
C6	59	66

3.6 Selection Operation

The selection process implements the idea of "the survival of the fittest". In the process of evolution, it is expected that the "better" chromosome will create a larger number of offspring and thus have a higher chance of surviving in the subsequent generation (Man *et al*, 1999). After selecting chromosomes from the solution pool for the mating pool, the selected chromosomes will be the parent chromosomes of the next generation.

As the selection process is basically arbitrary, all chromosomes may have a probability to be chosen. The stronger chromosomes will have higher probability for selection. This is based on the idea that only the stronger ones can survive, while the weaker ones will be eliminated. Moreover, every chromosome can appear more than once in the mating pool. The sum of the chromosomes selected is equal to the size of the solution pool initially defined.

The Roulette Wheel selection approach is one of the most common techniques used for proportionate selection mechanism (Davis 1987, Dawid 1996, Onwubolu 2001, Sakawa 2002). In Table 3.2, there are six chromosomes with the fitness value evaluated as 26, 73, 67, 65, 0 and 66 for chromosomes 1, 2, 3, 4, 5 and 6 respectively. The probability value (P_i) for chromosome i is calculated by Equation 3.2. For example, the probability (P_1) of chromosome 1 (0.087) is equal to 26.

$$\mathrm{Pr}_i = f_i / \sum_n f_n, \tag{3.2}$$

where

P_i – probability of chromosome *i*

After calculating the probabilities of all chromosomes, the chromosomes are all put in a Roulette wheel as shown in Figure 3.4. Each sector represents the probability of a chromosome being selected. As shown in Table 3.2, this random number lies between the cumulative values of the probability of the chromosomes. Therefore, to select a chromosome in this example, a number between 1 and 297

Table 3.2 Sample calculation of selection probability.

Chromosome	Fitness Value	Cumulative	Range	Probability
C1	26	26	$0 < x \leqq 26$	0.087
C2	73	99	$26 < x \leqq 99$	0.246
C3	67	166	$99 < x \leqq 166$	0.226
C4	65	231	$166 < x \leqq 231$	0.219
C5	0	231	$231 < x \leqq 231$	0
C6	66	297	$231 < x \leqq 297$	0.222
Total	297	–	–	1

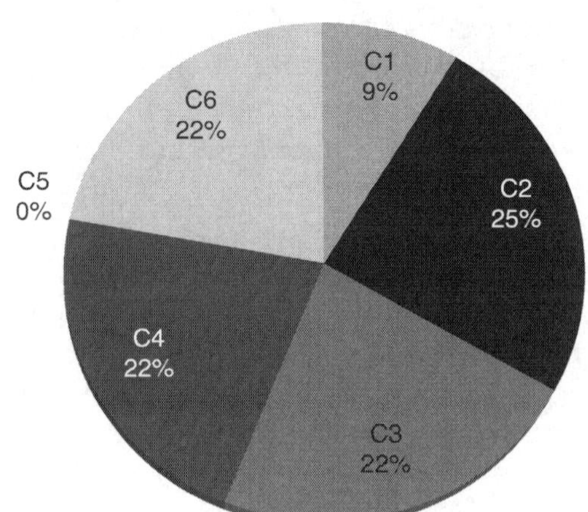

Figure 3.4 A sample of Roulette wheel of six chromosomes.

will be randomly generated. For example, if the random number generated is 100, then chromosome 3 will be selected.

3.7 Crossover Operations

Crossover is the exchange of genes among chromosomes in the mating pool in order to replace the weaker genes with the stronger genes from other chromosomes. A pair of selected parents will generate a pair of offspring as shown in Figure 3.5. The genes for exchanging will be selected randomly from the two parent chromosomes. For example, the eighth genes in parent 1 and 2 are selected and exchanged for generation of offspring.

3.8 Mutation Operations (k-opt Exchange)

One of the efficient exchange methodologies for local fine searching is k-opt exchange, proposed by Lin and Kernighan (1973), whose basic idea is to use a set of k links to replace the current set of k links in the routed processes. The size of k would critically affect the complexity of the heuristic. For larger k values, the heuristic tends to give better computational results, but the computational time will also increase.

The idea of mutation is inspired from the genetic mutation in nature. It changes the value of genes from the mating pool randomly and creates new potential solutions which are independent from the current chromosomes. A high mutation frequency will change the GA to some random searching algorithms (Dawid, 1996). However, mutation may destroy the good solutions, since those strong genes are replaced simultaneously (Al-Hakin, 2001). Figure 3.6 demonstrates the process of mutation. A gene will be

Before crossover:

Parent 1 : 1.3.1 - 1.2.1 - 1.1.1 - 2.3.2 - 1.3.3 - 1.2.2 - 1.1.2 - 2.1.3 - 2.2.3
Parent 2 : 1.1.1 - 1.3.1 - 1.2.1 - 2.3.2 - 1.2.2 - 1.2.3 - 1.1.2 - 1.3.3 - 1.1.3

After crossover:

Offspring 1 : 1.3.1 - 1.2.1 - 1.1.1 - 2.3.2 - 1.3.3 - 1.2.2 - 1.1.2 - 1.3.3 - 2.2.3
Offspring 2 : 1.1.1 - 1.3.1 - 1.2.1 - 2.3.2 - 1.2.2 - 1.2.3 - 1.1.2 - 2.1.3 - 1.1.3

Figure 3.5 Demonstration of crossover operation.

Parent 1:	1.3.1 - 1.2.1 - 1.1.1 - 2.3.2 - 1.3.3 - 1.2.2 - 1.1.2 - 2.1.3 - 2.2.3
Offspring 1:	1.3.1 - 1.2.1 - 1.2.3 - 2.3.2 - 1.3.3 - 1.2.2 - 1.1.2 - 2.1.3 - 2.2.3

Figure 3.6 Demonstration of mutation operation.

randomly selected and swapped as k-opt exchange in order to form an offspring of new potential solution.

In general, all chromosomes in the mating pool will be mutated. It can prevent the solution pool from saturating with the chromosomes of similar structure so that the solutions in the population will not fall into a local optimal range easily.

3.9 Conclusion

This chapter presents a hybrid GA for the integration of set up and process planning when minimizing the makespan of the processes. It explains the steps involved in applying the hybrid GA, including the formation of the initial pool, mating pool etc. In addition, different kinds of genetic operators involved are also explained including selection operator, crossover, operator and mutation operator.

References

Abdinnour-Helm, S., Network design in supply chain management, *International Journal of Agile Management Systems*, 1(2), 99–106, 1999.

Al-Hakin, L., An analogue genetic algorithm for solving job shop scheduling problems, *International Journal of Production Research*, 39(7), 1537–1548, 2001.

Allahverdi, A., Gupta, J.N. and Aldowaisan, T., A review of scheduling research involving setup considerations, *Omega*, 27, 219–239, 1999.

Aytug, H., Khouja, M. and Vergara, F.E., Use of genetic algorithms to solve production and operations management problems: a review, *International Journal of Production Research*, 41(17), 3955–4009, 2003.

Baker, K., *Introduction to Sequencing and Scheduling*. New York: John Wiley & Sons, 1974.

Burdett, R. L. and Kozan, E., Evolutionary algorithms for flowshop sequencing with non-unique jobs, *International Transactions in Operational Research*, 7, 401–418, 2000.

Cavalieri, S. and Gaiardelli, P., Hybrid genetic algorithms for a multiple-objective scheduling problem, *Journal of Intelligent Manufacturing*, 9, 361–367, 1998.

Das, S.R., Gupta, J.N.D. and Khumawala, B.M., A saving index heuristic algorithm for flowshop scheduling with sequence dependent set-up times. *J Opl Res Soc*, 46, 1365–1373, 1995.

Davis, L., Handbook of Genetic Algorithms, *New York: Van Nostrand Reinhold*, 1987.

Dawid, H., *Adaptive Learning by Genetic Algorithms Analytical Results and Applications to Economic Models, Lecture Notes in Economics and Mathematical* Systems 441, Springer, 1996.

Franca, P.M., Gendreau, M., Laporte, G. and Muller, F.M., A tabu search heuristic for the multiprocessor scheduling problem with sequence dependent setup times. *Int JProd Econ.* 43, 79–89, 1996.

Gagne, C., Price, W.L. and Gravel, M., Comparing an ACO algorithm with other heuristics for the single machine scheduling problem with sequence-dependent setup times, *Journal of the Operational Research Society*, 53, 895–906, 2002.

Glover, F., Tabu Search part I, *ORSA Journal on Computing*, 1(3), 190–206, 1989.

Godfrey, C., *Emerging optimization techniques in production planning and control.* London: Imperial College Press, 2002.

Goldberg, D.E., Genetic Algorithms in Search, Optimization and Machine Learning, Reading, MA: Addision-Wesley, 1989.

Gravel, M., Price, W.L. and Gagne, C., Scheduling jobs in a Alcan aluminium factory using a genetic algorithm, *Int JProd Res*, 38, 3031–304, 2000.

Holland, John H., *Adaptation in Natural and Artificial Systems, Ann Arbor, MI:University of Michigan Press*, 1975.

Janiak, A. and Portmann, M. C., Genetic algorithm for the permutation flow-shop scheduling problem with linear models of operations, *Annals of Operations Research*, 83, 95–114, 1998.

Kim, S., and Bobrowski, P., Impact of sequence dependent setup time on job shop scheduling performance, *International Journal of Production Research*, 32(7), 1503–1520, 1994.

Lin, S. and Kernighan, B.W. "An effective heuristic algorithm for the traveling salesman problem," Operation Research, vol. 21, pp. 498–516, 1973.

Man K.F., Tang K.S and Kwong S.. *Genetic algorithms: concepts and designs.* Springer, 1999.

Neppalli, V., Chen, C. and Gupta, J. N. D., Genetic algorithms for the two-stage bicriteria flowshop problem, *European Journal of Operational Research*, 95, 356–373, 1996.

Onwubolu, G.C and Mutingi, M., Optimizing the multiple constrained resources product mix problem using genetic algorithms, *International Journal of Production Research*, 39(9), 1897–1910, 2001.

Onwubolu, G. C. and Mutingi, M., Genetic algorithm for minimizing tardiness in flow-shop scheduling, *Production Planning and Control*, 10, 462–471, 1999.

Panwalkar, S., Dudek, R., and Smith, M., Sequencing research and the industrial scheduling problem, in: M. Beckmann, P. Goos, H. Zurich (Eds.), Symposium on the Theory of Scheduling and Its Applications, Springer-Verlag, New York. 29–38, 1973.

Pinedo, M., *Scheduling: theory, algorithms and systems.* Prentice-Hall: Englewood Cliffs, NJ, 1995.

Pinedo, M., Scheduling: Theory Algorithms, and systems. UpperSaddleRiver, N.J.: Prentice Hall, 2002.

Sakawa, M., Genetic algorithms and fuzzy multi-objective optimization, *Kluwer Academic Publisher*, 188–222, 2002.

Skylab, R. G. and Jeffrey, S.S., Algorithms for single machine total tardiness scheduling with sequence dependent setups, 2005.

Srikar, B.N. and Ghosh, S., A MILP model for the N-job, M-stage flowshop with sequence dependence set-up times, *International Journal of Production Research*, 24 (6), 1459–1474, 1986.

Tapan, P., *Multiobjective scheduling by genetic algorithms*. Boston: Kluwer Academic, 1999.

Vignaux, G.A. and Michalewicz, Z., A genetic algorithm for the linear transportation problem, *IEEE Transactions on systems, man, and cybernetics*, 21(2), 445–452, 1991.

Wight, W., *Production and Inventory Management in the Computer Age*. New York: Van Reinhold Company, 1984.

Wilbrecht, J.K. and Prescott, W.B., The influence of setup time on job performance, 16, B274–B280, 1969.

Wortman, D.B., *Managing capacity: getting the most from your firms assets*. Ind Eng 24: 47–49, 1992.

Yoon, S.-H. and Ventura, J. A., An application of genetic algorithms to lot streaming flow shop scheduling, *IIE Transactions*, 34, 779–787, 2002.

Zandieh, M., Fatemi Ghomi, S.M.T., and Moattar Husseini, S.M., An immune algorithm approach to hybrid flowshops scheduling with sequence with sequence-dependent set-up time. *Applied Mathematics and Computation*, 180(1), 111–127, 2006.

4

Design for Supply Chain with Product Development Issues Using Cellular Particle Swarm Optimization (CPSO) Technique

Vikas Kumar[1] and F.T.S. Chan[2]

[1]*Department of Management, Dublin City University Business School, Dublin, Republic of Ireland*

[2]*Department of Industrial & Systems Engineering, Hong Kong Polytechnic University, Hong Kong*

Abstract

Increasing global competitiveness has forced manufacturers to develop and manage families of products to remain competitive in the market while simultaneously increasing their profitability and market share. Manufacturers achieve this goal by aiming to design a variety of consumer appealing products while maintaining the production cost at a minimum level. Therefore product design and development issues have generated a lot of curiosity among researchers in the last few decades. The product family design concept incorporates the challenges of designing, developing and coordinating the design of multiple products. At the same time it also aims to maximize the commonality across a set of products without compromising their individual performances. The design for supply chain (DFSC) helps in the selection of an appropriate product family. In this research, to address the DFSC issues, a product platform approach has been proposed that integrates the component modularity as well as the function modularity in the product design. The optimization model proposed in this research for the product development and the supply chain design is based on a Generic Bill of Materials (GBOM) representation. Realizing the complexities associated with the design of supply chain this research uses a newly developed cellular particle swarm optimization (CPSO) technique to resolve the product development problem. This algorithm inherits the property of cellular automata (CA) and standard

Manoj Tiwari and Jenny A. Harding (eds.) Evolutionary Computing in Advanced Manufacturing, (51–76) © Scrivener Publishing LLC

particle swarm optimization (PSO). The main objective of this research is to maximize the market share while taking into account the customer view during the new product development. The performance of the CPSO algorithm has been compared with the GA and standard PSO on a case study and randomly generated data sets of increasing complexity are used to show the efficacy and robustness of the algorithm.

Keywords: Product development, supply chain design, particle swarm, cellular automata, GBOM

4.1 Introduction

In the present business environment the market has become very much user oriented giving customers multiple options to choose from. Therefore the customer has an option to choose similar kinds of products offered by different manufacturers whose products only vary by a small amount. Over time customers have developed a mindset of having a product that can serve multiple purposes such as a mobile phone that can be used as a camera, a music device, or a device to surf the internet apart from providing its basic functionality of making and receiving calls. Since these customer needs are being fulfilled by multiple players in the market, customers have become more demanding and often opt for a product based on a number features such as the functionality, colour, looks, user friendliness and several other features. To deal with this kind of situation the manufacturer puts a lot of emphasis on the process flexibility to enhance the overall scale efficiency of the manufacturing plant and imitate a high production dominant strategy. The overall cost of the product does not solely depend on the design or manufacturing in the plant, rather it depends on all the processes right from the design to until it reaches to the market. Therefore, addressing the supply chain issues in a manufacturing context without discussing product development issues seems to be incomplete. Therefore the Design for Supply Chain (DFSC) problem becomes a crucial issue for the manufacturers. DFSC problems deal with designing a supply chain and very often involve resolving some decisions regarding the product development.

In recent years, the focus of product development competitive advantage has been on time-to-market, overlapping problem solving, and multifunctional teamwork encompassed within the philosophy of concurrent engineering. Modularity compels a designer to create a higher level of independence between the components by redefining their interface details (Sanchez, 1997). Depending upon the intricacy of the technological development or production capacity, modules can be either function or component modules (Pahl and Beitz, 1988). The function modules become very important while designing a product that has several intricate functional requirements. Different models and modes of bills of materials such as planning bills (Braker, 1996), modular bills, kit bills (Oden *et al.*, 1993), and Generic Bill of Material (Jiao *et al.*, 2000) can be utilized in various design problems. Products can be designed in accordance with the existing customer diversity by designing products for different market segments with optimum profit and differentiation. This can be accomplished using a platform approach, leveraging the design cost between the lower and higher market segments.

The design and development of a product requires the coordination of many different skills and knowledge, and between the different elements of the manufacturing processes. Therefore, coordination between supply chain networks is an essential factor for overall enhanced performance. The delivery of products quicker to the customer while ensuring quality throughout the production and delivery phases at lower cost often gives firms competitive advantage over their rivals (Twigg, 1998). Graves and Tomlin, (2003) found that supply chains with higher values of process flexibility enhance the overall scale efficiency of the plants. The process flexibility imparted in the supply chain helps to imitate a high production dominant strategy. However, the production plants with the lower degrees of flexibility suffer more from the shortfalls. Blackhurst *et al.* (2004) point out that the supply chain faces uncertainties such as the uncertainty of its network operations in addition to the demand uncertainty. The supplier performance also plays a crucial role in a supply chain network. The supplier in the supply chain is evaluated in terms of not only the supply chain cost, but also the delivery performance and the quality assurance.

Liu *et al.*, (2000) pointed out that the performance evaluation of suppliers should be based on performance measures for all part types supplied to the purchasing company. This evaluation provides benchmarking data for reducing the number of suppliers, which in turn results in benefits including reduction in costs of parts and order processing, and better partnerships with suppliers. They applied Data Envelopment Analysis (DEA) to evaluate the suppliers. Jain *et al.* (2004) argued that a fuzzy set approach can be implemented to evaluate a suppliers' performance. Stewart (1995) argued that an increase in delivery performance is possible through a reduction in the lead time attribute. Realizing the importance of the product variety and the supply chain, Randall and Ulrich (2001) proposed some hypotheses which reflect that manufacturing plants with large scale efficiency and local nature, with respect to the target market, have least dominance of production cost or market mediation cost in altering the product varieties. Lamothe *et al.* (2005) proposed a mixed integer linear programming model to select a diverse product family with an optimum supply chain cost using the market segment design strategy, modular design strategy or the combination of both these strategies. These findings highlight the different aspects of supply chain management and suggest that supply chain is an integral part of the production process and one needs to handle it efficiently to improve the overall performance of the firm.

Realizing the growing significance of the design for supply chain and product development this paper addresses a multi-criteria DFSC problem dealing with a diverse product family selection and its supply chain design. The research considers a hybrid market segment and modularity technique for the diversification as well as the component modularity vis-à-vis functional modularity in the product design part of the problem. The research proposes, a comprehensive optimization framework considering various control factors such as process flexibility, production or market mediation dominance on product variety, supply chain cost, profit and product design complexity. The proposed framework inherits a number of decision variables such as which and how many items to choose, the existence, location and capacities of plants, the shipping channels to use, product performance levels, and product prices. A number of constraints have also been considered such as the demand and flow conservations, safety inventories, feasible product flows, resource lines, market segmentation, and product differentiation. Additionally, the framework integrates the optimization of the objective function and a few supply chain performance hypotheses. The framework imparts process flexibility to the supply chain to tackle the demand uncertainties to a remarkable extent. The optimization problem for DFSC in this research can be broadly divided into two parts,

1. Product family selection.
2. Design for supply chain.

The optimization model aims to meet the following three objectives, i.e.

- Minimize the supply chain cost,
- Maximize sales profit of end products and
- Minimize the product design complexity.

To resolve this kind of complex DFSC problem, the literature suggests a number of different constraint based algorithms however, these fail to generate real time solutions for large size problems. Evolutionary algorithms, which have emerged over the years, have come to the rescue of these issues, as they can efficiently handle complex optimization problems. This research uses a newly developed Cellular Particle Swarm Optimization (CPSO) technique. More specifically this research uses a version of CPSO proposed by Shi *et al.* (2010) known as the CPSO-outer algorithm. This algorithm is capable of searching many optimum points in parallel and reaches an optimal or near optimal solution with high probability in less computational time. A problem of a wiring harness supplier of an automated guided vehicle (AGV) manufacturer has been tested to experimentally evaluate the propositions. The main objective is to achieve a trade-off between the supply chain cost, sales profit and the design complexity.

4.2 Problem Formulation

This chapter proposes a model for selecting the optimal product family based on the Generic Bill of Materials (GBOM) model developed by Lamothe *et al.* (2006). In the model the logical and physical items are represented as nodes. The logical items in the model are the hypothetical ones and they can neither be manufactured in the plants nor be stored or shipped. Whereas, the physical items can be manufactured, stored and shipped. These nodes are connected with each other through AND and Ex-OR nodes. AND nodes are those which flourish over the contributions from each and every one of their children, whereas an Ex-OR node flourishes over the contribution from one and only one of its children. This chapter adopts a hybrid market-modular approach that has been adopted for the product development. Here the market is a logical item. The hybrid market-modular strategy helps to reach a compromise between high equipment costs, an excessive level of satisfaction, and high management costs (Lamothe *et al.*, 2006). In this approach, the market is first divided into several segments by fixing specific performance levels for each function. The market segmentation is shown in Figure 4.1 (a) and 4.1 (b) whereas the component modularization is shown in Figure 4.1 (c). The detailed explanation of the figure is presented in the work of Yadav *et al.* (2011). Different children choices at Ex-OR nodes result in the different combinations of product families. An optimal product family is selected based on the objective function.

In our model, we have assumed that the market is segmented into two segments (lower and higher). We also assume that consumers in a market segment are homogenous. Different products offered to the two consumer segments with different performance levels, share a common platform which is defined as components and subsystem assets shared across a product family (Krishnan and Gupta, 2001). A product platform reduces product development cost, imparts a greater degree of reuse and enhances the responsiveness of firms, but may incur over or under design cost of the lower and the higher market segment respectively. The mathematical modeling is described in detail by Yadav *et al.* (2011). The three criteria addressed in this paper are sales profit, supply chain cost and product design complexity.

4.2.1 Notations

Φ_x^{exist} = Fixed cost of existence of an item s

δ_s = Binary decision variable for the existence of an item s

Ψ_f^{open} = Fixed cost of opening a facility f

Ψ_{rf}^{open} = Fixed cost of opening a resource line of type r at facility f

χ_f = Binary decision variable for the existence of a facility f

ε_{rf} = Number of resource lines of type r opened at facility f

Π_{fe}^{path} = Fixed cost of building a shipping path between facility f and e

λ_{fe} = Binary decision variable for the existence of a shipping path between facilities f and e

Ω_{sf}^{req} = denotes the costs incurred in making an item s at a facility f

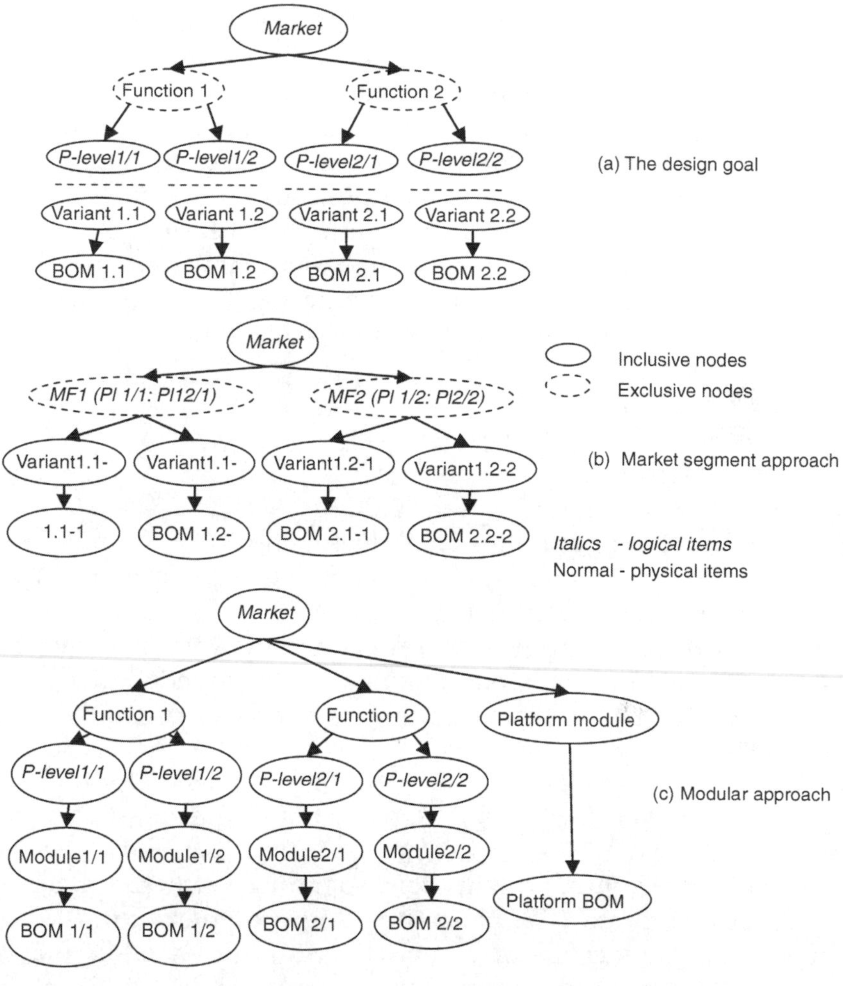

Figure 4.1 A tree representation of the Market and modular approaches.

N_{sft} = Decision variable corresponding to the net quantity of requirement relevant to an item s at a facility f

N_{max} = Maximum quantity of any item that can be manufactured, stored or shipped on a time period

Δ_{sf}^{req} = Variable cost in storing an item s at a facility f

H_{sft} = Quantity of a physical item s

ρ = Performance level of product in general

ρ_h = Performance level of the product target to the higher market segment (having contributions from platform components as well as non-platform components)
ρ_l = Performance level of the product target to the lower market segment having contributions from platform components as well as non-platform components)
ρ_p = Performance level of the common platform (i.e., performance level delivered through modular components)
I = Coefficient of integration, accounting for integration benefits of platform based product development
b = Constant
c = Constant
C = Set of all the physical items
k = Constant
p = Index for any item in the GBOM
η = Number of modules of all the functions
η_{max} = The maximum number of product the firm can produce
d_i = Certainty of ith function technology
θ_l = Number of modules at a functional coordination level l
u_l (' = Number of functions contained in 'i^{th}' function module
α = Coefficient of scale economy of non platform units

These three criteria are combined to form a measure for the supply chain effectiveness. The objective function 'F' can then be written as $F = w_1 SCCost - w_2 PRF + w_3 COMP_{des}$, where w_1, w_2 and w_3 are the magnitude of the three weights corresponding to the supply chain cost (SC Cost), profit (PRF) and the design complexity ($COMP_{des}$) respectively. Coefficients of w_1 and w_3 are positive, while that of w_2 is negative, because the intention is to minimize SC Cost and $COMP_{des}$, whereas PRF needs to be maximized. The expressions for SC Cost, PRF and $COMP_{des}$ as discussed above are given as follows;

$$\begin{aligned} SC\ Cost = & \sum_{s}^{S}\left(\Phi_{s}^{exist} \cdot \delta_{s}\right) + \sum_{f}^{F}\Psi_{f}^{open}\chi_{f} \\ & + \sum_{f}^{F}\sum_{e \neq f}^{F \cup F^{C}} \Pi_{fe}^{path}\lambda_{fe} + \sum_{r}^{R}\sum_{f}^{F}\Psi_{rf}^{open}\varepsilon_{rf} \\ & + \sum_{s}^{S}\sum_{f}^{F}\sum_{t}^{T}\left[\Omega_{sf}^{req}N_{sft} + \Delta_{sf}^{req}S_{sft} + \sum_{e \neq f}^{F \cup F^{C}}\Delta_{sf}^{req}H_{sfet}\right] \end{aligned} \tag{4.1}$$

$$PRF = \sum_{t}^{T}\left(\sum_{f}^{F} N_{s_{ENDL}ft}\left[p_l - c_{\circ}IK\left\{1-[(1-\alpha)/\eta_{\max}]\sum_{f}^{F} N_{s_{ENDL}ft}\right\}\rho_l^c\right]\right.$$
$$+\sum_{f}^{F} N_{s_{ENDH}ft}\left[p_h - IK\left\{1-[(1-\alpha)/\eta_{\max}]\sum_{f}^{F} N_{s_{ENDH}ft}\right\}\rho_h^c\right]$$
$$\left.-g(\rho_h^b)-g(\rho_l^b)-(h-2g)(\rho_p^b)\right) \quad (4.2)$$

The total fixed cost to develop the product family is $g(\rho_h^b-\rho_p^b)+g(\rho_l^b-\rho_p^b)=g(\rho_h^b)-g(\rho_l^b)+(h-2g)(\rho_p^b)$ where, '$h(\rho)$' is a polynomial function and $g(\rho)$ is not necessarily a polynomial function. The profit term 'PRF' is comprised of the sum of contributions from the individual market segments product minus the fixed cost of developing the platform based product family. The decision variables involved in this term are the total number of end products manufactured for each of the market segments, the performance levels and the selling prices. The performance level (currency per valuation of a market segment) of a member of the product family is assumed to be determined from the supply chain cost incurred in manufacturing a single piece during a given time period, i.e.

$$COMP_{des} = -1/\theta_l\sum_{i}^{\eta}(d_i \ln d_i + (1-d_i)\ln(1-d_i))$$
$$+-\sum_{i}^{\theta_l}((u_l(i)/\eta)\ln((u_l(i)/\eta) \quad (4.3)$$

The objective function is combined so as to give a best trade-off between the supply chain cost, sales profit and the design complexity, in building the functional topography. The objective function does restrict its decision variables due to a number of factors such as, the limitation imposed on the resources, market targeting and customer valuation, and more product differentiation than the non-platform approach. It also follows the policies of GBOM and avoids the cost incurred by the excess or unsafe inventory. So, the constrained non-linear programming DFSC problem is given as follows.

$$\text{Minimize } OBJ = w_1SC\ \text{Cost-w}_2PRF + w_3COMP_{des} \quad (4.4)$$

This objective function is subjected to a number of constraints and the mathematical expression of these constraints can be found in the work of Yadav *et al.* (2011). The GBOM constraints express the conditions to draw a consensus on the implementation of GBOM concepts. A few other constraints stipulate that a net requirement associated with an item compels it to exist and specifies that the selection of an item (without external demand) requires either the existence of one of its parents as an AND node or the existence of a relevant link to an Ex-OR node. Some other constraints considered are defined for all the nodes; the existence of an AND node relies on the existence of each and every one of its children; and some are defined only for Ex-OR nodes. There is also a constraint considered here that shows the existence of a link to an Ex-OR node which forces the existence of the concerned child and parent. One of the constraints reveals that the existence of an Ex-OR node 'p' compels the selection of one and only one relevant link. Whereas, one constraint considered in this work ensures that a net requirement on an Ex-OR item 'p' induces a gross requirement on its child item. Finally, the last constraint considered implies that whenever a link p-s exists, the net quantity required of 'p' requires the child items.

Apart from the factors incorporated into the mathematical model, several other factors are also considered in the framework to improve the performance of the supply chain, though not necessarily in the optimization. These factors are process flexibility, product and market variety, different flow types in the supply chain, market diversity parameter, and different order of activities for designing the robust supply chain framework. The detailed explanation on these factors can be found in Yadav *et al.* (2011). The next section elaborates a small simulated problem.

4.2.2 Simulated Problem

As stated earlier in the introduction section in this research a problem of a wiring harness supplier of an AGV manufacturer has been tested to experimentally evaluate the propositions The wiring harness supplier connects all the electrical and electronic components of the AGV such as connectors, relays, fuses, and computer (Yadav *et al.*, 2011). In order to simplify the calculations, we make the assumption that three main functional requirements need to be considered, these are the driving and carrying function (F1), the supporting function (F2) and the monitoring and controlling functions (F3). The consideration of the modular function architecture becomes more vital in the case of a highly automated product like an AGV. To interlink the optimum number of function modules in a determination problem with the GBOM, we assume that the latter is a combinatorial problem, in which three combinations are possible viz. F1, F2 and F3 in a single module; two modules, one with F1 and F3 and other with F2; and the three modules each comprising of one functional requirement (Figure 4.2). This can be done by setting the correlation level at l=2 (with the number of modules at this correlative level = 2). In some of the other nodes in the GBOM, P-level denotes the performance level corresponding to a functional module and C-module denotes the corresponding component module.

The proposed model is executed to resolve this problem in which two types of AGV are manufactured, one targeted to the lower market segment (Market frag.1) and other to the higher segment (Market frag.2). This is done by limiting the number of versions of computers. The effective performance level for each of the lower segment AGV and the higher segment AGV are to be determined by making children choices for the Ex-OR nodes. The GBOM indicates that the higher segment can be served at any one of the two possible performance levels, while the lower segment can be served at any one of the five possible values of performance levels. Overall nine product families are possible, from which the one having the best trade-off between the supply chain cost, sales

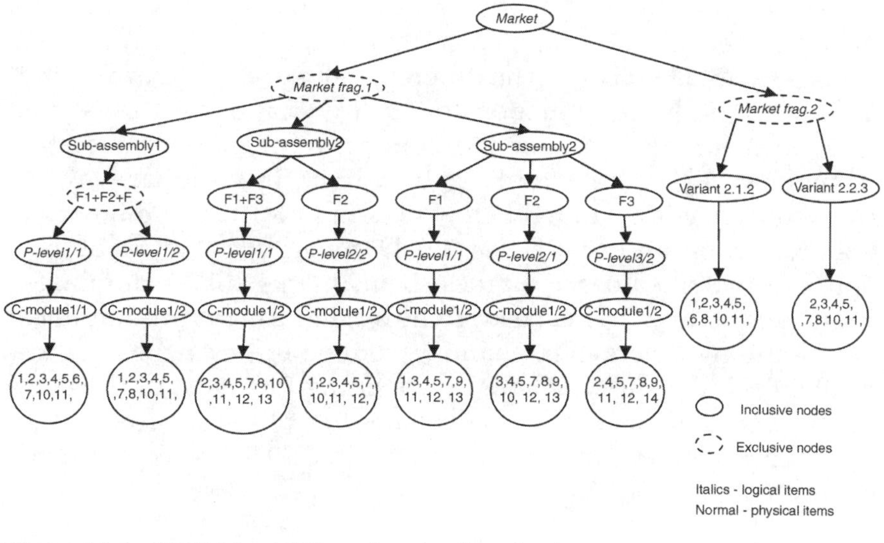

Figure 4.2 A GBOM for AGV product family selection.

profit and the design complexity (functional) is to be selected. The variants are materialized by assembling them from the basic components (from 4.1 to 4.10) and the computer components (4.11 to 4.15). Different combinations of these numbers in the GBOM indicate the bill of materials of the components of the AGV connected by the wiring harness supplier. These combinations are made to be located at the leaves of the GBOM. Different versions of computers are used for the lower and higher segments, though some are common along with other basic components. The common components include platform components manufactured on common platforms whereas the unique ones represent the non-platform components in the GBOM.

The simulated problem studied in this research consists of 7 manufacturing facilities (for basic components) and 2 synchronous facilities. Each facility is dedicated to a customer and the time period over which demand is made is taken to be 2. The mandatory data is simulated and the optimization is performed by executing the CPSO-outer algorithm. The problem studied also involves the planning of setting the locations for all the 7 manufacturing facilities to strengthen the product variety using the hypotheses proposed by Randall and Ulrich (2001). Afterwards all possible locations for opening sub-assembly, computer and synchronous facilities are marked and the scale efficiency for different combinations of facility location is evaluated. The sum of the distances between each of the different manufacturing facilities and the customer facilities are also calculated and, the mean of these sums is denoted by D_{mean}. The different values of D_{mean} corresponding to each of the combinations of facility locations are calculated and a benchmark is set for the dominance of market mediation by making D_{mean} bounded above by a limit. In a similar way, the scale efficiency is bounded below by a limit for the dominance of the production on product variety. Different possibilities of locations are evaluated based on these benchmarks, till the dominated strategy is escaped. This task can be accomplished through linear programming. Thereafter, chaining is done to impart the process flexibility to the supply chain by assuming that each of the items can be processed in at most three plants, to realize the process flexibility with a degree greater than or equal to 1. This is accomplished using equation 4.5.

$$\mu = \min \quad \{\omega(M)\} : L(M) < Q \tag{4.5}$$

Where, 'μ' is the degree of flexibility, '$w(M)$' denotes the excess capacity, 'Q' is the number of plants, and $L(M)$ is the set of plants that can process one or more product in the set M. After imparting the process flexibility, the production capacity of the AGV wiring harness supplier also changes. This task can be accomplished manually or through computer programming. Although, increasing the value leads to a higher investment cost, it also strengthens their ability to tackle the effects of the lost sales, in particular by attaining a flexibility measure greater than or equal to 1, which reduces the configuration losses drastically (Graves and Tomlin, 2003). Further, the final computation of the optimized value of the objective function (through the CPSO-outer algorithm discussed in the next section) and the corresponding decision variables, are exploited to realize the facilities, shipping paths, items selected and the net requirement of different items associated with each of the items at each facility or shipped from one facility to the other. Genuine flow types are utilized to emulate the shipping and the flow conservation of different items in the finally designed supply chain. The next section describes the Cellular Particle Swarm Optimization (CPSO-outer) algorithm adapted to resolve this problem in detail. A brief description of PSO is first provided and then the variant CPSO-outer has been explained.

4.2.3 Particle Swarm Algorithm (PSO)

Nature has plenty to offer and this has always fascinated researchers, making them explore the different activities carried out in nature that can assist in finding solutions to different problems. This inspiration and the understanding of nature has led to the development of a number of nature inspired algorithms also known as evolutionary algorithms such as Genetic Algorithm (GA), Ant Colony Optimization Algorithm (ACO), Bee Colony Optimization (BCO) and Artificial Immune System (AIS) based algorithms etc. to resolve complex optimization problems. Each of these algorithms is inspired by different activities followed in nature, for example; Ant Colony Algorithm emulates the real behaviour of ants and Bee Colony emulates the natural behaviour of bees while searching for the food around the hive. For large size design for supply chain problems, optimization techniques such as mixed integer linear programming might face oscillatory problems leading to a slower solution time. To overcome this problem, in this research a variant of the normal PSO developed by Shi *et al.* (2010) and known as the cellular particle swarm optimization (CPSO-outer) algorithm, has been applied.

Particle swarm optimization (PSO) is one of the biologically inspired evolutionary algorithms that have been derived from the idea of flocking of birds. Many examples could be extracted from the nature that demonstrates that social sharing of information among the individuals of a population may provide an evolutionary advantage. Kennedy and Eberhart (1995) first proposed PSO and since then it has achieved considerable attention in the global optimization areas. PSO originally intends to graphically mimic the elegant way in which swarms find their food sources and save themselves from predators (Eberhart and Kennedy 1995). It is a population-based stochastic optimization paradigm, in which each individual, termed as a *'particle'* from the population of the *'swarm'* changes their position with time and represents a potential solution. PSO in some ways resembles other existing Evolutionary Algorithms, such as Genetic Algorithm, but the difference lies in its definition in a social context rather than in a biological context. According to Eberhart and Shi (2001) PSO is based on simple concepts with the ease of implementation and computational efficacy.

The Particle Swarm Optimization (PSO) algorithm is motivated by the flocking of birds, and it works on the social behavioral interaction among the particles in the swarm. Taking inspiration from the cognitive behavior of the swarms this algorithm begins with the random initialization of a population of particles in the search space. These particles are assumed to be in the multidimensional space (D-dimensional) each having a position and velocity. The position and velocity of the particle demonstrates the particle's status in the multidimensional search space. Hence, in a PSO system particles move around in multiple directions in the search space, and the position of each particle is guided accordingly by the memory of their own best position, as well as of a neighbouring particle. One of the prominent features of flocking of the swarm is that the particles communicate their best positions to each other and accordingly they adjust their own position and velocity. Parsopoulos and Vrahatis (2002) proposed basically two main variants of the PSO algorithm:

- Global neighbourhood, where the best global position is communicated to all particles and updated immediately in the swarm

- Local neighbourhood, where each particle moves towards its best previous position and in the direction of the best particle in its restricted neighborhood.

In this research the global neighbourhood variant has been adapted because of the faster convergence. The position and velocity vectors of the ith particle can be represented as

$$x_i = (x_{i1}, x_{i2}, \ldots, x_{iD}) \tag{4.6}$$

$$p_i = (p_{i1}, p_{i2}, \ldots, p_{iD}) \tag{4.7}$$

The fittest particle among all the particles in the population is represented by

$$F = (f_1, f_2, f_D) \tag{4.8}$$

The velocity vector for the *i*th particle can be represented as

$$V_i = (V_{i1}, V_{i2}, \ldots V_{id}) \tag{4.9}$$

The updated velocity and position for the next fitness evaluation of each particle could be determined according to the following equations:

$$v_i^{t+1} = \omega v_i^t + c_1 R_1()\left(p_i^t - x_i^t\right) + c_2 R_2()\left(f_d^t - x_i^t\right) \tag{4.10}$$

$$x_i^{t+1} = x_i^t + v_i^{t+1} \tag{4.11}$$

Where t is the iteration number, $d = 1, 2 \ldots D$; $i = 1, 2 \ldots N$, and N is the size of the population (swarm). $c1$ and $c2$ are two positive values called acceleration constants, $R_1()$ and $R_2(\ )$ are two independent random numbers that uniformly distribute between 0 and 1 and are used to stochastically vary the relative pull of pi and f. The introduction of such random elements into the optimization is intended to simulate the slightly unpredictable component of natural swarm behavior. 'ω' is the inertial weight which helps to improve the performance of the particle swarm optimizer.

The equation (4.10) contains the three terms on the right hand side in which the inertial effect of the movement is represented by the first term. The memory of the individual and whole is referred by the second and third terms respectively. Basically, equation (4.10) is used to calculate the particle's new velocity which depends on its preceding velocity and the distances of its present position from both its own best past position and the group's best past position.

All the other particles follow the best available position and move closer to it, exploring the region more thoroughly in the process. According to Robinson and Rahmat-Samii (2004) equation (4.10) is the central constituent of the entire optimization. The stochastic tendency to return towards a particle's preceding best position is represented by the second term of the equation (4.10). The third term in equation (4.10) is referred to as the *social influence* term. The variable F keeps moving towards the best solution i.e. optimal solution found by the neighbor particles in the search space. A particle farther from the global best is more strongly pulled from its location, and moves more rapidly than a closer particle. The particle's velocity comes to a halt when it reaches the location of best fitness. This is the point from where the particles are pulled back in the opposite direction. The performances of the individual particles are evaluated by a predefined fitness function dependent on the problem during the evolution of the swarm. In the case of maximization of the fitness function Fitness *(xi)*, the individual best position of each particle *pi* and the global best position *f* are updated after each iteration using the following two equations, respectively:

$$p_i^{t+1} = \begin{cases} x_i^{t+1} : Fit\ (x_i^{t+1}) > Fit\ (p_i^{t+1}) \\ p_i^t : Fit\,(x_i^{t+1}) \leq Fit\,(p_i^{t+1}) \end{cases} \tag{4.12}$$

$$F^{t+1} = \arg\max Fit\,(p_i^{t+1}) \tag{4.13}$$

where 'Fit' refers to the fitness value for the respective iteration.

The pseudo code for the normal PSO is shown in Figure 4.3.

Although, the particle swarm optimization has a strong ability to search for the global optimal solution, it often struggles to escape from the local optima and it is important to prevent the premature convergence, especially when the complexity of the problem increases. Under such circumstances the performance of the normal PSO for finding global optimum sharply decreases. To overcome this problem of the normal PSO, Shi *et al.* (2010) proposed a new variant of the PSO known as Cellular Particle Swarm Optimization algorithm that inherits the qualities of the Cellular Automata (CA). CA is a discrete dynamical system model capable of producing complex phenomena (Schiff, 2007). The CA model generally simulates the macro-behavior by micro-dynamics, which are produced by the interaction of a population of individual cells that are connected in particular neighborhood structures. The next section elaborates the Cellular Particle Swarm Optimization (CPSO) algorithm in detail.

```
// Initialization
for i=1 to the swarm size do
        Randomly initialize X_id within the search range of (X_min, X_max);
        Randomly initialize V_i within the velocity range of (V_min, V_max);
P_i = X_i;
end for
Calculate the fitness value of each particle;
Indentify the best position P_best;
// Loop
While (stop criterion is not satisfied and k< max no. of iteration times) do
    for i=1 to the swarm size do
            v_i^{t+1} = ωv_i^t + C_1R_1 () (p_i^t − x_i^t) + C_2R_2 () (f_d^t − x_i^t)
            x_i^{t+1} = x_i^t + v_i^{t+1}
            p_i^{t+1} = p_i^t
            f_d^{t+1} = f_d^t
            Evaluate Fitness
            if fitness p_i^{t+1} < fitness (x_i^{t+1}) then
              Update f_d^{t+1};
            end if
            if fitness f_d^{t+1} < fitness (x_i^{t+1}) then
              Update f_d^{t+1};
          end if
    end for
end while
```

Figure 4.3 Pseudo Code for normal PSO.

4.2.4 Cellular Particle Swarm Optimization (CPSO) Algorithm

Nature inspired optimization algorithms have always fascinated researchers trying to resolve complex problems. However, a number of factors affect the performance of population based algorithms. The most common of these factors include the communication structure and the information inheriting and diffusion mechanisms, since both of these have a capability to further influence the co-operation of the population and self adaption of the individual (Shi *et al.*, 2010). Cellular automata have been rigorously used by researchers from different fields such as physics, biology and computer science. Shi *et al.* (2010) point out that the popularity of research on CA and its application is due to two prime reasons: firstly, the simplicity of the implementation of CA; and secondly because the research and application of CA does not necessarily call for rigorous mathematical reasoning. In CA homogeneously interconnected deterministic finite automata (cells) work synchronously at discrete time steps and each cell interacts locally with its neighbors. The collection of such cells then assembles to present enormous power to solve problems. This emerging property of CA can be utilized to model complex systems and hence was employed by Shi *et al.* (2010) in PSO.

CA and PSO come from different fields, however they have several similarities. Both the mechanisms are population based, although in PSO they are termed as particles and in CA they are known as cells. Secondly, in both the mechanisms every individual is either a cell (as in CA) or particle (as in PSO) and they interact with each other to transmit certain intrinsic information. The cells in CA, communicate with their neighbours and changes in a cell's state take into account the cell's current state and its neighbours' states. In PSO, each particle communicates with other particles and updates intrinsic information such as velocity, position, fitness, personal best position, global best position in PSO. Thirdly, in CA, a transition rule is used to govern the evolution of cells, while in PSO the two equations about velocity and position are used to update particles. And finally, both the CA and PSO run in discrete time steps (Shi *et al.*, 2010). These similarities between the two provide the possibility of integrating them together to form as a more powerful and efficient mechanism that can be used to resolve complex problems.

Shi *et al.* (2010) employed this idea and proposed a hybrid PSO known as the Cellular Particle Swarm Optimization (CPSO) algorithm. In order to enhance the performance of the interaction in the PSO, they employed the idea of CA in exploring the communication structure and information inheriting, and diffusing mechanisms of the swarm system of PSO. In their proposed CPSO, individuals can only exchange information within their neighbourhood. This helps to explore the search space due to the slow information diffusion through the population, promoting the preservation of diversity, and exploiting every cell's local information inside the neighbourhood. Shi *et al.* (2010) designed a mechanism sophisticatedly by integrating adapted CA and PSO so that concentration can be attained. The CA model for PSO is defined as follows;

a. cell: selected candidate solution (assuming there are U cells);
b. cell space: the set of all cells;

c. cell state: the particle's information of such respects as Pi, f_d and Xi at time k, denoted by $S_i^t (i = 1, 2, 3, ...U)$, and $S_i^t = [p_i^t, f_d^t, V_i^t, X_i^t]$;
d. Neighbourhood: a kind of topology based on lattice structure containing a set of particles, defined by $N(i) = \{i + \delta_1, i + \delta_2, ...i + \delta_m ....i + \delta_l\}$, where l is the size of the set,
m = 1, 2,. . . , l;
e. Transition rule:
$$S_i^{t+1} = f(S_i^t \cup S_{N(i)}^t) = f(S_i^t, S_{i+\delta_1}^t, S_{i+\delta_2}^t, ...S_{i+\delta_l}^t)$$
f. Discrete time step: iterations in PSO.

Under this framework Shi *et al.* (2010) proposed two versions of CPSO known as CPSO-inner and CPSO-outer. This research will employ the CPSO-outer to resolve the product development problem because of its superiority over the CPSO-inner (Shi *et al.*, 2010). The CPSO-outer algorithm is explained in the next section.

4.2.5 CPSO-outer Algorithm

Shi *et al.* (2010) proposed the CPSO-outer algorithm by extending a generalized-CA strategy to guide particles in order to enhance their optimization capability. Firstly, they assumed that the whole search space is considered as the cell space, and defined every potential candidate solution in the search space as a cell on the basis of the above assumption. To better understand the interaction behaviour of cells, they also considered that the cell space is cut by infinite virtual grids, and the cell space could not be subdivided any more. Every grid contains only one solution. In order to differentiate particles and potential candidate solutions, they defined a "smart-cell" as a cell that undergoes search behaviour, i.e. a "smart-cell" represents a "particle" of the Particle Swarm Optimization method. Then, a cell that is not "smart", is one that represents a candidate solution that is not sampled by any of the particles in the search space. Every smart-cell could construct its neighbourhood smartly by a neighbourhood function. The force of neighbours may draw the particle out of its previous trajectory. The ith particle's position x_i^t is defined as the cell state of this particle i.e. $S_i^t = x_i^t$. When implementing CPSO-outer, only smart-cells participate in the updating process. A smart-cell constructs its neighbourhood by a neighbourhood function given in Eq. (4.14).

$$N(i)=\begin{cases} X_i^t+\dfrac{fitness(f_d^t)}{fitness(x_i^t)}R_3{}^\circ V_i^t & fitness(X_i^t)\neq 0, fitness(f_d^t)\geq 0 \\ X_i^t+\left|\dfrac{fitness(x_i^t)}{fitness(f_d^t)}\right|R_3{}^\circ V_i^t & fitness(X_i^t)\neq 0, fitness(f_d^t)<0 \\ X_i^t+\left(\dfrac{e^{fitness(f_d^t)}}{e^{fitness(x_i^t)}}\right)^2 R_3{}^\circ V_i^t & fitness(X_i^t)=0, fitness(f_d^t)\geq 0 \\ X_i^t+\left(\dfrac{e^{fitness(f_d^t)}}{e^{fitness(x_i^t)}}\right)^2 R_3{}^\circ V_i^t & fitness(X_i^t)=0, fitness(f_d^t)<0 \end{cases} \tag{4.14}$$

Where R_3 is a $1\times d$ matrix composed by d uniform random numbers in [-1, 1], and "º" is the operation symbol of Hadamard product (Havil, 2003) which is named after French mathematician Jacques Hadamard and also known as the entry-wise product. R_3 as a direction coefficient is used to adjust the direction and distance of the neighbourhood. For the jth dimension (j = 1, 2,. . . ,d) of x_i^t, N(i) generates random points within

$$radius_j=\left\|\frac{fitness(f_d^t)}{fitness(X_i^t)}R_{3j}V_{ij}^t\right\|,\left\|\frac{fitness(X_i^t)}{fitness(f_d^t)}\right\|R_{3j}V_{ij}^t \ or \left\|\left(\frac{e^{fitness(f_d^t)}}{e^{fitness(x_i^t)}}\right)^2 R_{3j}V_{ij}^t\right\|$$

away from X_{ij}^t. Figure 4.4 shows an intuitive description for the neighbourhoods of the jth dimension and the kth dimension. l neighbours could be obtained by l different $R_3^{\delta m}$, namely, m = 1, 2, . . . , l. Particles in CPSO-outer can select neighbours independently and dynamically to communicate with according to the evaluation of its current position and global best position, and then conduct a local search around every smart-cell. Eq. (14) could be simplified as $N(i)=X_i^t+\xi_i^t+{}^\circ V_i^t$. For the jth dimension (j = 1, 2, . . ., d) of ξ_i^t, the range of ξ_{ij}^t could be small at early iterations when the difference of the fitness value of particle with that of the global best is relatively large. Then, along with the particles gradually converging to the global best, that difference reduces and leads to a relatively larger range of ξ_{ij}^t, and finally, when all particles converge to the equilibrium point, ξ_{ij}^t would be a uniform random number in [−1, 1]. The transition rule of CA in CPSO-outer is described as follows:

$$f(\phi) = \min\,(fitness(N(i)), fitness(N(i+\delta_1)), \ldots, fitness\,(N(i+\delta_m)), \ldots, fitness(N(i+\delta_m))) \tag{4.15}$$

$$\text{Where } \phi = \begin{cases} i & if \;\; f(\phi) = fitness(N(i)), \\ i+\delta_m & if \;\; f(\phi) = fitness(N(i+\delta_m)), \end{cases} \tag{4.16}$$

$$S_i^{t+1} = S_\phi^t \tag{4.17}$$

The transition rule gives particles the power to make a wise jump, and also helps in exploring the search space in a local competition neighbourhood and enhances the diversity of the swarm. Hence, the CPSO-outer algorithm has greater potential to search for a global optimum in the search space. The pseudo code of the algorithm is shown in Figure 4.5.

4.3 Computational Analysis and Result

After resolving the objective function using the CPSO-outer algorithm the value for each solution was converted into a fitness value. The near optimal/optimal fitness value after converging through the CPSO in 827 generations was found to be 164.5 whereas using

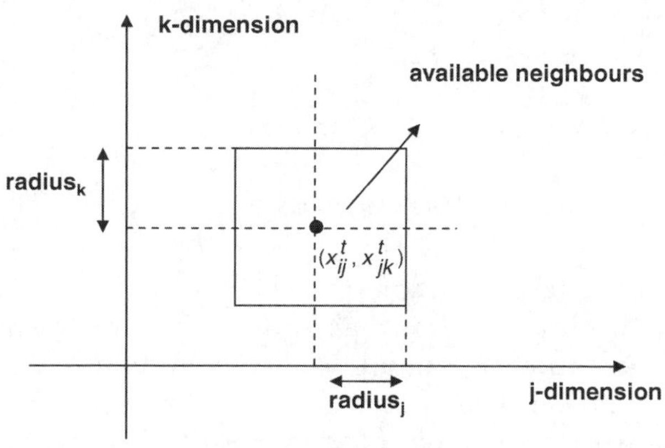

Figure 4.4 Neighbourhood for j-dimension and k-dimension of x_i^t [Shi *et al.*, 2010].

// Initialization
Randomly generate a number corresponding to each bit
Check whether it satisfies all the problem constraints
If it does not go to step 2
for i=1 to the swarm size do
    Randomly initialize X_{id} within the search range of (X_{min}, X_{max});
    Randomly initialize V_i within the velocity range of (V_{min}, V_{max});
$P_i = X_i$;
end for
Calculate the fitness value of each particle;
Indentify the best position f_d;
let initialize position be each particle's state;
// Loop
While (stop criterion is not satisfied and k< max no. of iteration times) do
  for i=1 to the swarm size do
    $v_i^{k+1} = \omega v_i^k + c_1 R_1 () \left(p_i^k - x_i^k\right) + c_2 R_2 () \left(f_d^k - x_i^k\right)$
    $S_i^{k+1} = S_i^k + v_i^{k+1}$
    $p_i^{k+1} = p_i^k$
    $f_d^{k+1} = f_d^k$
    Evaluate Fitness $\left(S_i^{k+1}\right)$
    Generate *l* neighbours using Eq. (14);
    Evaluate l neighbours;
   for a = 1 to the *l* do
     if fitness S_i^{k+1} < *fitness* ($S_{i+\delta_a}^{k+1}$) *then*
       Update S_i^{k+1};
     end if
   end for
     if fitness p_i^{k+1} < *fitness* (S_i^{k+1}) *then*
       Update p_i^{k+1};
     end if
     if fitness f_d^{k+1} < *fitness* (p_i^{k+1}) *then*
       Update f_d^{k+1};
     end if
   end for
end while

Figure 4.5 Pseudo Code for CPSO-Outer.

GA it was after 1000 generations (Yadav *et al.*, 2011) and by standard PSO the near optimal solution was found only after 922 generations. The final near optimal/optimal product family of AGV selected is shown in Figure 4.6. The optimum number of modules was found to be two, consisting of a module comprised of the driving and carrying function (F1), and the monitoring and controlling functions (F3), whilst the other one was comprised of the supporting function

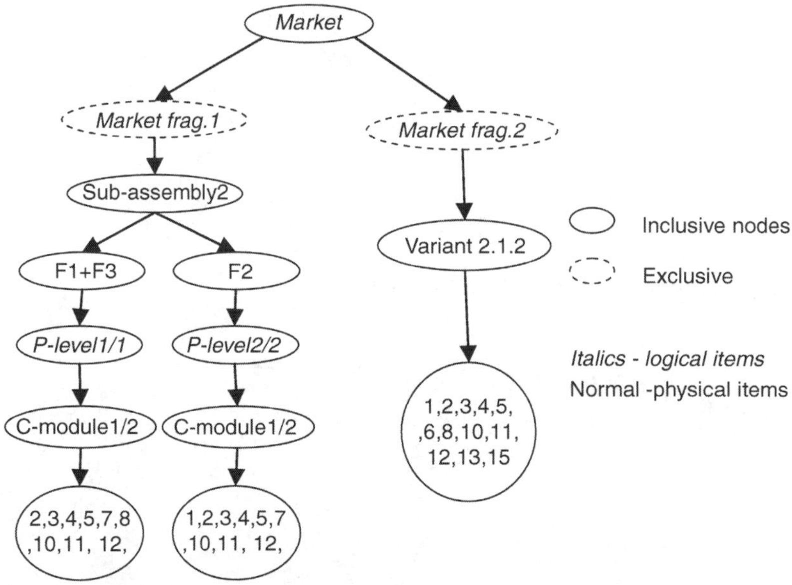

Figure 4.6 A Product Family Tree for the AGV.

(F2). The market diversity parameter was found to be 0.24, i.e. the optimal quantity of the products manufactured has low relative amount and the higher end product per unit difference valuation of the two market segments. For the supply chain, the binary decision variables corresponding to the existing facilities, shipping channels, etc. were obtained after optimization was exploited to design the supply chain. The designed supply chain is shown is Figure 4.7. The comparative convergence trend of the proposed algorithm with PSO is shown in Figure 4.8.

A computational analysis of the proposed DFSC problem was also carried out to address the product development issue, by studying the variation in the market diversity, attained by varying the coefficient of the scale of economy of non-platform units. The outcome showed that the value of the market diversity parameter increased when the value of the coefficient of scale of economy for non-platform units of AGV was increased. However, at very large non-platform economies of scale market, diversity suffered some falls indicating the lesser suitability of the platform based product development in ensuring the appropriate market diversity. In such situations the near optimal/optimal fitness value obtained decreases, resulting in an undesirable increase in the supply chain

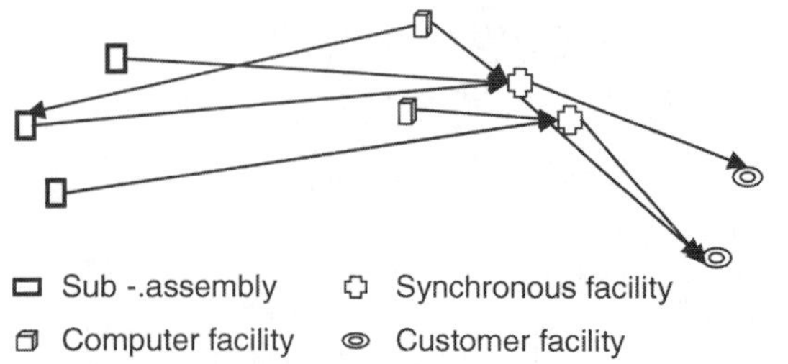

Figure 4.7 Designed Supply Chain for the best solution.

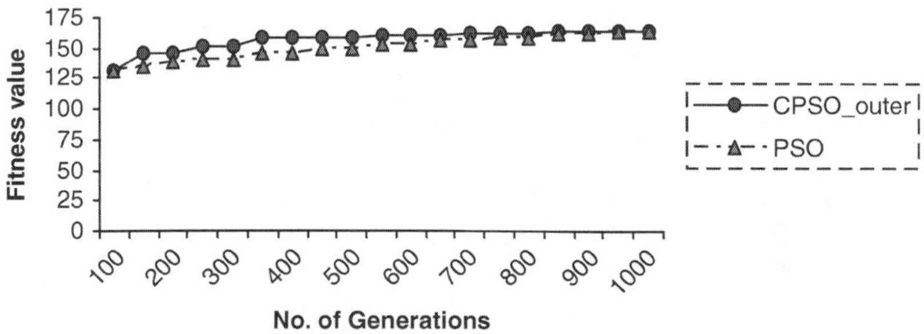

Figure 4.8 Comparative convergence trend of the proposed algorithm with PSO.

cost and decrease in the profit. The values of the market diversity for the three different values of the lower and higher AGV consumer segment valuation were also computed and the variation of the market diversity parameter was observed. Consequently, the most competent value of the market diversity was observed at the intermediate difference between the valuations of two consumer segments and intermediate scale of economies. However, promising results were obtained in the case of intermediate values for differences even at high non platform economies of scale, as compared to the results obtained for very small or large valuation difference at large economies of scale.

4.4 Conclusions

The present study aims to address the design for supply chain problem considering the product development issues. The design framework for supply chain was constructed considering several factors such as process flexibility, production planning and marketing dominance, and product variety. The proposed framework was aimed at minimizing the supply chain costs, the degree of design complexity and maximizing the sales profits. This research also proposes a platform based product development strategy to reduce the product development cost by using mixed market-modular strategies. The research employs a new variant of the PSO proposed by Shi *et al.* (2010) known as the CPSO-outer to resolve this complex problem. The efficacy of the algorithm was showed by comparing the results with GA and simple PSO. The results indicate that CPSO-outer outperformed the GA and PSO algorithms in terms of faster convergence. Thus, there is scope to use this algorithm to resolve complex manufacturing and supply chain problems.

Future research may involve testing the CPSO-outer algorithm on a number of other complex problems in product family design and manufacturing supply chain problems. The model can also be further extended to include time and cost functions. The model can also include the effect of financial costs (duties, taxes, exchange rates). The problem specifically related to product family design could also be addressed. The information flow plays a significant role in the success of a supply chain and in product development. Hence, the future work may also involve designing a multi-agent supply chain framework for better information exchange.

References

Blackhurst, J., Wu, T., and O'Grady, P., 2004, Network-based approach to modelling uncertainty in a supply chain, *International Journal of Production Research*, 42 (8), pp. 1639–1658.

Braker, S. P. W., 1996, Restructuring the bill of material for productivity: a strategic evaluation of product configuration, *International Journal of Production Economics*, 45 (1–3), pp. 251–260.

Eberhart, R. C. and Shi, Y., 2001, Particle swarm optimization: developments, applications and resources, Proc. congress on evolutionary computation 2001 IEEE service center, Piscataway, NJ. Seoul, Korea, 2001a.

Eberhart, R. C., and Kennedy, J., 1995, A New Optimizer Using Particles Swarm Theory, Proc. Sixth International Symposium on Micro Machine and Human Science (Nagoya, Japan), IEEE Service Center, Piscataway, NJ, pp. 39–43.

Graves, S. C. and Tomlin, B. T., 2003, Process Flexibility in Supply Chains, *Management Science*, 49 (7), pp. 907–919.

Havil, J., 2003, Gamma: Exploring Euler's Constant. Princeton, NJ: Princeton University Press.

Jain, V., Tiwari, M. K, and Chan F. T. S., 2004, Evaluation of Supplier performance using a fuzzy based approach, *Journal Manufacturing Technology Management*, 8, pp. 735–744.

Jiao, J., Tseng, M. M., Ma, Q., and Zou, Y., 2000, Generic bill-of-materials-and-operations for high-variety production management, *Concurrent Engineering*, 8 (4), pp. 297–321.

Krishnan, V., and Gupta, S., 2001, Appropriateness and impact of platform based product development, *Management Science*, 47 (1), pp. 52–68.

Lamothe J., Hadj-Hamou, K., and Aldanondo, M., 2006, An optimization model for selecting a product family and designing its supply chain, *European Journal of Operational Research*, 169 (3), pp. 1030–1047.

Lamothe, J., Hadj-Hamou, K. and Aldanondo, M., 2005, Product Family and Supply Chain Design: An integration of both points of view, Supply Chain Optimisation, *Applied Optimization*, 94 (II), pp. 175–190.

Liu. J., Ding, F.-Y., and Lall, V. 2000, Using data envelopment analysis to compare suppliers for supplier selection and performance improvement, *Supply Chain Management*, 5 (3/4), pp. 143–150.

Oden, H., Langenwalter, G., and Lucier, R., 1993, *Handbook of material and capacity requirements planning*. McGraw-Hill, London.

Pahl, G., and Beitz, W., 1988, *Engineering design: a systematic approach*, Springer, London.

Parsopoulos, K. E. and Vrahatis, M. N., 2002, Recent approaches to global optimization problems through Particle Swarm Optimization, *Natural Computing*, 1 (2–3), pp. 235–306.

Randall, T. and Ulrich, K., 2001, Product variety, supply chain structure, and firm performance: analysis of the U.S. bicycle industry, *Management Science*, 47 (12), pp. 1588–1604.

Robinson, J. and Rahmat-Samii, Y., 2004, Particle swarm optimization in electromagnetics, *IEEE Transactions on Antennas and Propagation*, 52 (2), pp. 397–407.

Sanchez, R. and Mahoney, J. T., 1996, Modularity, flexibility, and knowledge management in product and organization design, *Strategic Management Journal*, 17 (10), pp. 63–67.

Schiff, J. L., 2007, *Cellular Automata: A Discrete View of the World*, Wiley & Sons.

Shi, Y., Liu, H. Gao, L. and Zhang, G., 2010, Cellular particle swarm optimization, *Information Science*, doi:10.1016/j.ins.2010.05.025 (in press).

Stewart, G., 1995, Supply chain performance benchmarking study reveals keys to supply chain excellence, *Logistics Information Management*, 8 (2), pp. 38–443.

Twigg, D., 1998, Managing product development within a design, *International Journal of Operations & Production Management, 18* (5), pp. 508–524.

Yadav, S. R., Mishra, N., Kumar, V. and Twari, M. K., 2011, A Framework for Designing Robust Supply Chains considering Product Development Issues, *International Journal of Production Research* (http://www.informaworld.com/10.1080/00207543.2010.524258).

5

Genetic Algorithms with Chromosome Differentiation (GACD) Based Approach for Process Plan Selection Problems

Nishikant Mishra[1] **and Vikas Kumar**[2]

[1]*School of Management and Business, Aberystwyth University, Aberystwyth, UK*
[2]*Department of Management, Dublin City University Business School, Dublin, Republic of Ireland*

Abstract
The changing global scenario and increased competitiveness has enforced manufacturers to optimize their processes to sustain their position in the market. Manufacturing firms rely on an effective process plan to efficiently utilize their manufacturing resources. Therefore, process plan selection is a crucial and often very complex task. The complexity further increases when there are alternative machines, setups, and processes during manufacturing operations. This research discusses process plan selection problems using Genetic Algorithms with Chromosome Differentiation (GACD) optimization technique. Comparative results on a case study as well as on randomly generated datasets of increasing complexity confirm that the proposed algorithm achieves an improved balance between exploration and exploitation, and has a better ability to escape from the local minima than other efficient meta-heuristic approaches.

Keywords: Process plan selection problem, genetic algorithm, chromosome differentiation

5.1 Introduction

Nowadays the manufacturing sector is flooded with new technologies that have changed the production processes and capabilities of

Manoj Tiwari and Jenny A. Harding (eds.) Evolutionary Computing in Advanced Manufacturing, (77–94) © Scrivener Publishing LLC

firms to produce differentiated products using single multi-function machines. The new machines are flexible, capable of handling multiple tools and performing multiple tasks and operations, and fixtures are competent to hold different types of parts. This has on one hand made manufacturing processes simpler and on the other hand it has increased their complexity. Since different processes can now be handled by multiple available options, this has increased the complexity of selecting the appropriate options for different processes. Therefore, manufacturing firms very often struggle to find the best suited process plan while manufacturing a part or group of parts together to achieve cost advantage as well as to save the total production time. Manufacturing of different products or part type needs the completion of the several design tasks prior to the commencement of the manufacturing. Moreover, the geometrical and tolerance restrictions among several features of the part type poses significant challenges while arranging setups and performing machining operations. To simplify these processes one needs to have different processing routes consisting of several alternative machines, tools and fixtures. The process plan selection problem deals with these kinds of issues and aims to provide a path so that different machining and other key operations can be performed efficiently. There are other factors such as production volume and loading capacities of the manufacturing system that also affect the selection of a process plan.

The main aim of the process plan selection problem is to select the best alternative from the available options so that the product can be manufactured from raw material to the finished product efficiently. Generally, different parts of a product are manufactured first and then assembled together to form the final product. The assembly parts of the products are manufactured independently and during the manufacturing more than one machine, fixture and tools etc. are used. Different operations performed on different parts may be executed on the same set of machines and the same or group of fixtures can be used to hold the different parts. Parts are usually manufactured in different batches and the size of these batches varies. The manufacturing of these batches are not mixed i.e. the manufacturing of the next batch starts only after the manufacturing of the previous batch is over. Process plan selection is a NP hard problem and different researchers have attempted to resolve the problem using knowledge based systems, heuristic and meta-heuristic, and agent based systems.

Tiwari and Vidyarthi (1998) considered the similarity measures among the process plans of the parts and developed four different kinds of algorithm to resolve the process plan selection problem in an automated manufacturing system. Kusiak and Finke (1988) attempted to resolve the process plan selection problem using a new heuristic model. The main objective of their work was to minimize the cost and the number of tools and fixtures used in manufacturing the product. In order to achieve this objective in minimum computational time they used a graph theoretical formulation method coupled with integer programming. Bhaskaran (1990) aimed at easing the planning and scheduling and further reducing the manufacturing cost by formulating an intransigent cost model to minimize the number of steps, total time and dissimilarity between the two process plans. In the real world, many aspects of manufacturing process related information can be imprecise or vague, such as machining time and tool changing time. In order to address these kinds of problems Zhang and Hunag (1994) used a fuzzy based approach. In their approach, all the alternative process plans for each part type were generated first and then were combined together. In the process plan selection problem, other objective functions may concern minimizing the material handling time, total processing time, transportation time and material handling cost etc. and these objectives have been highlighted by many researchers such as Seo and Egbelu (1996) and Rebaine and Strusevich (1999).

In order to better utilize the resources efficiently using the different available process plans in circumstances when there is simultaneous formation of part families and machine groups, Rajamani *et al.* (1990) used the Integer programming model. Brandimarte and Calderini (1995) proposed a two-level hierarchical Tabu search algorithm to resolve the joint process plan selection problem. Jain and Elmaraghy (1997) applied a genetic algorithm model to the single process plan scheduling problem and showed the efficacy of GA in finding the optimal/near optimal solutions. Ming and Mak (2000) proposed a hybrid Hopfield Neural Network and Genetic Algorithm based approach to select the optimal process plan. Chang and Chen (2002) used a dynamic programming based process planning selection strategy in a computer integrated manufacturing (CIM) environment simultaneously considering the use of workstations to efficiently resolve the stage-type network problem. A pragmatic approach for optimal process and partner selection in a virtual enterprise was proposed by Feng and Yamashiro (2003). Using this approach they were able to efficiently reduce the manufacturing cost, improve the product quality, and shorten the lead-time. The integrated process planning and selection problem has also been discussed by several researchers such as Tonshoff *et al.* (1989), Khoshnevis and Chen (1990) and Zhang and Mullar (1994). The next section describes the problem formulation part of this research.

5.2 Problem Formulation

In a manufacturing scenario whenever an order arrives, the manufacturing unit makes a number of decisions, such as decisions related to the requirements of the raw materials and requirements of the different part types for the assembly purpose. The manufacturing unit also generates a CAD/CAM diagram of the different part types. Once the diagram is ready, the next task is to define the different operations that need to be performed on the part types. According to the materials and the operations required, different machines are then selected for further machining operations. Thereafter, the tool and fixture allocation for different operations are defined. This leads to the generation of alternative process plans for carrying out these operations.

As discussed earlier in this chapter, the process plan selection problem deals with the selection of the best alternative from the available options so that the product can be manufactured efficiently and this involves the selection of machines, cutting tools, fixtures, setups etc. The process plan problem selects one process plan for each part type from the available alternative plans. Figure 5.1 shows a process plan selection problem adopted from Dashora *et al.* (2006) where it is clearly evident that there are a number of alternatives available for a particular part type. For example, part type 2 has three different process plan alternatives. However, the increase in the part types and their alternatives significantly enhances the computational complexity and thus the problem can be resolved by using different available optimization methods. In this paper this complexity has been resolved using a hybrid GACD algorithm.

The process plan selection problem considered in this chapter is aimed at minimizing the cost taking into account the cost associated

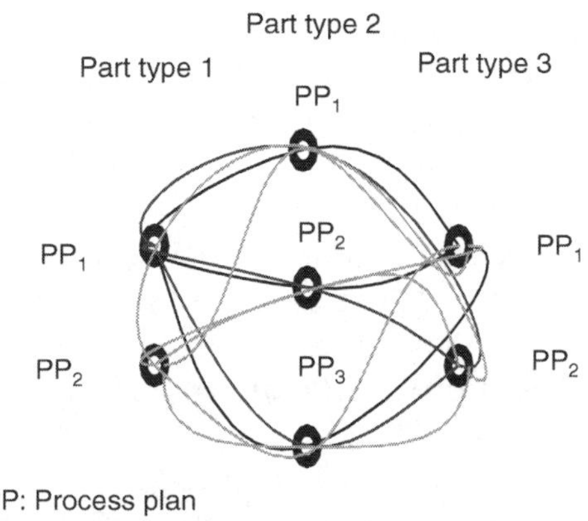

Figure 5.1 A simple PPS problem.

with processing time and cost of each processing step. Hence, the objective function can be represented as (Dashora *et al.* 2006):

$$Cost = \alpha \times \sum_{i \in Tabu_1} (Ct_{pt} \times PT_i + CS_s \times SP_i) + \beta \times \sum_{i,j \in_1 edge\ in\ Tabu_1} DIST\ [i][j] \tag{5.1}$$

Where, α, β are the constants

Ct_{pt} = Cost associated with processing time

PT_i = Processing time of plan 'i'

CS_s = Set-up cost

SP_i = Set of processing steps

5.3 Genetic Algorithm with *Chromosome Differentiation*

5.3.1 Overview of GA

GAs are probabilistic search and optimization techniques that simulate the process of evolution by taking a collection of chromosomes called a population of solutions and thereafter applying various biologically inspired genetic operators like selection, crossover, and mutation to achieve an optimal or near optimal solution (Goldberg *et al.* (1989), Holland (1975)). Each member of the population is evaluated according to some fitness measures, and the fitter solutions are used for reproduction. New offspring of individuals are generated and poor solutions are replaced. The cycle of evaluation, selection, and reproduction is continued until a stopping criterion related to search time or quality of current best solution is met (Meloni *et al.* (2003) and Chen *et al.* (2001)). From the research contributions in the broad area of GA's applications, it is found that in most applications, the number of generations required for a conventional GA to achieve optimal/near optimal solutions is substantially high. Therefore, especially on large-size complex combinatorial optimization problems, such as those considered here, it is not only desirable but rather indispensable to seek for some evolution of the GA so as to make it more efficient (reduce search time) and reliable (reach hi-quality solutions in a repeatable way). Keeping in mind these primary requirements, we have devised the GACD to introduce and maintain a high level of population diversity without excessively slowing the convergence process. This in turn results in faster information interchange among chromosomes, and therefore ensures a better ability to escape local optima.

5.3.2 Genetic Algorithm Incorporating *Chromosome Differentiation*

In the proposed algorithm, on the basis of sexual differentiation genotypes (chromosomes) are divided into two classes i.e. *Male* (M) and *Female* (F), thereby giving rise to two populations, a Male Population (MP) and a Female Population (FP). In addition, these populations are made dissimilar artificially, and the two populations are generated in a way that maximizes the Hamming Distance between individuals in the two classes. Crossover is only allowed between individuals from distinct populations, and this introduces a greater degree of diversity and simultaneously leads to greater exploration in the search space. Selection is applied over the entire population so as to exploit the information gained so far.

5.3.3 Description of GA with Chromosome Differentiation

The detailed descriptions of the various parameters of GACD are discussed in the next subsections.

Population Initialization: The construction of a chromosome in GACD is delineated in Figure 5.2.

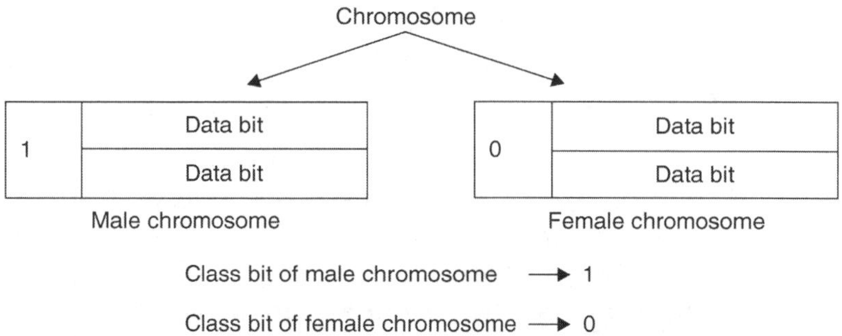

Figure 5.2 Framework of chromosome in GA having chromosome differentiation.

The data bits are represented as '∂' individual bits here and are used to encode the parameters of the problem. In addition, the first bits of a chromosome are called class bits, as these are used to indicate whether the class of the chromosome is M or F. Two separate populations, one consisting of M chromosomes (M population) and other containing the F chromosomes (F population) are maintained over each generation. The total population (TP) = MP + FP, and initially MP = FP = TP/2, but the sizes of these two populations, MP and FP, may vary in different generations. The MP is initialized first; the data bits for each M chromosome are assigned 1. The FP is then initialized, and data bits are generated for F chromosome, by maximizing the Hamming Distance between these two classes of populations. The class bits of each F chromosome are set to 0. In this research, the chromosomes must contain both sequencing (the order in which operations are considered) and assignment (operation-to-machine) information. Such information can be easily organized in a two-dimensional array of bits, which can be immediately interpreted as a bi-dimensional chromosome, whose first row determines operation priorities, and second one specifies the machines where subsequent operations are to be performed. For a generic pair of chromosomes $Y_1, Y_2 \in$ INI_POP, where INI_POP is the initial population, the *Hamming Distance* HD(Y_1, Y_2) represents a straightforward metric to measure the overall difference between the two chromosomes. In our case, the Hamming Distance (HD) between the Male Population (MP) and Female Population (FP) is denoted as:

$$HD(mf, fp) = \sum_i \sum_j HD(Y_{i,} Y_j) \quad \forall Y_i \in MP \text{ and } Y_j \in FP \qquad (5.2)$$

Here, HD(y_1, y_2) is calculated as:

$$HD\left(Y_i, Y_j\right) = \sum_{k=1}^{\partial} \left| Z_{i,j} - Z_{j,k} \right| + \sum_{k=1}^{\partial} \left| MT_{i,k} - MT_{j,k} \right| \tag{5.3}$$

Where, $Z_{i,k}$ is i^{th} bit of kth chromosome and $MT_{i,k}$ is process time of ith part of kth chromosome.

Consequently, the Average Hamming Distance (AHD) of a population size n is defined as (Bandyopadhyay and Pal (1998)):

$$AHD(mp, fp) = \frac{\sum_{i=1}^{n/2} \sum_{j=1}^{n/2} HD\left(Y_i, Y_j\right)}{{}^{n}C_2} \tag{5.4}$$

To generate the female population corresponding to each male chromosome, two different criteria (complementary functions) for the first and second rows are used. The methodology adopted to generate the female population from the male population is illustrated in Figure 5.3.

Fitness evaluation:- Only the '∂' data bits of each chromosome are used for the fitness computation. Class bits do not contribute to the calculation of the fitness function, and therefore do not affect the solution.

Selection:- Next generation members are selected from all chromosomes in the enlarged sampling space, based on their fitness function. The best chromosomes are therefore preserved, and the characteristics of elitism are therefore incorporated in both Male and Female populations.

Crossover:- Crossover is applied with the probability C_p between the participating *M* and *F* chromosomes. Each parent contributes one class bit to the child. In the present context, various types of crossover operators are applied on the data bits. Crossover is applied until the following conditions are satisfied.

1. There are no traces of chromosomes remaining in the mating pool.
2. There are either *M* or *F* chromosomes remaining in the mating pool.

In case of condition (1), the crossover procedure terminates, while for condition (2) the remaining *M* or *F* chromosomes are

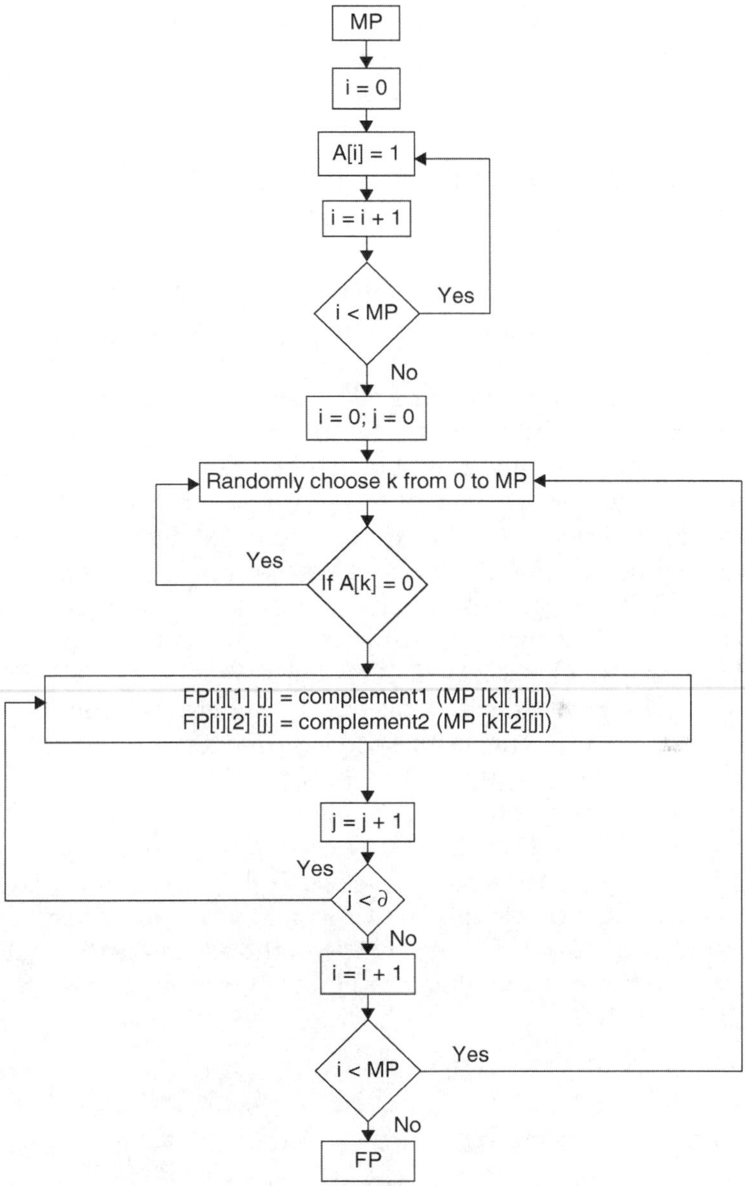

Figure 5.3 Flow chart displaying generation of female population from male.

mated with the best *F* or *M* chromosome. If at the initial stage, the mating pool contains chromosomes of only one class, the crossover procedure is discontinued.

Mutation:- After crossover, mutation is generally applied on the chromosomes with mutation probability Mp (Srinivas *et al.* (1994)). Mutation operators are not applied on the class bits.

5.4 GACD Based Solution Methodology to Process Plan Selection Problem

The major advantages of GACD are its flexibility and ability to adapt to the changing optimization criteria and constraints. It can be modeled in the desired form according to the problem complexity. As in all evolutionary algorithms, factors such as representation of individuals, encoding methods, Initial M and F population, selection and scaling approaches, and the choice of genetic operators have a tremendous influence on the performance of the proposed heuristics. To explain the basic peculiarities of the GACD, a small example is considered. The detailed description of the considered example is given below.

The problem can be described as follows. The problem consists of 5 part types. The corresponding process plans for the respective part types are; P1 = {1, 2} P2 = {3,4} P3 = {5, 6, 7} P4 = {8, 9, 10} P5 = {11, 12}. Table 5.1 shows the incidence matrix where rows indicate corresponding manufacturing resources (machines, fixtures, tools etc.), and columns give alternative process plans.

The processing time for alternative plans are given as: PrT′={17, 21, 12, 11, 8, 9, 18, 13, 17, 8, 12, 20}. The set of corresponding processing steps is represented as St′={6, 8, 5, 4, 3, 4, 7, 4, 8, 5, 6, 8}. The cost associated with processing time Ct = 1 $/unit time. Cost of each processing step Cs = 1 $/step. The weight assigned to machines, tools, and fixtures to calculate the Hamming distance is as follows (Dashora *et al.* 2006);

Machine weight	Wm	= 3.
Fixture weight	Wf	= 2.
Cutting tool weight	Wc	=1.

The weights were assigned empirically. The objective of this problem is to minimize the cost associated with the process plans given by the following relation:

$$\text{Cost} = A \times \sum_{i \in \text{Tabu}} (Ct \times PrT[i] + Cs \times St'[i]) + B \times \sum_{i,j \in_1 \text{edge in Tabu}_1} \text{DIST}[i][j];$$

Table 5.1 Incidence Matrix.

	Part 1		Part 2		Part 3			Part 4			Part 5	
m_1	1			1		1		1				
m_2		1		1	1		1			1	1	1
m_3	1		1				1		1			
f_1						1	1	1	1		1	
f_2	1	1	1								1	1
f_3			1		1						1	
f_4				1			1				1	
f_5							1			1	1	
t_1	1	1		1	1			1				
t_2		1				1		1		1		
t_3	1		1	1						1		
t_4				1	1	1		1		1		
t_5	1	1					1	1				
t_6							1				1	
t_7			1						1	1	1	1

m : Machines.
f : Fixtures.
t : Cutting tools.

Where, A, B are the constants, which, based on past experiences and consulting the relevant literature, have been taken to be one.

Encoding: Encoding the chromosome of a solution and its representation are key issues in the implementation of GAs. This includes the encoding of the class bit and the encoding of the data bit. During the last two decades, various string encoding techniques have been developed such as real encoding, integer encoding, permutation and matrix encoding etc. In this research, and integer encoding scheme has been used for the data bit and a binary encoding scheme has been adopted for the class bit.

Population Initialization: As mentioned in Section 5.3.2., the Male chromosomes are first generated randomly, and thereafter the female chromosomes are generated corresponding to these male chromosomes, by considering Hamming Distance criteria.

Fitness Computation: During each generation, chromosomes are evaluated using some measures of fitness. In this research, the fitness value is calculated according to an operation sequence and selection of an appropriate machine, whilst taking into account all the technological constraints. A mathematical model of the objective functions is shown in equation. 5.1.

Selection: Selection is performed over all the population disregarding the class information. A scaling approach helps to sustain the steady selection pressure in the population and prevent the premature convergence of the population to a suboptimal solution. In this research, various scaling approaches have been tested, including Dynamic Linear Scaling, Sigma Truncation, Power Low Scaling, Logarithmic Scaling, Windowing, Normalizing and Boltzmann Selection. Computational experiments reveal that the performance of a proportional selection based "Roulette Wheel Strategy" with embedded features of the "Boltzmann Selection" scheme works more effectively than others. For chromosome C, and fitness ψ_c, the scaling function (S_c) may be defined as follows:

$$S_c = e^{\psi_c / T} \tag{5.5}$$

The selection pressure is low when the control parameter T is high. Now the selection probability can interpreted as

$$p_c = \frac{S_c}{\sum_{c=1}^{pop\_size} S_c} \tag{5.6}$$

$$f_c = pop\_size \times p_c \tag{5.7}$$

Where, f_c is the improved roulette wheel count and POP_SIZE is population size. The best individual always survives into the next generation so as to enable the GACD to converge faster.

Crossover: In GACD, crossover is applied until the condition mentioned above is satisfied. Crossover is a process in which one Male and one Female chromosome recombine to produce two chromosomes (Male or Female or both). Various types of standard crossover operators are most commonly and frequently used by researchers. This includes, Partially-Mapped Crossover (PMX), Enhanced Edge Recombination (EER), Order Crossover (OX), Uniform Order Based Crossover (UOX). Each of the aforementioned crossover operations has its own advantages. The PMX retains the absolute position within the permutations, and requires more generations to converge towards the best solution (Oliver *et al.* (1997). As in other GAs (Whitley *et al.* (1989) and Whitley *et al.* (1991)) the EER operator is computed on every individual of every generation, and thus introduces an additional overhead that slows down the GACD. UOX and OX are arguably the best crossover techniques in terms of preserving and exploiting the important characteristics of the parents (each position in the offspring is filled with elements from the same position in one of the parents (Grefenstette *et al.* (1995)). In our tests, the UOX and OX give results that are only slightly worse than those of the PMX but it requires significantly fewer generations. This result has stimulated the search for an improved operator, referred to as Hybrid Mapped Crossover (HMX), which seems to combine the positive features of the considered operators. The basic operations of HMX can be summarized as follows:

Step 1. Select a Male and Female chromosome for crossover operation.
Step 2. Randomly select an aforementioned crossover operator.
Step 3. Execute the selected crossover operator on the first row of data bit.
Step 4. Randomly select two operations and change their machine from existing alternatives.

Mutation: After crossover, strings are normally subjected to mutation. This introduces some extra variability, provides and maintains diversity in a population and explores the search space. In this research, mutation is applied only on the data bit. During the past decades, several mutation operators have been proposed for permutation representation, such as inversion, insertion, displacement and reciprocal exchange mutation. In this research, a Heuristic based method is applied. It is designed with the neighborhood technique in order to produce an improved offspring. The mechanism of the mutation operator is discussed below.
Procedure: Heuristic based mutation

Step 1. Select a chromosome for mutation operation.
Step 2. Pick up 'n' genes at random.
Step 3. Considering all possible permutations of the first row of selected genes, neighborhood offsprings are generated.
Step 4. Randomly select two operations and change their machine from existing alternatives.
Step 5. Evaluate all neighbors and select the best one as offspring.

5.4.1 Selection of GACD's Parameter

Parameter setting is a time-consuming issue but crucial for any GA application. Michalewicz (1994) described the determination of appropriate GA's parameters as an art rather than a science. The main parameters for our GACD are population size (POP_SIZE), maximum number of generations (MAX_GEN), initial male population (INIM_POP), initial female population (INIF_POP), crossover probability (C_p), mutation probability (M_p). Following the practice of combining general guidelines available in literature with preliminary configuration experiments, we have obtained the configuration for the GACD used in this paper as follows:

1. *Population Size factor (PSF)*: GA practice recommends that the population size varies as a multiple of chromosome length (CL).

   $$\text{POP\_SIZE} = \text{PSF} \times \text{CL} \ (\text{PSF} = 1,2)$$

2. *Crossover and Mutation probability*: A moderately large value of C_p (0.5–0.9) and a small value of M_p (0.05–0.2) are commonly employed in GAs. Increasing values of C_p and M_p promote exploration at the expense of exploitation. Our preliminary experiments confirm that values within these ranges can provide satisfactory results. The specific values of these parameters in our numerical results are provided in the next section.

5.5 Numerical Experiments

In this section, we summarize the extensive campaign of numerical investigations performed to validate the effectiveness of the proposed ideas. To demonstrate the efficacy and robustness of the proposed model and the quality of the solutions found with the GACD, illustrative examples are taken from Dashora *et al.* (2006). As mentioned in section (5.4.1) the selection of GACD's control parameters is a fundamental process to achieve a satisfactory evolution of the search. In this research work, intensive experiments have been carried out to determine the suitable values for the parameters governing the execution of GACD (initial male population, initial female population, crossover rate, mutation rate, crossover type, mutation type and selection strategies, etc.) Various crossover operators, which are used for planning, scheduling, and control problems were tested on a sample problem described in the section 5.4.1. The proposed GACD based heuristic applied on the process plan selection problem has been coded in C^{++} and the program was run on an IBM PC with Pentium IV CPU at 1.7 GHz.

After executing all the steps of the algorithm, the best tour cost obtained by the GACD algorithm for the test problem described in section 5.4 is 182.00. A comprehensive analysis of the proposed approach is shown in Table 5.2. The convergence trend of GACD, GA and ACO algorithms are shown in Figure 5.4. Further, to show the robustness of the proposed approach it is also tested on various problems taken from the literature Dashora *et al.* (2006). Optimal/suboptimal solutions were achieved in each case and the results are shown in Table 5.3. In some cases the objective function values are the same however, the GACD algorithm leads to quick convergence to reach the optimal / near optimal solution. ACO takes around 900 iterations and GA takes 950 iterations to converge however, GACD only take 850 iterations, which is much faster than the other two.

Table 5.2 Convergence trends of different algorithms.

	GACD	GA	ACO
Number of iterations	82	103	100

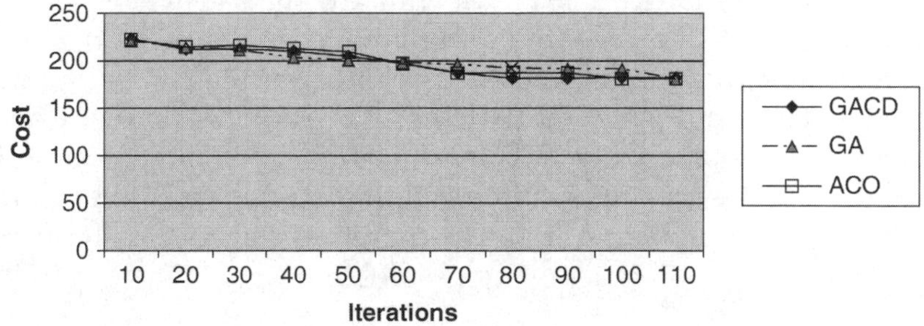

Figure 5.4 Convergence trends of GACD, GA and ACO algorithms.
GACD=Genetic Algorithms with Chromosome Differentiation.
GA: Genetic Algorithm.
ACO: Ant Colony Optimization.

Table 5.3 Comparative study of different algorithms.

	Example of Bhaskaran (1990)	**Example of Kusiak and Finke (1988)**	**Example of Zhang and Huang (1994)**
Applied methodology	Objective function value	Objective function value	Objective function value
Kusiak and Finke (1988)	34.5	30.8	59
Bhaskaran (1990)	33.1	30.8	61
Zhang and Huang (1994)	32.5	30.8	61
Dashora *et al.* (2006)	32.5	30.8	59
Proposed GACD approach	32.5	30.8	59

5.6 Conclusions

Nowadays a major challenge for researchers is to find a way to achieve maximum utilization of resources while manufacturing the product in the minimum time. To accomplish this, a task flexible manufacturing system has gained considerable success. The success of flexible manufacturing systems depends on the successful selection of the process plan.

In this chapter the GACD algorithm has been used for the process plan selection problem. The GACD inherits the features of Genetic Algorithms with chromosome differentiation. This helps the algorithm to quickly converge towards the optimal or near optimal solution. The result showed that the GACD algorithm converges much faster than the ACO and GA algorithms. Thus the result indicates that the GACD algorithm would be very useful in resolving large real world problems. In future therefore, the GACD algorithm needs to be tested on more complex problems.

References

Bandyopadhyay. S. and Pal S.K., 1998, Incorporating chromosome differentiation in genetic algorithms, *Information Science*, 104, pp.293–319.

Bhaskaran, K., 1990, Process plan selection. *International Journal of Production Research*, 28 (8), 1527–1539.

Brandimarte, P. and Calderini, M., 1995, A hierarchical bicriterion appraoch to integrated process plan selection and job-shop scheduling. *International Journal of Production Research* 33 (1995), 161–181.

Chen, A.H., Fu. L.C., Lin, M.F, and Huang, A. C., 2001, Petri-net and GA-based approach to modeling, scheduling, and performance evolution for water fabrication, *IEEE Trans. Robotics Automat*, 17, 619–636.

Chang, H.C. and Chen, F.F., 2002, A Dynamic Programming based process planning selection strategy considering utilization of machines, *International Journal of Advanced Manufacturing Technology*, 19(2002), 97–105.

Dashora, Y., Kumar, S., Tiwari, M. K. and Shankar, R., 2006, Ant colony optimization to select best process plan in automated manufacturing environment, *Proceedings of the Institution of Mechanical Engineers, Part B: Journal of Engineering Manufacture*, 220, 1457–1472.

Feng, D.Z., and Yamashiro, M., 2003, A Pragmatic Approach for Optimal Selection of Plant Specific Process Plans in a Virtual Enterprise. *Production Planning & Control*, 14(6), 562–570.

Goldberg, D., Kog, E. E. B. and Deb,K., 1989, *Messy Genetic Algorithm: Motivation Analysis and First Result*. Complex System Publication Inc., 493–530. 1989.

Grefenstette, J., Gopal, R., Rosmaita, B. and Gucht D. V, 1995, Genetic Algorithms for the TSP. In Grefenstette, J. J. (ed.) *Proceedings of the First International Conference on Genetic Algorithms and Their Applications*. Hillsdale, New Jersey: Lawrence Erlbaum, 160–165.

Holland, H.H., 1975, *Adaptation in Natural and Artificial Systems*. Detroit MI: University of Michigan Press.

Jain, A. K. and Elmaraghy, H.A., 1997, Single process plan scheduling with genetic algorithm, *Production Planning and Control*, 8 (4), 363–376.

Khoshnevis, B., and Chen, Q., 1990, Integration of process planning and scheduling functions. *Journal of Intelligent Manufacturing*, 1, 165–176.

Kusiak, A., and Finke, G., 1988,Selection of process plans in automated manufacturing systems. *IEEE Transactions of Robotics and Automation*, 4(4), 397–402.

Meloni, C., Naso, D. and Turchiano, B., 2003, Multi-Objective Genetic Algorithm for a class of sequencing problems in manufacturing environments, *IEEE International Conference on Systems, Man & Cybernetics*, Washington, D.C., USA.

Michalewicz, Z, 1994, *Genetic Algorithm + Data Structure = Evolution Programs*. New York: Springer Verlag.

Ming, X.G., and Mak, K.L., 2000, A hybrid Hopfield network-genetic algorithm approach to optimal process plan selection, *International Journal of Production Research*, 38(8), 1823–1839.

Oliver, I. M., Smith, D. J. and Holland, J. R. C., 1987, A Study of Permutation Crossover Operators on the TSP. In Grefenstette, J. J. (ed.) *Genetic Algorithms and Their Applications: Proceedings of the Second International Conference*, Hillsdale, New Jersey: Lawrence Erlbaum. 224–230.

Rajamani, D., Singh, N., and Aneja, Y. P., 1990, Integrated design of cellular manufacturing systems in the presence of alternative process plans. *International Journal of Production Research* , 28 (8), 1541–1554.

Rebaine D, Strusevich V, 1999, Two-machine open shop scheduling with special transportation times. *The Journal of the Operational Research Society*, 50(7), 756–764.

Seo, Y. and Egbelu, PJ, 1996, 'Process Plan Selection Based on Product Mix and Production Volume,' *International Journal of Production Research*, 34(9), 2639–2655.

Srinivas, M. and Patnaik, L.M, 1994, Adaptive probabilities of crossover and mutation in Genetic Algorithm. *IEEE Transaction on system man and cybernetics*, 24(4), 655–667.

Tiwari, M.K., and Vidyarthi, N.K., 1998, An integrated approach to solving the process plan selection problem in an automated manufacturing system, *International Journal of Production Research*, 36(8), 2167–2184.

Tonshoff, H.K., Beckendorff, U., and Andres, N., 1989, Flexplan – a concept of intelligent planning and scheduling. Proceedings of CIRP International Workshop on Computer Aided Process Planning, Hanover University, Germany.

Whitley, D., Starkweather, T. and Shaner, D., 1991, The Traveling Salesman and Sequence Scheduling: Quality Solutions Using Genetic Edge Recombination. In Davis, L. (ed.) *Handbook of Genetic Algorithms*, New York: Van Nostrand Reinhold. 350–372.

Whitley, D., Starkweather, T. and Fuquay D., 1989, Scheduling Problems and Travelling Salesman: The Genetic Edge Recombination Operator. In Schaffer, J. (ed.) *Proceedings on the Third International Conference on Genetic Algorithms*, Los Altos, CA:Morgan Kaufmann Publishers, 133–140.

Zhang, H.C., and Huang, S.H., 1994, A fuzzy approach to process plan selection. *International Journal of Production Research*, 32(6), 1265–1279.

Zhang, H.C., and Mullar, S., 1994, An integrated model of process planning and production scheduling. *International Journal of Computer Integrated Manufacturing*, 7(6), 356–364.

6

Operation Allocation in Flexible Manufacturing System Using Immune Algorithm

Mayank K. Pandey

Department of Mechanical Engineering, University of Alberta, Canada

Abstract

Flexible manufacturing systems (FMS) are suited to the niche manufacturers who desire increased flexibility and efficiency from their automated systems to stay competitive. FMS provide manufacturers with machine as well as routing flexibility. Machine flexibility provides opportunities to change the order of operations to be performed on a part and change in the product type. Routing flexibility enables the use of different machines for the same operation, change in the ordered amount and change in the capacity of the system. Machine loading problems, which include job selection and operation allocation problems, have widely been studied because of their key role in determining the overall efficiency of the manufacturing system. The complexity of the machine loading problems is so high that even a small scale problem takes a large computational time. Evolutionary algorithms are apposite tools to deal with such problems. They can also provide good solutions to large scale problems. Specifically, algorithms which have self learning and memory acquisition capability, are good for solving machine loading problems because of their complex characteristics. Immune algorithms (IA) possess the properties of evolutionary algorithms and hence are suitable to use for machine loading problems. In this chapter different issues related to the machine loading problem have been discussed and a successful implementation of the immune algorithm or 'Artificial Immune System' (AIS) on a machine loading problem is delineated.

Keywords: Machine loading problem, artificial immune system, hypermutation, Gaussian and Cauchy mutation

Manoj Tiwari and Jenny A. Harding (eds.) Evolutionary Computing in Advanced Manufacturing, (95–122) © Scrivener Publishing LLC

6.1 Introduction

To survive in the current global competition, every manufacturing company needs to follow an effective production schedule such that maximum utilization of available resources is possible. One of the major issues in such time effective production is related to the optimal loading decisions. This becomes more critical in the complex manufacturing environments with the customer specific products, high end technologies, multiple machines and job types etc. In such cases manual approximation of the best possible loading operation is less likely to give a good outcome. The machine loading problem is one of the six post-release decisions of a flexible manufacturing system which involves computational complexity and high level of variability (Stecke, 1983). A typical FMS consists of 5–25 Numerical Control Machines, a central storage system and an automated material handling system. All these components are controlled by a computer system. A FMS system is designed to combine the efficiency of a high production transfer line and the flexibility of a job shop, to best suit the batch production of a mid volume and mid variety of products. However, its effectiveness is directly related to its design and operation strategies. Generally two types of decisions are associated with the FMS operations, i.e. pre-release decisions and post release decisions. Pre-release decisions involve pre-arrangement of jobs and tools whereas job routing and sequencing decisions come under post-release decisions. Different post release decisions are as follows:

- Machine grouping
- Production ratio estimate
- Batching of jobs
- Pallets and fixture allocation
- Job selection
- Operations and tool allocation among machines

The Machine loading problem in a random FMS environment has been widely studied. It aims to minimize the system unbalance and maximize the throughput under different tooling and machining constraints. In general, a machine loading problem is defined as (Prakash *et al.* 2008):

> *"Let there be J jobs each needing assignment to a set of machines. Each job is characterized by a set of operations that can be performed*

on a definite set of alternate machines. Let there be M machines in the system, each of which has a definite processing time and fixed tool slots. Every operation of a job requires a pre-specified processing time, fixed tool slot and a set of alternate machines. The problem is to find a unique sequence of jobs so that system unbalance can be minimized and throughput can be maximized by satisfying various technological constraints such as available machining time and tool slots"

In the literature, above objectives are adopted to reflect the system efficiency and the system utilization. The machine loading problem, discussed in the context of FMS, usually consist of several *NC* machines (typically 5–25 in number), a storage rack and an automated material handling system. The combination of job selection and operation allocation problems in machine loading makes it *NP-hard* in nature (Shanker and Srinivasulu, 1989). A subset of jobs is selected for immediate production and corresponding operations are assigned to the available machines. Researchers have further partitioned the machine loading problem into three parts as job sequence determination, operation allocation on machines and job reallocation (Swarnkar and Tiwari, 2004). The grouping of the resources, selection of part-mix and aggregate planning work as input for the resource scheduling. Planning of dynamic operations are also important. Hence, it is obvious that the loading decisions are an important link between operational and strategic level decisions. Many researchers have addressed these problems in a sequential manner. The solution approaches adopted by researchers can be classified into the following five domains:

- Mathematical programming: Lashkari *et al* (1987) formulated a 0–1 integer program to solve the operation allocation problem with planning aspects of the refixturing and limited tool availability. Sarin and Chen (1987) developed a mathematical model to determine the routing of the parts and allocation of the tools to minimize the machining costs. Shankar and Srinivasulu (1989) developed a two-stage branch and backtrack procedure with the objective of maximizing the assigned workload. Sawik (1990) proposed a hierarchical decision structure which includes: part type selection, machine loading, part input sequencing and operation. Liang and Dutta (1992) developed an integrated approach to solve the job selection, tool allocation and machine loading problem, but their approach was limited to only small sized problems. Guerrero *et al.* (1999) considered the entire production route instead of individual operations and used this approach to solve the tool allocation problem. He also presented an extensive review of work done in FMS using mathematical modeling. Das *et al.* (2009) formulated an integer programming model to group the parts and to address the operation-sequencing problem. Zeballos *et al.* (2010) used an integrated constraint programming (CP) model where the dimensionality of the problem is reduced by using two sets of two-index variables to capture the manufacturing activities rather than one set of four indexes.

- Multi-criteria decision making: Kumar *et al.* (1990) used a max-min approach to find the best compromise solution while considering five different objectives namely: machine processing time, machine to machine movement, equal size machine grouping, unequal size machine grouping, tool slot availability and operation priorities. Chen and Askin (1990) compared different machine loading problems on three FMS environments and concluded that the overall system performance consideration is better than optimizing individual operations. Kim and Yano (1997) minimized the maximum percentage overload (relative to the optimal continuous workload allocation) across machine groups and found it to be a key objective. Their results also indicated that reducing the number of machine groups and balancing the workloads among the machines help to reduce makespan.
- Heuristic oriented: Stecke and Soldberg (1981) investigated sixteen dispatching rules which allow an idle machine to select its next operation from those parts available according to some priority. Marino and Ding (1993) proposed two heuristic methods to balance workloads and meet the due dates. The goal of the 1st heuristic is to achieve a workload balance whereas the second heuristic reduces the number of late part types. Chen and Chung (1996) suggested several alternative

measures for the assessment of machine flexibility and routing flexibility. Tiwari *et al.* (1997) proposed Petri net based modeling of the loading problem and suggested a heuristic to calculate system unbalance and throughput. Sarma *et al.* (2002) presented a Tabu search based heuristic to solve the problem.

- Simulation based: Jain *et al.* (1989) used forward and backward simulation to generate schedules for automotive plants. They used heuristics designed by experts to govern the simulation. Sabuncuoglu and Hommertzheim (1992) analyzed the relative performances of machine and automated guided vehicle (AGV) scheduling rules with a FMS simulation model. Basnet and Mize (1995) used discrete event simulation to evaluate decision alternatives. They used Expert systems to make operating decisions, to choose heuristics and to make operating policies.
- Artificial Intelligence (AI) based: Kim and Kim (2005) used a symbiotic algorithm to solve the loading problem. Tiwari *et al.* (2006) used a variant of simulated annealing to solve the machine loading problem. Prakash *et al.* (2008) and Khilwani *et al.* (2008) used a modified version of the immune algorithm to solve operation allocation and job selection problems, and their work is described in this chapter. Biswas and Mahapatra (2008) used particle swarm optimization to solve the machine loading problem. Yogeswaran *et al.* (2009) used a hybrid algorithm comprising of GA and simulated annealing. Mandal *et al.* (2010) solved a dynamic machine loading problem using GA, SA and AIS and found that performance of AIS is better than the other two algorithms.

Due to the inherent complexity of the machine loading problem, optimization approaches like linear programming or dynamic programming become unreasonable to use with large problem sizes. The complexity of the machine loading problem can be so high that even for 8–10 jobs there can be more than 10^8 possible operation machine allocation combinations. The huge solution space makes it difficult to reach the optimal solution and reduces the convergence speed to optimal or near optimal solutions. In recent times, nature inspired evolutionary algorithms have been successfully used to solve this complex combinatorial problem as mentioned above in the AI based methods like Genetic Algorithm (GA), Ant Colony optimization (ACO) and Immune algorithm (IA). Mandal *et al.* (2010) found that the performance of the Immune algorithm, which is also called the Artificial Immune System (AIS) is better than GA and Simulated Annealing (SA). Due to self organizing and learning capability of AIS, it has been widely used in real world applications like pattern recognition, data utilization, machine learning and evolutionary computation etc. Castro and Zuben (2005) found that the existing immune algorithm was not interactive and required more computational time to reach the optimal/near optimal point. Considering this problem, Prakash *et al.* (2008) proposed a modified immune algorithm to solve the loading problem. In this chapter their work has been referred to present the modified immune algorithm and its application to an operation allocation and job selection problem. Hypermutation is the key step in AIS which determines the solution quality. With better solution quality it is more likely that new offspring will survive in the next evolution. The new hypermutation operation included:

1. Gaussian mutation strategy for higher affinity antibodies.
2. Cauchy mutation strategy for lower affinity antibodies.
3. Elitist based immune memory to preserve the superior antibodies present in the clonal pool and to pass its traits to the next generation.

Additionally, chaotic sequence for initialization and roulette wheel selection were used to amplify the effectiveness of the algorithm. They solved ten standard data sets for the machine loading problem and compared the performance of their proposed algorithm with existing heuristic and conventional clonal algorithms.

6.2 Machine Loading Problem

The FMS under consideration consists of multi functional NC machines which can have varying processing requirements for the same job. Each job can have different processing requirements. The allocation of operations of selected jobs from a pool of jobs under different tooling and machining constraints must be carried out while solving machine loading problem. Let there be a machine loading problem consisting eight jobs and four machines (Tables 6.1 and 6.2 Shanker and Srinivasulu, 1989). The eight jobs can be sequenced in 8 ways. For any of these options there exist $1\times2\times 2\times1\times2\times6\times9\times6=2,592$ operation machine allocation combinations. Hence the total possible allocations for jobs are $8!\times2592$. Obviously, with the increase in the problem size, the possible number of combinations and thus the solution space also increases. Evolutionary algorithms are appropriate for such environments where computational cost is very high compared to getting an optimal or near optimal solution.

Table 6.1 Description of problem #1.

Job No.	Batch Size	Operation Number	Unit Processing Time	Machine Number	Tool Slot Needed
1	8	1	18	3	1
2	9	1	25	1,4	1
		2	24	4	1
		3	22	2	1
3	13	1	26	4,1	2
		2	11	3	3
4	6	1	14	3	1
		2	19	4	1
5	9	1	22	2,3	2
		2	25	2	1
6	10	1	16	4	1
		2	7	4,2,3	1
		3	21	2,1	1
7	12	1	19	3,2,4	1
		2	13	2,3,1	1
		3	23	4	3
8	13	1	25	1,2,3	1
		2	7	2,1	1
		3	24	1	3

Table 6.2 Data related to problem given in Table 6.1.

Machine Number	Available Time	Available Tool Slot
1	480	5
2	480	5
3	480	5
4	480	5

To solve the problem, the following objectives were considered:

1. Maximizing throughout: Throughput is defined as the summation of the batch size of selected part-types. Maximization of throughput increases the productivity of the manufacturing systems. In order to maximize the throughput, machines are allowed to work over the specified processing limit.
2. Minimizing system imbalance: System unbalance (SU) is the summation of the remaining time (idle time) on all available machines. By minimizing system unbalance, the utilization level of machines can be increased.

If system imbalance is zero (minimum imbalance), perfect utilization of a machine is ensured. It is desirable to consider both under utilized and over utilized times to calculate imbalance. The upper bound of the throughput for a time duration is equal to the sum of the batch sizes of all the jobs needing to be assigned. The following assumptions are considered to solve the problem:

- Processing requirements of a job are know a priori
- Tool sharing or duplication is not permitted
- Loading/unloading and transportation time is negligible
- Any machine can not perform more than one operation of the job at a time
- Any particular operation on a job does not require more than one machine
- Many machines can not perform more than one operation on a job at a time

6.2.1 Problem Formulation

6.2.1.1 *Subscript/Superscript*

j	job index, $1 \le j \le J$
o	operation index, $1 \le o \le O$
m	machine index, $1 \le m \le M$

6.2.1.2 *Decision Variables*

$$\alpha_j = \begin{cases} 1 & \text{if job } j \text{ is selected,} \\ 0 & \text{otherwise} \end{cases}$$

$$\beta_{jom} = \begin{cases} 1 & \text{if operation o of job } j \text{ is performed on machine m,} \\ 0 & \text{otherwise} \end{cases}$$

6.2.1.3 *Notations*

SU	system unbalance
TH	throughput
BS_j	batch size of job
SU_{max}	maximum system unbalance (= $M \times$ *Planning Horizon)*
OTM_m	over utilized time on machine m
UTM_m	under utilized time on machine m
O_j	set of operations for job j
M_{jo}	set of machines for performing o^{th} operation of job j
$M_{jom}^{t\_av}$	available time on machine m for performing o^{th} operation of job j
$M_{jom}^{t\_req}$	required time by machine m for performing o^{th} operation of job j
$M_{jom}^{t\_rem}$	remaining time on machine m for performing o^{th} operation of job j
$M_{jom}^{tool\_av}$	available tool slots on machine m for performing o^{th} operation of job j
$M_{jom}^{tool\_req}$	required tool slots by machine m for performing o^{th} operation of job j

$M_{jom}^{tool\_rem}$ remaining tool slots on machine m for performing o^{th} operation of job j

6.2.1.4 *Imbalance Coefficient*

$$C_I = \frac{\mathrm{SU}_{\max} - \sum_{m=1}^{M} \mathrm{OTM}_{\mathrm{m}} + \mathrm{UTM}_{\mathrm{m}}}{\mathrm{SU}_{\max}} \tag{6.1}$$

6.2.1.5 *Throughput Coefficient:*

$$\mathrm{C}_{TH} = \frac{\sum_{j=1}^{J} \mathrm{BS}_j \times \alpha_j}{\sum_{j=1}^{J} \mathrm{BS}_j} \tag{6.2}$$

For both of the above coefficients:

$0 \le C_I \le 1$ and $0 \le C_{TH} \le 1$

For a perfectly balanced work station $C_I = 1$ and completion of all jobs within a given timeframe gives $C_{TH} = 1$. Hence, the objective is to maximize both C_I and C_{TH}.

The weighted average of both objectives is used in the literature to find the final objective function. If the associated weights for C_I and C_{TH} are w_I and w_{TH}, respectively, then the final objective function is defined as:

Maximize

$$C = \frac{w_I \times C_I + w_{TH} \times C_{TH}}{(w_I + w_{TH})}$$

$$= \frac{w_I \times \left(\dfrac{\mathrm{SU}_{\max} - \sum_{m=1}^{M} \mathrm{OTM}_{\mathrm{m}} + \mathrm{UTM}_{\mathrm{m}}}{\mathrm{SU}_{\max}} \right) + w_{TH} \times \left(\dfrac{\sum_{j=1}^{J} \mathrm{BS}_j \times \alpha_j}{\sum_{j=1}^{J} \mathrm{BS}_j} \right)}{(w_I + w_{TH})} \tag{6.3}$$

The above objective function is subjected to the following constraints:

a. Machining Constraints:

$$\sum_{m=1}^{M} \mathrm{OTM}_{\mathrm{m}} + \mathrm{UTM}_{\mathrm{m}} \geq 0 \tag{6.4}$$

This constraint ensures that the sum of remaining time (over utilized and under utilized time) after allocation of all jobs remains positive. The minimum value reached by this constraint is zero.

$$\sum_{o=1}^{O_j} M_{jom}^{t\_req} \beta_{jom} \leq M_{jom}^{t\_av}, \ \forall j \tag{6.5}$$

The above constraint implies that the time available on any machine should always be equal to the time required by that machine to perform the operation of jobs.

$$\sum_{o \in M_{jo}} \beta_{jom} \leq 1, \quad \forall j, \forall O_j \tag{6.6}$$

Constraint (6) exhibits that only one machine is required from a given set of machines for performing optional operations on a job.

$$\sum_{m=1}^{M} M_{jom}^{t\_rem} \beta_{jom} \geq 0 \tag{6.7}$$

Constraint (7) enforces that sum of remaining time on all machines after completion of a particular operation of a job must be a positive quantity or at least zero.

b. Tooling Constraints:

$$\sum_{o=1}^{O_j} M_{jom}^{tool\_req} \beta_{jom} \leq M_{jom}^{tool\_av} \tag{6.8}$$

The above constraint shows that the tool required by any machine to perform an operation on the jobs must be within the available tool slots on that machine.

$$M_{jom}^{tool\_rem} \beta_{jom} \geq 0 \tag{6.9}$$

This dictates that the remaining tool slot on any machine after completion of a job must be greater or equal to zero.

c. Processing Constraint:

$$\sum_{m=1}^{M}\sum_{o=1}^{Oj}\beta_{jom}=\alpha_j\times O_j \tag{6.10}$$

The above constraint implies that all operations on a job must be completed sequentially before considering the next job. It is obvious from the above formulation that a machine loading problem has to satisfy many constraints simultaneously and search for solutions in a large search space. Considering the self-learning and memory capacity of the immune algorithm, it is a good AI tool to solve such problems. In the next section, a detailed discussion of the biological and computational immune system is provided. Further, works available in the literature to solve machine loading problems with the modified immune algorithm are elaborated.

6.3 Solution Methodology

6.3.1 Introduction to Immune System and Analogy to Immune Algorithm

The immune algorithm is based on the biological immune system. The immune system is a complex arrangement of cells, molecules, and organs. It protects organisms against infections. One of the major roles of an immune system is to recognize and eliminate disease-causing agents known as *pathogens*. Molecules called *antigens* which are a small portion of pathogens help an immune system to recognize pathogens. There are two types of antigens; self and non-self. Non-self-antigens are disease causing elements, whereas self-antigens are harmless to the body. *B-cells*, and *T-cells* are two major groups of immune cells. They depend on their protein molecules called *antibodies* to recognize antigens. A B-cell recognizes the antigen when its antibodies come into contact with an antigen of complementary shape (Dasgupta, 1999). The B-cell, thus stimulates the antigen.

The clonal selection principle explicates the immune system response when an antigenic pattern is recognized by the B-cell.

Affinity is the term used to define the success of an antibody to connect and neutralize the antigen. The process of clonal selection is shown in Figure 6.1.

When a B-cell encounters an antigenic pattern with a suitable affinity threshold, it proliferates and differentiates into both memory cells and effectors cells (Ada and Nossel, 1987). Generating B-cells are simulated by affinity to proliferate and mature into antibody secreting cells. These cells then divide themselves into clones. These properties are called proliferation and differentiation. Hypermutation and receptor editing are other two important characteristics of the immune system. Hypermutation involves mutation of inferior bodies at a higher rate than the antibodies with higher antigenic affinity. This phenomenon is called receptor editing (Tiwari *et al.* 2005).

Generally, the clonal selection principle is utilized to design an immune algorithm. The similarity between the biological immune system and the computational immune system is shown in Table 6.3.

Initially the immune system was used to maintain diversity in a genetic algorithm (GA). Later on it was developed to use itself as an optimization tool. The first use was reported by Cutello and Nicosia (2002). They proposed the Simple Immune Algorithm (SIA) to analyze and predict the dynamics of an algorithm.

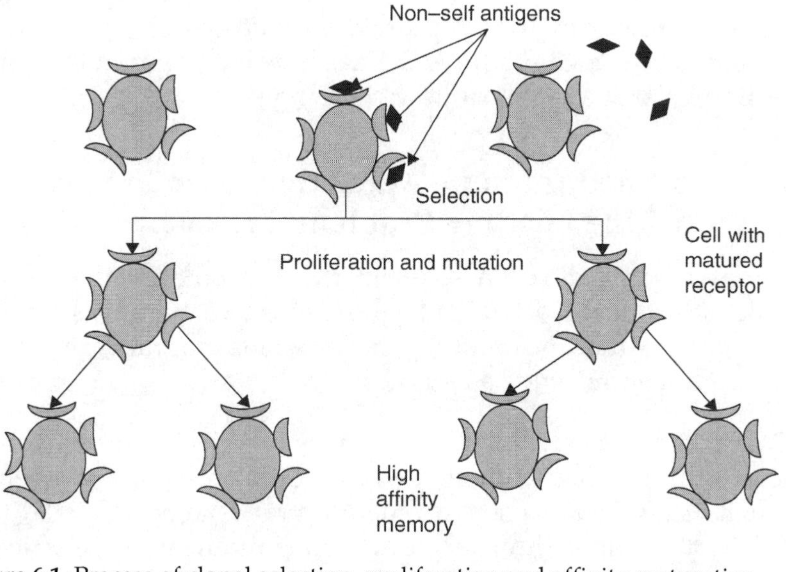

Figure 6.1 Process of clonal selection, proliferation and affinity maturation.

Table 6.3 Similarity between biological and computational immune systems.

Biological Immune System	Computational Immune System
Antigen	Objective Function
Antibody	Solution
Affinity between antibody and antigen	Utility values of solution
Hyper-mutation	Hyper-mutation
Colonization	Colonization

Castro and Zuben (2005) developed the clonal algorithm (CLONALG) by manipulating the hypermutation and cloning properties. A fitness value was used in their work for proportional cloning and counter-proportional hypermutation. Later on, many researchers followed CLONALG and solved various kinds of problems like multiobjective optimization, machine learning, pattern selection, portfolio selection etc. Tiwari *et al.* (2005) proposed another algorithm related to clonal selection, called the Psycho-Clonal algorithm. This algorithm has specific characteristics like various levels of immune memory, needs and affinity maturation. Prakash *et al.* (2008) used a modified immune algorithm to solve the machine loading problem. Their algorithm is detailed in the following section.

6.3.2 Modified Immune Algorithm Used to Solve Machine Loading Problem (Prakash *et al.* 2008)

Their proposed algorithm reduced the computational time and searches the solution domain in a more effective manner. They proposed various modifications to the previous immune algorithm. Their modifications were as follows:

1. Chaotic sequence: Chaotic sequences are the random number generator that is characterized by its ergodic and stochastic behavior. Such behavior was used to generate a random number at different parts of the algorithm, like population initialization, selection of offspring and introducing random changes in the population.

Random number generators (RNG) are found to give slow convergence. One probable reason is that it follows a normal distribution and random numbers close to the peak in the distribution have a higher probability of occurrence. The following chaotic random number generator was used (Yao and Liu, 1999):

$$N(t+1)=R*N(t)*(N(t-1)) \tag{6.11}$$

A graphical presentation of a one-dimensional logistic fitness function is shown in Figure 6.2. In the above equation *N(t)* is the value of the chaotic random number for the tth iteration and it is the bifurcation parameter of the system. In the above figure value $N(0)=0.1$, and $R=4$ is considered. It is clear that the spread spectrum characteristics of logistic mapping can be utilized in place of RNG.

2. Selection: The quality of antibodies i.e. quality of population members are evaluated based on their antigenic affinity i.e. utility (fitness) value of population. Selection is then performed based on the affinity value to proliferate and produce clones. A roulette wheel selection methodology was used for proliferation of the high affinity antibodies (Gen and Cheng, 1997) in place of the deterministic selection rule. Hence, the probability of getting stuck at a local optima owing to the deterministic rule was reduced. More diversity was maintained in the clonal pool. In roulette wheel selection

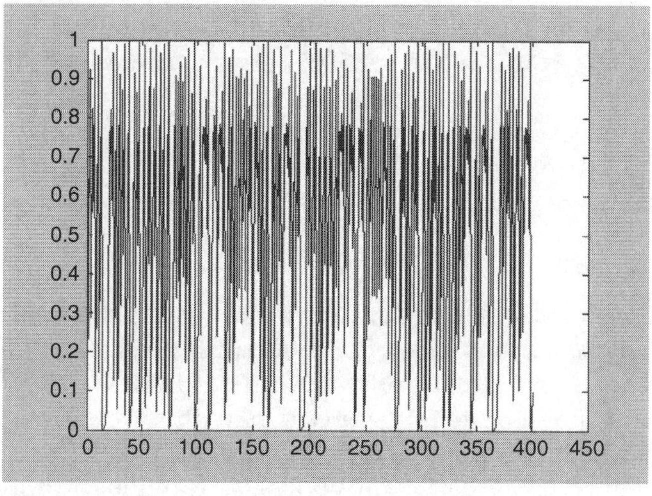

Figure 6.2 Chaotic function of logistic mapping.

antibodies are ranked according to their fitness value. Defining f_i as the ith individual of the population, then the selection probability of the kth individual (antibody) is given as:

$$p_k = \frac{f_k}{\sum_{i=1}^{pop\_size} f_i} \tag{6.12}$$

3. Proliferation: The antibodies with higher affinities are proliferated by cloning i.e. cloned independently and proportionally to their antigenic affinity. The antibodies are selected in higher affinity order. If say k antibodies are selected, the generated number of clones is given by:

$$N_{clones} = \sum_{i=1}^{k} N(i) = \sum_{i=i}^{k} round\left(\frac{\lambda \cdot k}{i}\right) \tag{6.13}$$

where N is the total number of clones to be generated for selected antibodies, $N(i)$ is the number of clones generated for the ith antibody and λ is a user defined multiplication factor.

4. Variation operators: The hypermutation and receptor editing mechanism is responsible for introducing variation in the antibodies. In hypermutation, antibodies with inferior affinity are selected and hypermutated at higher rates. This is similar to mutation in a normal GA except that in a GA, solutions with superior quality are selected rather than solutions with inferior quality. One point mutation was suggested by George and Garey (1999) to generate random changes in the clones. This generates new antibodies by changing one or more genes. For binary bits 0 is replaced by 1 or vice versa; this is called flipping. In other than binary bits swapping of bits is performed to introduce change (Figure 6.3).

Antibodies are mutated based on the hypermutation rate given by:

$$\alpha = \exp(-\mu \cdot f) \tag{6.14}$$

here, μ is the control factor for decay and f is the antigenic affinity. It is evident that antibodies with lower affinities will be mutated at a higher rate than antibodies with higher affinities.

It is important to note that antibodies with lower affinity require mutation which can make larger increases in their affinity so that they can survive. Similarly, antibodies with higher affinity require a relatively smaller increase in their affinity to be able to survive. Also, different step sizes for mutation will ensure that individual

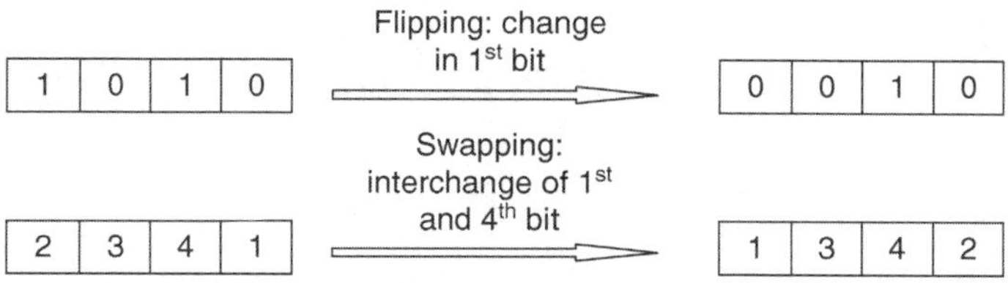

Figure 6.3 Mutation performed on binary and integer bits.

antibodies are not trapped in local minima due to the fact that lower affinity antibodies no longer tend to lower affinity and higher affinity antibodies do not follow larger step size which can lead to premature convergence. Obviously, different mutation schemes are required for antibodies with lower and higher affinities. Considering this, a parallel mutation was proposed which follows two kind of mutation strategy:

1. Gaussian mutation strategy for small mutation step.
2. Cauchy mutation strategy for large mutation step.

These point mutation (PM) strategies are applied in parallel to antibodies on the basis of their affinity. It is based on the selection probability p_k. After calculating the selection probability for each individual in the population, the point mutation is applied as follows:

For each individual in population
Generate a random number 'r' = [0,1]

$$PM = \begin{cases} \text{Gaussian mutation} & \text{if } r < p_k \\ \text{Cauchi mutation} & \text{otherwise} \end{cases} \tag{6.15}$$

If an antibody consists of $x_1, x_2, \ldots, x_n$ particles and the x_kth particle is selected for mutation then a new particle after Gaussian mutation is:

$$x_{k\_gaussian}^{new} = x_k + N(\mu, \sigma) \tag{6.16}$$

where $N(\mu, \sigma)$ is a normally distributed random number following the Gaussian (Normal) distribution:

$$f(x) = \frac{1}{\sigma\sqrt{2\pi}} \exp\left[\frac{1}{2}\left(\frac{x-\mu}{\sigma}\right)^2\right], \quad -\infty < x < \infty \tag{6.17}$$

Usually, $N(0,1)$ is used to generate the random number.

Similarly for Cauchy distribution, a new particle is generated as follows:

$$x_{k\_cauchi}^{new} = x_k + \eta \cdot \beta_k \tag{6.18}$$

where η is the correction step and β_k is the Cauchy random number which follows the one dimensional centered at origin as follows:

$$f(x) = \frac{t}{\pi\left(t^2 + x^2\right)}, \quad -\infty < x < \infty \tag{6.19}$$

The difference in the Gaussian and Cauchy distributions is shown in Figure 6.4. It is important to choose different control parameters appropriately to get faster convergence.

5. Immune memory: After proliferation and hypermutation, the clonal pool (progenies) are evaluated on the basis of their affinity and only a set of high affinity antibodies are stored in *immune memory*. Only these high affinity antibodies were used to generate new populations for the next generation. This memory is updated every iteration. Based

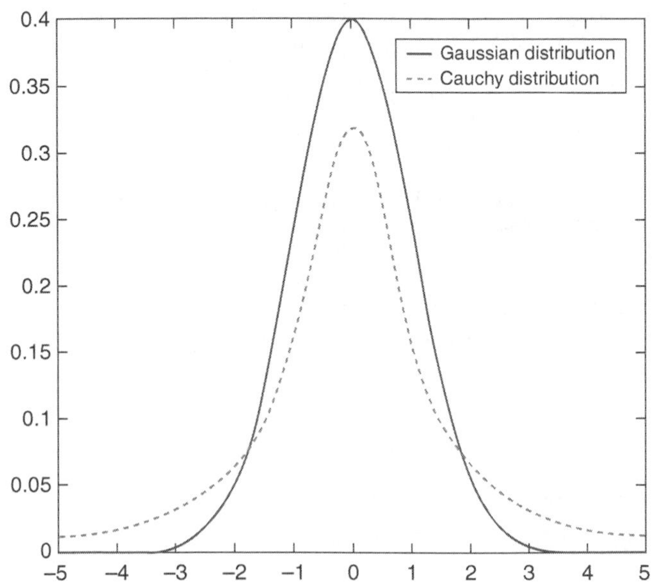

Figure 6.4 Plot of Gaussian and Cauchy distribution with standard deviation 1, mean 0 and t = 1. (Prakash *et al.* 2008).

on the number of candidate antibodies, some of the best available antibodies are carried on to the next generation and the rest of the antibodies are selected randomly (See Prakash *et al.* 2008 for details).

6.3.3 Fast Clonal Algorithm (Khilwani *et al.*, 2008)

Their paper mainly explains the pros and cons of the conventional immune algorithm and modifies various features of the canonical immune algorithm like the hypermutation mechanism, clonal expansion, immune memory, initialization and selection. The authors have developed a new variant and tested their algorithm on 23 benchmark functions as well as on the machine loading problem. Various issues related to the clonal algorithm are now discussed in detail. They have explained the use of a chaotic sequence for population generation with different representation schema like binary coding, decimal coding and permutation coding. They also proved that the number of clones is directly proportional to the rank assigned to it. They also listed following properties of variation operators and provided its proof:

1. Hypermutation guides the system towards local optima, while receptor editing helps to avoid local optima.
2. Point mutation is unable to satisfy the necessary condition of hypermutation.

Because of the second property a new mutation operator was required. They also used Gaussian mutation and Cauchy mutation as explained in the previous section. The phenomenon of step size used in mutation with binary, decimal and permutation coding was also explained. They also calculated the complexity of their algorithm based on De Castro and Zuben (2005).

6.4 Implementing Immune Algorithm for Machine Loading Problem

For both of the above articles, the benchmark setting for the machine loading problem includes four workstations (each with 480 min of available machine time) and five slots of the tools to perform the operation of jobs. Usually, operations can be classified into two types, essential and optional. Essential operations can be performed by a particular machine using a definite processing time. Optional operations can be performed on any of the given machines in the same or different processing times and tool slots (Table 6.1). The implementation of the modified immune algorithm on the machine loading problem is explained in the following steps:

Step 1: Generate the new population. Set the immune memory to zero in the beginning
Step 2: For each generation evaluate the affinity of each antibody and select $m \leq \textit{population size}$ based on the affinity value.
Step 3: Perform cloning (selected antibodies are cloned)
Step 4: Perform hypermutation (Cloned antibodies are hypermutated using parallel mutation) and evaluate the fitness of the hypermutated particle
Step 5: Update the immune memory set (based on the selected high affinity antibodies from the hypermutated particles)
Step 6: Update the population (Based on the updated immune memory size)
Go to step 2
Follow step 2 to step 6 till termination criterion is satisfied.

6.5 Computational Result

In this section, the results obtained by different researchers using the modified immune algorithm are compared. Prakash *et al.* (2008) have given detailed results for all 10 benchmark loading functions. Their detailed results are given in Table 6.4. In the table, $\textit{Job\_Rejected}_{\text{NSU}}$ implies that a job was rejected due to negative system imbalance and $\textit{Job\_Rejected}_{\text{TSC}}$ shows that a job was rejected due to tool slot constraints. They also tested the algorithm for some benchmark test functions and compared it with the Simple Immune Algorithm and the Clonal Algorithm. They found that the modified immune algorithm has better convergence than previously reported algorithms. They also compared the system unbalance and throughput obtained by the modified immune algorithm with other peer works in the machine loading problem, namely Shankar and Srinivasulu (1989), Mukhopadhyay *et al.* (1992) and Tiwari *et al.* (1997). Khilwani *et al.* (2008) have also compared their algorithm with the results of the aforesaid work in machine loading. Hence, we have summarized all these result in Table 6.5.

Table 6.4 Operation machine allocation combination for 10 benchmark machine loading problem (Prakash *et al.*, 2008).

Job #	Op#1	Op#2	Op#3	Remarks
	Machine # to be Assigned for this Operation			
Problem #1				
5	2	2	–	S
7	4	1	4	S
4	3	4	–	S
3	1	3		S
2	*	*	*	$Job\_Rejected_{NSU}$
6	*	*	*	$Job\_Rejected_{NSU}$
8	*	*	*	$Job\_Rejected_{TSC}$
1	3	–	–	S
Problem #2				
3	1	–	–	S
4	3	–	–	S
5	4	1	1	S
1	2	–	–	S
6	2	–	–	S
2	*	*	*	$Job\_Rejected_{NSU}$
Problem #3				
4	1	–	–	S
5	4	2	–	S
2	2	–	–	S
3	4	3	–	S
1	*	*	*	$Job\_Rejected_{NSU}$
Problem #4				
1	2	1	4	S
2	4	3	–	S
3	4	–	–	S
4	3	3	2	S
5	1	–	–	S

(Continued)

Table 6.4 Operation machine allocation combination for 10 benchmark machine loading problem (Prakash *et al.*, 2008). (Continued)

Job #	Op#1	Op#2	Op#3	Remarks
	Machine # to be Assigned for this Operation			
Problem #5				
2	1			S
1	2	2		S
4	4	2	3	S
6	1			S
3	*	*	*	$Job\_Rejected_{NSU}$
5	3	–	–	S
Problem #6				
1	4	–	–	S
4	3	–	–	S
6	2	–	–	S
2	1	3	–	S
5	4	1	–	S
3	*	*	*	$Job\_Rejected_{NSU}$
Problem #7				
1	1	–	–	S
5	3	–	–	S
2	4	2	3	S
3	1	–	–	S
6	2	2	–	S
4	*	*	*	$Job\_Rejected_{NSU}$
Problem #8				
7	4	2	–	S
5	1	3	–	S
4	3	4	–	S
6	2	2	1	S
3	*	*	*	$Job\_Rejected_{NSU}$
1	*	*	*	$Job\_Rejected_{NSU}$
2	*	*	*	$Job\_Rejected_{NSU}$

(Continued)

Table 6.4 Operation machine allocation combination for 10 benchmark machine loading problem (Prakash *et al.*, 2008). (Continued)

Job #	Op#1	Op#2	Op#3	Remarks
	Machine # to be Assigned for this Operation			
Problem #9				
1	3	1	–	S
2	4	–	–	S
3	2	–	–	S
4	1	3	–	S
5	4	3	–	S
6	4	1	–	S
7	2	1	–	S
Problem #10				
1	3	1	2	S
6	3	1	–	S
5	2	4	4	S
2	3	4	1	S
3	1	2	–	S
4	*	*	*	$Job\_Rejected_{TSC}$

It is clear from Figure 6.5 and Figure 6.6 that the modified immune based algorithm consistently performs better than a heuristic approach on almost all machine loading problems. Figure 6.5 gives a comparison of the best known heuristics in the literature with the results obtained from the modified immune algorithms. These algorithms have least system unbalance in most of the cases (Table 6.5). It is also found that the maximum throughput is achievable with these algorithms in almost all problems. Obviously, these immune based algorithms have outperformed the heuristic approaches. Based on the above results, it can be said that modified immune algorithm can be used in other real world computationally complex problems.

6.6 Conclusion

Machine loading in a flexible manufacturing system is a complex real world problem where system resources are to be used in an

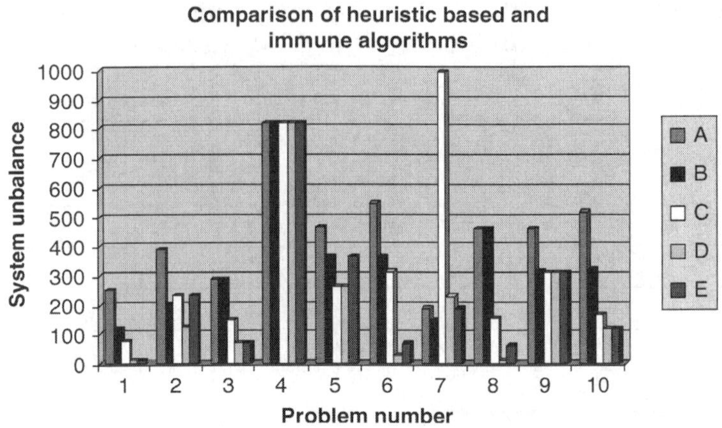

Figure 6.5 Comparison of System Unbalance; A: Shankar and Srinivasulu (1989), B: Mukhopadhyay *et al.* (1992), C: Tiwari *et al.* (1997), D: Prakash *et al* (2008), E: Khilwai *et al.* (2008).

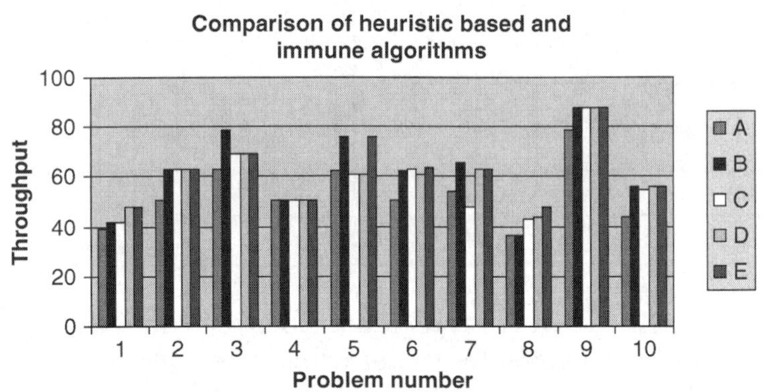

Figure 6.6 Comparison of Throughput; A: Shankar and Srinivasulu (1989), B: Mukhopadhyay *et al.* (1992), C: Tiwari *et al.* (1997), D: Prakash *et al* (2008), E: Khilwai *et al.* (2008).

optimal manner to maximize the productivity. There are different methods available in the literature to solve this problem like mathematical programming, heuristic approach and artificial intelligence etc. However, whatever method is used to solve the problem, it needs to achieve an optimum/near optimum solution in a reduced computational time. Artificial intelligence based approaches give good computational results in real time when compared to other methods, although conventional AI methods like GA, SA and AIS etc., are flawed with slow convergence and local minima

Table 6.5 Comparison of Heuristic based and Immune algorithms.

Prob. No.	No. of jobs	A		B		C		D		E	
		SU	TH	SU	TH	SU	TH	SU	TH	SU	TH
1	8	253	39	122	42	76	42	14	48	14	48
2	6	388	51	202	63	234	63	124	63	234	63
3	5	288	63	286	79	152	69	72	69	72	69
4	5	819	51	819	51	819	51	819	51	819	51
5	6	467	62	364	76	264	61	264	61	364	76
6	6	548	51	365	62	314	63	28	61	69	64
7	6	189	54	147	66	996	48	231	63	189	63
8	7	459	36	459	36	158	43	13	44	63	48
9	7	462	79	315	88	309	88	309	88	309	88
10	6	518	44	320	56	166	55	122	56	122	56

A: Shankar and Srinivasulu (1989), B: Mukhopadhyay *et al.* (1992), C: Tiwari *et al.* (1997), D: Prakash *et al* (2008), E: Khilwai *et al.* (2008).

entrapment. Considering these drawbacks, a new variant of the immune algorithm was used by researchers. Prakash *et al.* (2008) and Khilwani *et al.* (2008) have used a modified immune algorithm to solve the complex machine loading problem. Their work found that the modified immune algorithm is not only better than the conventional immune algorithm but it also outperforms heuristic based approaches while solving loading problems. Hypermutation, random number generation, population generation, and selection are key issues which are addressed in their work and modified to get faster convergence and avoid local entrapment. The selection of variation parameters is important to get better convergence. The efficient use of the modified immune algorithm in the loading problem encourages its use to solve other *NP*-hard problems.

References

Ada GL and Nossel GJV. The clonal selection theory. *Scientific American* 1987. 257(2), 50–57.

Basnet, C. and Mize, J. H. A rule based object-oriented framework for operating flexible manufacturing systems. *International Journal of Production Research,* 1995, 33(5), 1417–1431.

Biswas, S., and Mahapatra, S.S. Modified particle swarm optimization for solving machine-loading problems in flexible manufacturing systems. *International Journal of Advanced Manufacturing Technology*. 2008, 39 (9–10), 931–942.

Chen, I. J., and Chung, C. H. An examination of flexibility measurements and performance of flexible manufacturing systems. *International Journal of Production Research*. 1996, 34(2), 379–394.

Chen Y, Askin RG. A Multi-objective evaluation of flexible manufacturing system loading heuristics. *International Journal of Production Research* 1990, 28(5), 895–911.

Cutello V, Nicosia G. An immunological approach to combinatorial optimization problems. *Lecture Notes on Computing Science*. 2002, 2527, 361–370.

Das, K., Baki, M.F., and Li, X. Optimization of operation and changeover time for production planning and scheduling in a flexible manufacturing system. *Computers and Industrial Engineering*. 2009, 56 (1), 283–293.

Dasgupta, D. *Artificial Immune Systems and Their Applications*, Ed. Berlin, Germany: Springer-Verlag, 1999.

De Castro LN, and Zuben V. *Recent development in biologically inspired computing*. Idea Group Publishing; 2005.

Gen M, Cheng R. *Genetic algorithm and engineering design*. John Wiley & Sons; New York, 1997.

George AJT, Grey D. Receptor editing during affinity maturation. *Immunology Today*. 1999. 20(4), 196.

Guerrero *et al.*, 1999 F. Guerrero, S. Lozano, T. Koltai and J. Larraneta, Machine loading and part type selection in flexible manufacturing systems, *International Journal of Production Research* **37** (1999), pp. 1303–1317.

Jain, S., Barber,K., and Osterfeld,D. Expert simulation for online scheduling. *Proceedings of the 1989 Winter Simulation Conference*. 1989, pp. 930–935.

Khilwani, N., Prakash, A., Tiwari, M. K., & Shankar, R. Fast Clonal Algorithm. *Engineering Applications of Artificial Intelligence*. 2008, 21(1), 106–128.

Kim, J. Y., & Kim, Y. K. Multileveled symbiotic evolutionary algorithm: Application to FMS loading problems. *Artificial Intelligence*. 2005, 22, 233–249.

Kumar, P., Singh, N., and Tiwari, N. K. Multicriteria analysis of the loading problem in flexible manufacturing system using Max-Min approach. *International Journal of Production Research*, 1990, 2(2), 13–23.

Lashkari R.S., Dutta S.P. and Padhe A.M, A new formulation of operation allocation problem in flexible manufacturing system and computational experience. *Int J Prod Res*. 1987, 25, 1267–1283.

Liang, M., and Dutta, S. P., Combined part-selection, load sharing and machine loading problem in hybrid manufacturing system, *International Journal of Production Research*, 1992, 30(10), 2335–2350.

Mandal, S.K., Pandey, M. K. and Tiwari, M. K. Incorporating dynamism in traditional machine loading problem: an AI based optimization approach. *International Journal of Production Research*, 2010, 48 (12), 3535–3559.

Morino, A. A., and Ding, F. Y. Heuristics for FMS-loading and part-type selection problems. *International Journal of Flexible Manufacturing System*. 1993, 5, 287–300.

Mukhopadhyay SK, Singh MK, Srivastava R. FMS loading: a simulated annealing approach. *International Journal of Production Research* 1998, 36(6), 1629–1647.

Prakash, A. Khilwani, N. Tiwari, M.K. and Cohen, Y. Modified immune algorithm for job selection and operation allocation in flexible manufacturing systems. *Advances in engineering software*. 2008, 39, 219–232.

Prakash, A. Tiwari, M.K. and Shankar, R., Optimal job sequence determination and operation machine allocation in flexible manufacturing systems An approach using adaptive hierarchical ant colony algorithm". *Journal of Intelligent Manufacturing*. 2008,19, 161–173.

Sabuncuoglu, I. and Hommertzheim, L. D. Experimental Investigation of FMS machine and AGV scheduling rules against the mean flow time criterion. *International Journal of Production Research*. 1992, 30(7), 1617–1635.

Sarin, S.C. and Chen C.S., The machine loading and tool allocation problem in fexible manufacturing system. *International Journal of Production Research*, 1987, 27(7), 1081–1094.

Sarma, U. M. B. S., Kant, S., Rai, R., and Tiwari, M. K. A solution to machine loading problem in FMS using Tabu search based algorithm. *International Journal of Computer Integrated Manufacturing*. 2002, 15(4), 285–295.

Sawik, T. Modeling and scheduling of flexible manufacturing system, *European Journal of Operation Research*, 1990, 107, 656–668.

Shankar K. and Srinivasulu A., Some selection methodologies for the loading problem in flexible manufacturing system. *Int J Prod Res*. 1989, 27:6, 1019–1034.

Stecke K. Formulation and solution of non-linear integer production planning problems for flexible manufacturing system. *Management Science* 1983, 29, 272–88.

Swarnkar, Rahul, & Tiwari, M. K., Modeling machine loading problem of FMSs and its solution methodology using a hybrid tabu search and simulated annealing-based heuristic approach. *Robotics and Computer-integrated Manufacturing*. 2004, 20, 199–209.

Stecke, K. E., & Solberg, J. J. Loading and control policies for flexible manufacturing systems. *International Journal of Production Research*. 1981, 19, 481–490.

Tiwari,M.K., Hazarika, B.,Vidyarthi,N.K., Jaggi, P.,& Mukopadhyay, S. K. A heuristic solution approach to the machine loading problem of FMS and its Petri net model. *International Journal of Production Research*. 1997, 35(8), 2269–2284.

Tiwari MK, Prakash, Kumar A, and Mileham AR. Determination of an optimal assembly sequence using the Psychoclonal Algorithm. *Journal of Engineering Manufacture* 2005, 219, 137–149.

Tiwari, M. K., Kumar, Sanjeev, Kumar, Shashi, Prakash, & Shankar, R. Solving part type selection and operation allocation problems in A FMS: An approach using constraint based fast simulated annealing algorithm. *IEEE Systems Man and Cybernatics part-A: Systems and Humans*. 2006, 36(6), 1170–1184.

Yao X, Liu Y. Evolutionary programming made faster. *IEEE Transactions on Evolutionary Computation* 1999. 3(2), 82–102.

Yogeswaran, M., Ponnambalam, S.G., and Tiwari, M.K. An efficient hybrid evolutionary heuristic using genetic algorithm and simulated annealing algorithm to solve machine loading problem in FMS. *International Journal of Production Research*. 2009, 47 (19), 5421–5448.

Zeballos, L.J., Quiroga, O.D. and Henning, G.P. A constraint programming model for the scheduling of flexible manufacturing systems with machine and tool limitations. *Engineering Applications of Artificial Intelligence*. 2010, 23 (2), 229–248.

7

Tool Selection in FMS A Hybrid SA-Tabu Algorithm Based Approach

Nitesh Khilwani[1], J. A. Harding[1] and Nishikant Mishra[2]

[1]*Wolfson School of Mechanical and Manufacturing Engineering, Loughborough University, UK*
[2]*School of Management and Business, Aberystwyth University, Aberystwyth, UK*

Abstract
A major advantage of Flexible Manufacturing Systems (FMS) is that a family of related parts can be processed simultaneously using sets of computer controlled workstations, capable of performing a number of different operations and interconnected by automated material handling devices. Flexible manufacturing industry is marked by the coexistence of several resources with varying capabilities, capacities and functions. In such systems the selection of appropriate resources (e.g. tool selection, tool management or tool loading) for machines and allocation of appropriate resources to them is one of the major decision problems. This problem is generally handled by two approaches, i.e. by minimizing the total manufacturing cost or by minimizing the total manufacturing time. Both these approaches lead to NP-hard problems and thus need the application of heuristic techniques or AI based optimization tools to achieve an optimal or near optimal solution. In this chapter, a hybrid algorithm named Simulated Annealing-Tabu is utilized to solve this computationally complex problem. This hybrid algorithm exploits the features of tabu search and simulated annealing to ensure convergence at a faster rate.

Keywords: Flexible manufacturing system, tool selection, Tabu search and simulated annealing

Manoj Tiwari and Jenny A. Harding (eds.) Evolutionary Computing in Advanced Manufacturing, (123–150) © Scrivener Publishing LLC

7.1 Introduction

Today, manufacturing industries are experiencing a remarkable challenge from the consumer market to ensure high productivity in less time and at lower operating cost. Industries are interested in improving their competitive edge in the global market place by realigning their organizational structure and their competitive strategies. However, the product cost is no longer the predominant factor affecting the manufacturers' perception. Other competitive factors such as flexibility, quality, efficient delivery, and customer satisfaction are attracting equal attention. Manufacturing industries are striving to achieve these capabilities through automation, robotics, and other innovative concepts such as just-in-time (JIT), quick response manufacturing (QRM), etc. (Suri 1998).

Flexible manufacturing is a concept that allows manufacturing systems to operate under highly customized production requirements. Issues such as reduction of inventories and market-response time to meet customer demands, high flexibility to adapt to changes in the market, reduction of the cost of products and services to increase market shares, etc. have made it almost obligatory for many firms to switch over to flexible manufacturing systems (FMSs) as a viable means to accomplish the above requirements while producing consistently good-quality and cost-effective products. FMS is actually an automated set of numerically controlled machine tools and material handling systems (Mishra *et al*, 2006).

In recent years, the concept of FMS has emerged as a viable answer to the problems of flexibility and efficiency. However, proper preplanning is a key condition for FMS success in dealing with problems of poor utilization of equipment and tools, and costly and frequent setups. Thus the difficult task facing the production planner is the optimal selection of machines and tools and the assignment of part operations to the selected machines.

This chapter presents a mathematical model for tool selection problems and a heuristic is employed to solve these problems. The main objective is to minimize the operational-cost in the process of tool selection in an FMS. In order to solve this complex NP hard problem, a hybrid heuristic Simulated Annealing-Tabu is utilized in this chapter (Mishra *et al*, 2006). The hybrid heuristic inherits its traits from Simulated Annealing and Tabu search. The hybrid heuristic is applied to a benchmark dataset and the results are compared with the two base algorithms, i.e. SA and Tabu (Swarnkar *et al.*, 2004).

The performance of the proposed SA-tabu based algorithm is effective and reliable for solving intractable problems.

The rest of this chapter is arranged as follows: Section 2 provides a survey of literature on tool selection problems. In section 3 the problem environment is discussed and modelled, whilst section 4, gives the search algorithms and their specific features. Section 5 contains the general framework of the hybrid heuristic. Section 6 deals with computational experience. The solution obtained is also compared with the solutions, obtained by other evolutionally techniques. The superiority of hybrid SA-tabu algorithm is proven over SA and Tabu. Finally, the chapter concludes in Section 7 with final remarks and scope for future work.

7.2 Literature Survey

There have been many contributions to the literature on tool selection in FMS. Stecke (1985) has mentioned four rungs of decision-making for FMS: designing, planning, scheduling, and control. Kusiak *et al* (1985), Stecke and Morin (1985), Van Looveren *et al.* (1986), and Singhal *et al.* (1987) have all addressed the relationships between various decisions and their hierarchies in FMS. Assuming that the design decisions have already been taken care of, efforts are made in this chapter to emphasize the decisions pertaining to planning problems. According to Stecke (1983), FMS planning problems mainly comprise of five decisions:

1. *Part type selection:* Selecting the set of part types to be manufactured simultaneously;
2. *Production ratio determination:* Setting the production ratio for selected part types;
3. *Resource (pallets/fixtures) allocation:* Allocating the limited pallets and fixtures to part types;
4. *Machine grouping:* Partitioning machines into groups of identically tooled machines;
5. *Tool selection and operation allocation:* Loading the machine groups by assigning part operations and the required tools.

Van Looveren *et al.* (1986) refer to steps (1)–(3) as batching, and to steps (4) and (5) as loading. However, if the production ratio is already set, and resource allocation and machine grouping are neglected, the FMS planning problem can be considered to comprise primarily of two critical tasks:

1. *Part type selection (PTS);*
2. *Tool selection and operation allocation.*

These two problems are inter-related, as they follow common restrictions such as tool-magazine capacity, tool-machine compatibility and available machining time. The decision pertaining to the tool selection and operation allocation problem has been considered as a tactical-level planning decision that acts as a link between the strategic and operating level decisions in manufacturing. It receives input from the preceding decision levels, such as part mix selection, resource grouping, aggregate planning, etc., and transfers it to the succeeding decision levels of resource scheduling and dynamic operation planning and control. This is why this problem has been rigorously pursued in the recent past. The different approaches adopted by different researchers to solve tool selection and operation allocation in FMS can be widely classified as:

- Mathematical programming approach (Stecke 1983, Shanker and Tzen 1985, Sarin and Chen 1987, Sawik 1988, 1990, 1996, Escudero 1989, Liang and Dutta 1992);
- Multicriterion decision-making (Ammons *et al.* 1985, Shantikumar and Yao 1987, 1988, Chen and Askin 1990, Kumar *et al.* 1990, Kim and Yano 1994, Tiwari *et al.* 1997);
- Simulation-based approach (Choi and Malstorm 1988, Jain *et al.* 1989, Gympah and Meredith 1996);
- Heuristic-based approach (Shanker and Srinavasulu 1989, Mukhopadhyay *et al.* 1991, 1992, 1998, Moreno and Ding 1993, Mukhopadhyay and Tiwari 1995).

The different models developed by different researchers to formulate the tool selection and operation allocation can be classified as:

- mixed integer programming (MIP) model (Stecke 1983, Shanker and Tzen 1985, Stecke and Kim 1988, Sawik 2000);

- bin-packing model (Afentakis *et al.* 1989);
- discrete generalized network model (Ram *et al.* 1990);
- block angular structure model (Kouvelis and Lee 1991);
- goal programming model (O'Grady and Menon 1987, Moreno and Ding 1989, Rai *et al.* 2002);
- Petri net model (Tiwari *et al.* 1997);
- 0–1 integer linear programming model (Atmani and Lashkari 1998).

Although mixed integer programming (MIP) is the most popular approach for modelling these problems, excessive computational time and infeasibilities hamper its application. Consequently, a common approach is to model the problem with MIP and then solve it by some heuristics approach. Many of these heuristics are again based on mathematical programming tools such as branch and bound (Berrada and Stecke 1986, Shanker and Srinavasulu 1989), Langrarian relaxation (Sarin and Chen 1987, Modi and Shanker 1994), etc. and still require a substantial amount of computing time. In recent years, some modern heuristic techniques such as tabu search, simulated annealing, genetic algorithms, neural networks, etc., have been developed through emulation of certain natural phenomena that can be easily exploited to solve various combinatorial optimization problems. The main advantage of these metaheuristics is their ability to escape local optima in order to get as close as possible to the global optima.

In this chapter, the problem of machine tool selection has been considered for an FMS and a hybrid heuristic Simulated Annealing-Tabu is applied to solve this problem.

7.3 Problem Formulation

Flexible manufacturing industries are marked by the coexistence of several resources with different capabilities, capacities and functions. The selection of appropriate resources such as machines, cutting tools, fixtures, tool access directions etc, is a primary problem in FMS. In this chapter, a real world problem of machine tool selection of an FMS is considered. The FMS assumed for this problem is as follows (Li *et al* 2005, Zhang *et al* 1997):

The FMS has a set of multifunction CNC machines labelled as $m = \{1,2...,M\}$ with different tool magazine capacities, operating costs, maximum available horse power, and tool loading, replacing and changing times. There are T non-identical tools $t = \{1,2...,T\}$ with different tool access directions $w = \{1,2..., W\}$ and a set of P part types $p = \{1,2...,P\}$ that are to be processed by the system. The operations associated with the part type p are represented by the indices $s = \{1,2..., S(p)\}$. Each operation of the part type p requires a specific tool t and may be performed on more than one machine.

The objective is to determine the optimal set of machine and tool combinations for the set of part types. The problem of machine tool selection attempted here includes selection of machines and tools, assignment of selected tools to the machines, and allocation of operations of each part type to the selected machine tool combination. The entire process of selection and allocation includes different costs such as setup cost, machining cost, tool cost, machine change cost, tool change cost, etc. The mathematical formulation of these costs is given as follows (Zhang *et al* 1997):

1. *Machine Cost* (MC): Machining cost is the cost of machines required to perform all the machining operations. It is defined as:

$$MC = \sum_{p=1}^{P} \sum_{s=1}^{S(p)} \sum_{m=1}^{M} MCI_m \times x_{psm}$$

   where, MCI_m is the machine cost index for machine m and $x_{psm} = 1$ if operation s of part p is performed on machine m.
2. *Tool cost* (TC): Tool cost is the cost of tools required to perform all the machining operations. It is defined as:

$$TC = \sum_{p=1}^{P} \sum_{s=1}^{S(p)} \sum_{t=1}^{T} TCI_t \times y_{pst}$$

   where, TCI_t is the tool cost index for tool t and $y_{pst} = 1$ if operation s of part p is performed using tool t.
3. *Machine change cost* (MCC): Machine change cost is considered when there is a change of machine between two operations. It is the cost required to change the machine and all its components in a particular way. It is defined as:

$$MCC = MCCI \times \sum_{p=1}^{P} \sum_{s=1}^{S(p)-1} \sum_{m=1}^{M} \Omega\left(x_{psm}, x_{p(s+1)m}\right)$$

where, *MCCI* is the machine change cost index, and $\Omega\left(x_{psm}, x_{p(s+1)m}\right) = 1$ if both operations s and s+1 are performed on the different machine, else 0.

4. *Set-up change cost* (SCC): Set-up change cost is considered when two adjacent operations are performed on the same machine having different tool access directions (TAD). It is defined as:

$$SCC = SCCI \times \sum_{p=1}^{P} \sum_{s=1}^{S(p)-1} \sum_{m=1}^{M} \sum_{w=1}^{W} \left[1 - \Omega\left(x_{psm}, x_{p(s+1)m}\right)\right] \times \Omega\left(Z_{psw}, Z_{p(s+1)w}\right)$$

where, *SCCI* is the setup change cost index, and $\Omega\left(z_{psw}, x_{pw(s+1)}\right) = 1$ if both operations use different TAD but on the same machine, else 0.

5. *Tool change cost* (TCC): Tool change cost is considered when there is a change of tool on the same machine. Tool change cost between tool *s* and tool *s*+1 occurs when operations are performed on the machine M_m but with different tools:

$$TCC = TCCI \times \sum_{p=1}^{P} \sum_{s=1}^{S(p)-1} \sum_{m=1}^{M} \sum_{t=1}^{T} \left[1 - \Omega\left(x_{psm}, x_{p(s+1)m}\right)\right] \times \Omega\left(y_{pst}, y_{p(s+1)t}\right)$$

where, *TCCI* is the tool change cost index, and $\Omega\left(y_{pst}, y_{p(s+1)t}\right) = 1$ if both operations use different tools but are performed on the same machine, else 0.

6. *Overall Processing Cost*: Overall processing cost is the sum of Machine Cost (MC), Tool Cost (TC), Machine Change Cost (MCC), Tool Change Cost (TCC) and Setup Change Cost (SCC) and it is defined as:

$$OPC = MC + TC + MCC + TCC + SCC$$

Therefore, the objective of the tool selection problem is to minimize the overall processing cost. The random search algorithm used for solving this problem is discussed in the next sections. The two algorithms used in this heuristic are provided in the next section, prior to going into the detailed implementation of the SA-Tabu heuristic for the tool selection problem.

7.4 Background on SA-Tabu Heuristic

Due to the combinatorial nature of the tool selection problem, heuristic techniques are the most promising approach for real life industrial problems. A hybrid Tabu SA algorithm is introduced in this chapter to solve the problem of tool selection in FMS. This heuristic is employed to search for the optimal/near optimal solution by way of systematic reasoning, instead of time-consuming trial and error procedures. Prior to introducing the hybrid Tabu SA algorithm, the basic concepts that lead to the development of the proposed heuristic are presented in this section. As stated earlier the proposed technique employs concepts derived from Tabu search and the Simulated Annealing Algorithm.

7.4.1 Simulated Annealing

Simulated annealing (SA) is a random search optimization technique proposed by Kirkpatrick *et al* (1983). SA draws its inspiration from the physical annealing of solid metal. In annealing, a metal is brought to its lowest energy state by first heating it to a very high temperature (usually recrystallization temperature of metal) and then cooling at a very slow rate, to a very low temperature. The heat causes the atoms to become unstuck from their initial positions and wander randomly through states of higher energy; the slow cooling gives them more chances of finding configurations with lower internal energy than the initial one.

By analogy with the above physical process, the SA algorithm starts with a random solution and iteratively selects a different solution based on the objective value of the solution and an assumed parameter, known as transition probability (TP). The process accepts not only those solutions, which improve the objective function, but also those solutions, which do not improve the objective function on the basis of the transition probability. The value of TP is considered high at the start and is decreased gradually at a predefined speed. Thus, the selection of solutions is almost random in the initial phase of the algorithm and becomes selective at the end as TP becomes close to 0.

The generic flow of the SA algorithm is given below:

1. Generate an initial solution *Sol*
2. Get an initial temperature T, where $T > 0$,
3. Execute the following steps till the temperature does not freeze:

   Step 1. Perform the following loop K times

   I. Select a random neighbour solution Sol_p of *Sol*.
   II. Calculate $\Delta = F\,(Sol) - F\,(Sol_p)$
   III. If $\Delta > 0$, except the new solution $Sol = Sol_p$
   IV. If $\Delta \leq 0$, select the solution with a probability, $P(\Delta, T)$
   V. If $F\,(Sol) \geq F(Sol_{best})$, then $Sol_{best} = Sol$

   Step 2. Reduce the temperature, T.

4. Sol_{best} is the best solution.

The main features of SA that make this algorithm more sophisticated are the perturbation annealing schedule and transition probability. Perturbation generates a new solution, the annealing schedule controls the initial temperature, final temperature and rate of cooling while the transition probability helps the heuristic to escape from local optima. However, the SA algorithm with all its advantages also has some demerits; such as it requires a large number of iterations to generate an optimal or near optimal solution. In addition, SA has no concept of a short-term memory list of prohibited neighbouring solutions as in the tabu search algorithm and hence the possibility of revisiting a solution is increased. These two drawbacks posed by SA increase the number of iterations required and thus a longer computational time is required to generate the global optima solution.

7.4.2 Tabu Search

Tabu search, originally proposed by Glover (Glover 1990), is regarded as a higher level heuristic that can be used for solving various computationally complex problems. The main feature of tabu search is the tabu-list, a short-term memory used to store the solutions visited in the recent past and avoid them in further iterations. This memory structure helps the search to overcome the problem of being trapped in a local optima and thereby saves computational time. Many successful implementations of tabu search, for obtaining optimal or near optimal solution of problems pertaining to process planning, scheduling, set partitioning and loading etc, are reported in the literature (Glover 1990, Taillerd 1990, Aljaber *et al.* 1997, McMullen 1998, Swarnakar *et al* 2004).

Tabu search uses a local or neighbourhood search procedure to iteratively move from one solution to another better solution in the neighbourhood, until some stopping criterion has been satisfied. The search begins with an initial feasible solution and iterates to find a better solution by evaluating a large pool of neighbourhood solutions. This method is characterized by its inherent simplicity, high adaptability, and a short term memory structure called the tabu-list. The tabu search also allows backtracking to a previous solution that may lead to a better solution, and this is known as aspiration. These two features, tabu-list and aspiration, make tabu search a powerful tool for solving various complex optimization problems. A generic flow of the tabu search algorithm for a maximization problem is as follows:

1. Create an initial solution *Sol*
2. Initialize tabu list $Tl = NULL$
3. Perform the following loop *N* times

    Step 1. Perturb *Sol* and get Sol_p
    Step 2. I $F(Sol_p) > F(Sol)$ for $F(Sol_p) >$ *Aspiration level*

    Then $Sol = Sol_p$

    $Tl = Sol$

    Step 3. If $F(Sol) \geq F(Sol_{best})$

    Then $Sol_{best} = Sol$
4. Sol_{best} is the desired solution.

The main drawback associated with tabu search is cycling, in which it follows the same path unless a tabu neighbour exists. Alternatively, it can be said that a loop will be encountered if the search moves to a previously visited solution that has not been tabu for the last two iterations.

7.4.3 Simulated Annealing-Tabu

As evident from the above discussion, the two algorithms have complementary differences (Swarnkar *et al* 2004), i.e.

- Tabu search makes use of a tabu list, which is a short-term memory of recently visited solutions, to escape local optima, but it has a deterministic nature and cannot avoid cycling.
- SA has a stochastic characteristic and hence can avoid cycles but the rate of improvement of the solution is very slow. Further, SA does not keep any track of previously visited solutions, and thus it is always possible for the search to return to such a solution again.

The limitations of both the algorithms have forced the operational research community to conceive a new technique, which combines the beauties associated with the aforementioned algorithms. Similar sort of work has been carried out by Zolfaghari and Liang (1999), to solve the group scheduling and machine speed selection problems using a SA-tabu algorithm. The hybrid tabu search and simulated annealing based approach retains all the benefits of tabu and SA searches. The details of this hybrid technique applied to the tool selection problem are provided in the next section.

7.5 Implementation of Tabu-Simulated Annealing

This section discusses the implementation of the hybrid SA-Tabu algorithm on the tool selection problem discussed in Section 7.3.

7.5.1 Notations Used in SA-Tabu Heuristic

Sol	The current string of operations
Sol_p	The string from the current string
Sol_{best}	The best string encountered so far.
$F(Sol)$	Overall processing cost for the string *Sol*
ΔS	Change in function *F* due to perturbation over *F*(*Sol*)

Δs_{best}	Change in function F due to perturbation over $[F(Sol]_{best})$
Reject	Variable that stores the number of solutions rejected so far, for its value
i	Number of iterations in which the temperature is reduced
R	Random number between 0 and 1
T	Temperature
T_0	Initial temperature
TL	Tabu list, a set of currently visited solutions
Generate (Sol_p, Sol)	Subroutine for generating neighbouring string from string *Sol* randomly
A	Aspiration level
$P(\Delta, T)$	Transition probability function

7.5.2 Steps of the Hybrid SA-Tabu Heuristic

Step 1: Initialize a feasible string *Sol* and evaluate the total processing cost $F(Sol)$ for *Sol*.

Step 2: Initalize $T = T_0 = 200$; $i = 1$; $TL = \varnothing\varnothing\varnothing$; $Reject = 0$; $Sol_{best} = Sol$

Step 3: Generate(Sol_p, Sol); i.e. generate new sequence from the existing one by randomly selecting different machine and tool for any random operation.

Step 4: If $Sol_p \in TL$

Then goto step 5
Else goto Step 6

Step 5: If $[F(Sol]_p) \geq A$

Then goto step 6
Else goto Step 3

Step 6: $\Delta S = F(Sol) - F(Sol_p)$

If $(\Delta S \leq 0)$
Then goto step 7
Else goto step 10

Step 7: Assign $Sol = Sol_p$; $F(Sol) = F(Sol_p)$

Include Sol_p in the tabulist
$TL \leftarrow Sol_p$
Update aspiration level
$A \leftarrow F(Sol_p)$

Step 8: $\Delta S_{best} = F(Sol_{best}) - F(Sol_p)$
If $\Delta S_{best} \leq 0$
Then goto step 9
Else goto step 13

Step 9: Assign $Sol_{best} = Sol_p$; $F(Sol_{best}) = F(Sol_p)$
$Reject \leftarrow 0$
Goto step 13

Step 10: Compute

$$P = \exp\left(-\frac{\Delta S}{T}\right)$$

Sample a random variable R in the range of (0,1)
If ($P < R$)
Then goto step 12
Else goto step 11

Step 11: Assign $Sol = Sol_p$; $F(Sol) = F(Sol_p)$
Include Sol_p in the tabulist
$TL \leftarrow Sol_p$
Update aspiration
$A \leftarrow F(Sol_p)$

Step 12: $Reject = Reject + 1$
If $Reject \geq 3$
Then goto step 14
Else goto step 13

Step 13: $i = i + 1$
Change the value of T

$$T = \frac{T_0}{1 + \ln i}$$

If $i \geq 15$
Then goto step 14
Else goto step 3

Step 14: Freeze
Sol_{best}

7.5.3 Representation

An appropriate representation of the candidate solution plays a key role in maintaining the search effectiveness of any algorithm.

There are a number of encoding schemes proposed by different researchers, e.g. binary coding by Holland (1975), permutation encoding scheme by Goldberg and Lingle (1985), adjacent coding by Grefenstette *et al.* (1985), matrix-based encoding scheme by Sawaka *et al.* (1996), etc. However, for any application, it is necessary to perform a careful analysis to ensure an appropriate representation of solutions, together with meaningful and problem-specific genetic operators.

The tool selection problem mainly consists of two parts, selection of machine and selection of tool for a particular operation. Therefore, a typical solution to the problem is represented by two sets of strings, i.e. representing the machine selected for each operation of all parts and tool allocation to each operation of all part types. Figure 7.1 represents the structure of the string.

7.5.4 Search Parameters

The hybrid SA-tabu heuristic is to be applied on a minimization problem considered in this chapter. The heuristic starts with a random string generated using the above step and perturbation to generate new strings. The performance of the heuristic is governed by various parameters, such as:

1. *Neighbourhood generation (Perturbation)*: The method used in this chapter to generate a new string is taken from Sridhar and Rajendran (1993). The adjacent interchanged method generates a random number between 1 and total number of operations and interchanges the type of machine and tool with the available machines and tools randomly.
2. *Transition probability*: When a random string is generated, its function value is evaluated. If the function value

Operation Number	1	2	3	4	5	6	7	8	9	10
Machine Number	**2**	**4**	**1**	**6**	**8**	**7**	**10**	**3**	**5**	**9**
Tool Number	**5**	**6**	**1**	**4**	**7**	**9**	**2**	**8**	**3**	**10**

Figure 7.1 Solution String for 10 operations, where 1st row represents machine number for operations and 2nd row represents tool number for that operation.

is better than the old string, it is accepted, otherwise it is accepted only if the transition probability is higher than a uniform random number. For every new string that is inferior to the old string, the transition probability P is calculated as:

$$P = \exp\left(-\frac{\Delta S}{T}\right)$$

where, ΔS is the difference of function values of current solution and the neighbouring solution and T is the temperature.

The probability function has a higher value for small ΔS at high temperature (T), and a lower value for larger ΔS at lower temperature. The search initially starts with a high temperature, so there will be a better chance of escaping local optima. As the temperature freezes with the number of iterations, the transition probability approaches to zero. Thus, the search is no longer able to escape local optima and the neighbouring solution is accepted only if it shows improved results. In case the transition from the current solution to the neighbouring solution is rejected, another solution in the neighbourhood is selected and evaluated.

3. *Annealing schedule*: The temperature decline is performed using a function known as the annealing schedule. In this chapter, the following annealing schedule is used (Swarnkar *et al* 2004):

$$T = \frac{T_0}{1 + \ln i}$$

In this annealing schedule, the cooling rate is high in the beginning and low at the end. This schedule is adopted keeping in mind the fact that the chances of getting better results are higher in the beginning, therefore, there is not much use of a higher transition probability to escape the local optima. After a few iterations, an attempt has been made to prevent the temperature from decreasing abruptly, which may result in frequent rejection of a perturbed solution.

4. *Tabulist*: The Tabu list is used to check whether the solution generated is latterly visited or not and thus it helps in restricting the algorithm from revisiting the pre-visited solutions. This feature of hybrid SA-tabu reduces the computational time of the algorithm to reach the global optima.
5. *Aspiration level*: This is a tabu search variable denoted by A, which restricts the search from being trapped at a solution surrounded by tabu neighbours. If a solution has a neighbourhood of only tabu solutions, the one with an objective function value greater than the aspiration level is chosen to explore further.
6. *Reject*: In tabu search, a solution is often rejected on the basis of probability considerations. After a certain limit of rejection, it is assumed that no superior solution exists in the neighbourhood and the search has reached a near optimal/optimal solution i.e. the variable Reject stores the number of rejections of perturbed solutions. The parameter Reject increases by one, whenever a solution is not selected by probability consideration. If it reaches a predetermined fixed value, the search procedure is stopped or started again depending on the other criteria.
7. *Stopping criteria*: To stop the search procedure from roaming in the solution space, two criteria are incorporated in this chapter.

    a. Number of iterations: During an experiment, it is found that as long as K = 15 i.e. equivalent temperature falls to 54 'C , any further reduction in temperature would not be useful because at low temperatures the possibility of accepting an inferior solution is very small and the results obtained are virtually indistinguishable from the optimal solution.
    b. Reject variable: When Reject acquires a predetermined fixed value (3 in this chapter) then no optimal or near optimal solution has been achieved during the last four iterations, therefore the probability of obtaining any better solution is small and the search procedure is stopped from roaming in the solution space.

7.6 Test Cases

To demonstrate the efficacy and robustness of the proposed model and analyze the capability of the SA-tabu heuristic, a test case has been adopted from the literature. A multi-constrained data set with increasing complexity has been considered as a benchmark. The test case maps the scenario described by Zhang *et al* (1997) for the single plant FMS. In this case, the manufacturing unit produces a multi-feature prismatic part having 19 different design features. The information related to the operations, machines, tools, and Tool Access Direction (TAD) is shown in Table 7.1. From Table 7.1, it is evident that 23 operations are performed on a part type using the set of available tools, machines and TAD.

Table 7.1 Available machines, tools and TAD for different operations for the case study 1 (Adopted from Zhang (1997)).

Feature-ID	Operations no	Type of operation	Available TAD	Available machine	Available tool
F01	OP01	Drilling	+Z, –Z	M01, M02, M03	T01
F02	OP02	Drilling	+Z, –Z	M01, M02, M03	T01
F03	OP03	Milling	+Z, –Z	M02, M03	T08
F04	OP04	Milling	+Y, –Z	M02, M03	T05, T06
F05	OP05	Milling	+Y	M02, M03	T05, T06
F06	OP06	Milling	+Y	M02, M03	T05, T06
F07	OP07	Drilling	+Z, –Z	M01, M02, M03	T02
F07	OP08	Reaming	+Z, –Z	M01, M02, M03	T03
F07	OP09	Boring	+Z, –Z	M03, M04	T04
F08	OP10	Drilling	+Z, –Z	M01, M02, M03	T01

(Continued)

Table 7.1 Available machines, tools and TAD for different operations for the case study 1 (Adopted from Zhang (1997)). (Continued)

Feature-ID	Operations no	Type of operation	Available TAD	Available machine	Available tool
F09	OP11	Drilling	+Z, –Z	M01, M02, M03	T02
F09	OP12	Ramming	+Z, –Z	M01, M02, M03	T03
F09	OP13	Boring	+Z, –Z	M03, M04	T04
F10	OP14	Milling	+X	M02, M03	T05, T06
F11	OP15	Drilling	–Z	M01, M02, M03	T01
F12	OP16	Drilling	–Z	M01, M02, M03	T01
F13	OP17	Milling	–Y, –Z	M02, M03	T05, T06
F14	OP18	Milling	–Y, –Z	M02, M03	T05, T06
F15	OP19	Drilling	+Z, –Z	M01, M02, M03	T01
F16	OP20	Drilling	+Z, –Z	M01, M02, M03	T01
F17	OP21	Milling	–Y	M02, M03	T05
F18	OP22	Drilling	–Y	M01, M02, M03	T06
F19	OP23	Drilling	–Y	M01, M02, M03	T01

There are eight different types of tools (T01-T08) assigned on four different types of machines (M01-M03), that can perform three different types of TAD. The cost index of machine (MCCI), tool (TCCI), and setup (SCCI) changes are 300, 10, and 90 respectively. The other cost indexes such as machine and tool cost index are listed in the Table 7.2. The precedence relation between the

Table 7.2 Cost Index of different operations, machines and tools for the case study 1 (Adopted from Zhang (1997)).

Machine/ Tool ID	Type	Cost Indices
M01	Press drill	10
M02	Vertical milling	35
M03	Vertical CNC milling	60
M04	Boring machine	50
T001	Drill d 0.2	03
T002	Drill d 1.2	03
T003	Reamer	08
T004	Boring tool	15
T005	Milling cutter 1	10
T006	Milling cutter 2	15
T007	Slot cutter	10
T008	Chamfer tool	10

Machine Cost Index (MCI): 300.
Tool Cost Index (TCI): 10.
Set up Change Cost Index (SCCI): 90.

operations due to different technological rules such as filterability, tolerance factor, operation stages requirement for machining is listed in Table 7.3.

The two basic algorithms SA and Tabu and the hybrid SA-Tabu heuristic are implemented to get optimal solutions for this test case. The optimal solution with overall processing cost calculated for the test case is provided in Table 7.4. From Table 7.4, it is evident that the performance of the hybrid heuristic is superior to the two basic algorithms. The hybrid heuristic gives a better solution than the other two algorithms in 40 generations. The best solution obtained from the approaches is provided in Table 7.5–7.7. The table provides the operation sequence, operation number, machine type, tool type and total processing cost, along with the total number of tool changes, machine changes and TAD changes.

Table 7.3 Precedence relation between different operations for the case study 1 (Adopted from Zhang (1997)).

	01	**02**	**03**	**04**	**05**	**06**	**07**	**08**	**09**	**10**	**11**	**12**	**13**	**14**	**15**	**16**	**17**	**18**	**19**	**20**	**21**	**22**	**23**
01	0	1	1	0	0	0	0	0	0	0	0	0	0	0	0	0	0	0	0	0	0	0	0
02	0	0	1	0	0	0	0	0	0	0	0	0	0	0	0	0	0	0	0	0	0	0	0
03	0	0	0	0	0	0	0	0	0	0	0	0	0	0	0	0	0	0	0	0	0	0	0
04	0	0	1	0	0	0	0	0	0	0	0	0	0	0	0	0	0	0	0	0	0	0	0
05	0	0	0	1	0	1	0	0	0	0	0	0	0	0	1	1	1	1	1	1	1	1	1
06	0	0	0	1	0	0	0	0	0	0	0	0	0	0	1	1	1	1	1	1	1	1	1
07	0	0	0	0	0	0	0	1	1	1	1	1	1	0	0	0	0	0	0	0	0	0	0
08	0	0	0	0	0	0	0	0	1	1	1	1	1	0	0	0	0	0	0	0	0	0	0
09	0	0	0	0	0	0	0	0	0	1	1	1	1	0	0	0	0	0	0	0	0	0	0
10	0	0	0	0	0	0	0	0	0	0	0	0	0	0	0	0	0	0	0	0	0	0	0
11	0	0	0	0	0	0	0	0	0	1	0	1	1	0	0	0	0	0	0	0	0	0	0
12	0	0	0	0	0	0	0	0	0	1	0	0	1	0	0	0	0	0	0	0	0	0	0
13	0	0	0	0	0	0	0	0	0	1	0	0	0	0	0	0	0	0	0	0	0	0	0

(Continued)

Table 7.3 Precedence relation between different operations for the case study 1 (Adopted from Zhang (1997)). (Continued)

	01	02	03	04	05	06	07	08	09	10	11	12	13	14	15	16	17	18	19	20	21	22	23
14	0	0	1	0	0	0	0	0	0	0	0	0	0	0	0	0	1	1	1	1	1	1	1
15	0	0	0	0	0	0	0	0	0	0	0	0	0	0	0	1	0	0	0	0	0	0	0
16	0	0	0	0	0	0	0	0	0	0	0	0	0	0	0	0	0	0	0	0	0	0	0
17	0	0	0	0	0	0	0	0	0	0	0	0	0	0	0	0	0	0	0	0	0	0	0
18	0	0	0	0	0	0	0	0	0	0	0	0	0	0	0	0	1	0	1	0	0	0	0
19	0	0	0	0	0	0	0	0	0	0	0	0	0	0	0	0	0	0	0	0	0	0	0
20	0	0	0	0	0	0	0	0	0	0	0	0	0	0	0	0	0	0	0	0	0	0	0
21	1	1	1	0	0	0	1	1	1	1	1	1	1	0	0	0	1	1	0	0	0	1	1
22	0	0	0	0	0	0	0	0	0	0	0	0	0	0	0	0	0	0	0	0	0	0	1
23	0	0	0	0	0	0	0	0	0	0	0	0	0	0	0	0	0	0	0	0	0	0	0

Table 7.4 Computational result for the test problem.

Criteria	Total production cost	Number of generation
Hybrid SA-Tabu	1739	40
Tabu	1749	45
SA	1742	8000

The main observations pertaining to the application of the three approaches are as follows:

1. The proposed SA-tabu approach is stochastic in nature and its search scheme prevents the entrapment of a solution at local optimum points.
2. The deterministic nature of tabu search often leads to effective avoidance of cycling of search path by the probabilistic nature of the hybrid algorithm.
3. Due to the short-term memory or tabu list incorporated in the new algorithm the number of solution revisits are significantly reduced when compared to those in the SA approach. In this way the computational time needed to achieve a near-optimum solution is considerably reduced.

7.7 Conclusion

In this chapter, the tool selection problem of FMS has been addressed using a hybrid heuristic called the simulated annealing-tabu. The problem discussed is a well-known computationally complex problem having a huge search space. Thus to achieve an optimal / near optimal solution with respect to a set of objective functions, an exhaustive search algorithm is not applicable in a real world scenario and therefore it becomes essential to use a random search technique. However, application of the proposed solution methodology is restricted to a single test case with a single part type and multiple operations. This research can be further extended by considering more complex test cases. Further, a few more objective functions namely, measures of flexibilities associated with machines, material handling, etc can be considered in future research. The hybrid heuristic can also be extended to similar problems is such areas as sequencing, resource allocation, and multi-objective loading and scheduling problems.

Table 7.5 Final operation sequence obtained by SA.

	1	**2**	**3**	**4**	**5**	**6**	**7**	**8**	**9**	**10**	**11**	**12**	**13**	**14**	**15**	**16**	**17**	**18**	**19**	**20**	**21**	**22**	**23**
Oper	14	05	06	04	21	22	23	20	01	18	15	16	19	02	17	07	08	03	09	11	12	13	10
Mach	02	02	02	02	02	02	02	02	02	02	02	02	02	02	02	02	02	02	03	03	03	03	03
Tool	05	05	05	05	05	01	01	01	01	01	01	01	01	01	07	02	03	08	04	02	03	04	01
TAD	+x	+y	+y	+y	−y	−y	−y	−z	−z	−z	−z	−z	−z	−z	−z	−z	−z	−z	−z	−z	−z	−z	−z

The total operation cost: 1742.
Number of tool change: 09.
Number of machine change: 01.
Number of TAD change: 03.

Table 7.6 Final operation sequence by Tabu Search.

	1	**2**	**3**	**4**	**5**	**6**	**7**	**8**	**9**	**10**	**11**	**12**	**13**	**14**	**15**	**16**	**17**	**18**	**19**	**20**	**21**	**22**	**23**
PID	14	05	6	04	21	18	22	23	20	01	15	16	19	02	17	07	08	03	09	11	12	13	10
MID	02	02	02	02	02	02	02	02	02	02	02	02	02	02	02	02	02	02	03	03	03	03	03
TID	05	05	05	06	06	06	01	01	01	01	01	01	01	01	07	02	03	08	04	02	03	04	01
TAD	+x	+y	+y	+y	−y	−y	−y	−y	−z	−z	−z	−z	−z	−z	−z	−z	−z	−z	−z	−z	−z	−z	−z

The total operation cost: 1749.
Number of tool change: 09.
Number of machine change: 01.
Number of TAD change: 03.

Table 7.7 Final operation sequence by Hybrid SA-Tabu.

	1	2	3	4	5	6	7	8	9	10	11	12	13	14	15	16	17	18	19	20	21	22	23
PID	14	05	06	04	21	18	17	22	23	20	01	15	16	19	02	07	08	03	09	11	12	13	10
MID	02	02	02	02	02	02	02	02	02	02	02	02	02	02	02	02	02	02	03	03	03	03	03
TID	05	05	05	05	05	05	01	01	01	01	01	01	01	01	01	02	03	08	04	02	03	04	01
TAD	+x	+y	+y	+y	−y	−y	−y	−y	−y	−z	−z	−z	−z	−z	−z	−z	−z	−z	−z	−z	−z	−z	−z

The total operation cost: 1739.
Number of tool change: 08.
Number of machine change: 01.
Number of TAD change: 03.

References

Afentakis, P., Solomon, MM and Miller, RA The part type selection problem. Proceedings of the Third ORSA/TIMS Conference on Flexible Manufacturing Systems Cambridge, MA pp. 141–146.

Aljaber, N., Wonjang, B. and Chen, C. L., 1997, A Tabu search approach to the cell formation. *Computers in Industrial Engineering*, 32(1), 169–185.

Ammons, JC , Lofgren, CB and McGinnis, A. (1985) A large scale machine loading problem in flexible assembly. *Ann. Oper. Res.* 3, pp. 319–322.

Atmani, A. and Lashkari, RS (1998) A model of machine-tool selection and operation allocation in FMS. *Int. J. Prod. Res.* 36:5, pp. 1339–1349.

Berrada, M. and Stecke, KE (1986) A branch and bound approach for machine load balancing in flexible manufacturing systems. *Manage. Sci.* 32, pp. 1316–1335.

Chen, YJ and Askin, RG (1990) A multiobjective evaluation of flexible manufacturing system loading heuristics. *Int. J. Prod. Res.* 25, pp. 895–911.

Choi, RH and Malstorm, EM (1988) Evaluation of traditional work scheduling rules in a FMS with a physical simulator. *J. Manuf. Syst.* 7, pp. 3032–3045.

Escudero, LF (1989) An exact algorithm for part input sequencing and scheduling with wide constraints in FMS. *Int. J. Flexible Manuf.* Syst. 1, pp. 143–174.

Glover F. (1990) Tabu search: a tutorial. *Interfaces*, 20(4):74–94.

Goldberg, D.E. and Lingle, R., Alleles, loci and the traveling salesman problem, in *Proceedings of the First International Conference on Genetic Algorithm and their Applications*, Hillsdale, NJ, 1985, pp. 931–940.

Grefenstette, J.J., Gopal, R., Rormatia, B. and Vangucht, D., Genetic algorithm for travelling salesman problem, in *Proceedings of the First International Conference on Genetic Algorithms and their Applications*, Hillsdale, NJ, 1985.

Gympah, KA and Meredith, JR (1996) A simulation study of FMS tool allocation procedures. *J. Manuf. Syst.* 15, pp. 419–431.

Holland, J.H. (editor), *Adaptation in Natural and Artificial System*, 1975 (University of Michigan Press: Detroit, MI).

Jain, S. , Barber, K. and Osterfeld, D. Expert simulation for on-line scheduling. *Proceedings of the 1989 Winter Simulation Conference Washington DC* pp. 930–935.

Kim, YD and Yano, CA (1994) A new branch and bound algorithm for loading problem in flexible manufacturing systems. *Int. J. Flexible Manuf. Syst.* 6, pp. 361–382.

Kirkpatrick F, Gelatte CD, Vecchi MP (1983) Optimization by simulated annealing. *Science*; 220:671–780.

Kouvelis, P. and Lee, HL (1991) Block angular structures and the loading problem in flexible manufacturing system. *Oper. Res.* 39, pp. 666–676.

Kumar, P., Tiwari, NK and Singh, N. (1990) Joint consideration of grouping and loading problem in a flexible manufacturing system. *Int. J. Prod. Res.* 28:7, pp. 1345–1356.

Kusiak, A. Raouf, A. and Ahmed, SH (eds) (1985) Loading models in flexible manufacturing systems. *Manufacturing Research and Technology*—1 Elsevier, Amsterdam — edited by.

Li, L., Fuh, J.Y.H., Zhang, Y.F and Nee, A . Y. C. (2005) Application of genetic algorithm to computer-aided process planning in distributed manufacturing environment. *Robotics and computer integrated manufacturing*, 21, pp. 568–578.

Liang, M. and Dutta, SP (1992) Combined part-selection, load-sharing and machine-loading problem in flexible manufacturing system. *Int. J. Prod. Res.* 30:10, pp. 2335–2350.

Mcmullen, P. R., 1998, JIT sequencing for mixed-model assembly lines with setups using Tabu search. *Production Planning and Control*, 9(5), 504–510.

Mishra, S., Prakash, Tiwari, M. K., Lashkari, R. S., (2006)A fuzzy goal-programming model of machine-tool selection and operation allocation problem in FMS: a quick converging simulated annealing-based approach, *International Journal of Production Research*, Vol. 44, No. 1, 43–76.

Modi, BK and Shanker, K. (1994) Models and solution approaches for part movement minimization and load balancing in FMS with machine, tool process plan flexibilities. *Int. J. Prod. Res.* 33:7, pp. 1791–1816.

Moreno, AA and Ding, I. (1989) Goal oriented heuristics for the FMS loading (and part type selection) problems. *Proceedings of the Third ORSA/TIMS Conference on Flexible Manufacturing System* pp. 105–110.

Mukhopadhyay, SK, Maiti, B. and Garg, S. (1991) Heuristic solution to the scheduling problem in flexible manufacturing system. *Int. J. Prod. Res.* 29, pp. 2003–2024.

Mukhopadhyay, SK, Midha, S. and Krishna, VA (1992) A heuristic procedure for loading problem in flexible manufacturing system. *Int. J. Prod. Res.* 30, pp. 2213–2228.

Mukhopadhyay, SK, Singh, MK and Srivastava, R. (1998) FMS machine loading: a simulated annealing approach. *Int. J. Prod. Res.* 36, pp. 1526–1547.

Mukhopadhyay, SK and Tiwari, MK Solving machine loading problem of FMS by conjoint measurement. *Proceedings of the 13th International Conference on Production Research Jerusalem pp.* 74–76,1995.

O'Grady, PJ and Menon, U. (1987) Loading a flexible manufacturing system. *Int. J. Prod. Res.* 25, pp. 1053–1068.

Rai, R., Kameshwaran, S. and Tiwari, MK (2002) Machine-tool selection and operation allocation in FMS: solving a fuzzy goal-programming model using a genetic algorithm. *Int. J. Prod. Res.* 40:3, pp. 641–665.

Ram, B., Sarin, SC and Chen, CS (1990) A model and solution approach for the machine loading and tool allocation problem in FMS. *Int. J. Prod. Res.* 28:4, pp. 637–645.

Sarin, SC and Chen, CS (1987) The machine loading and tool allocation problem in flexible manufacturing system. *Int. J. Prod. Res.* 25:7, pp. 1081–1094.

Sawaka, M., Kato, K. and Muri, T., Flexible scheduling in a machining center through genetic algorithm. *Comput. Ind. Eng.*, 1996, 30(4), 931–940.

Sawik, TJ (1988) Modeling and scheduling a batch-type production on identical machines. *Eur. J. Oper. Res.* 35, pp. 393–400.

Sawik, TJ (1990) Modeling and scheduling of flexible manufacturing system. *Eur. J. Oper. Res.* 45, pp. 177–190.

Sawik, TJ (1996) A multilevel machine and vehicle scheduling in a flexible manufacturing system. *Math. Comput. Model.* 23, pp. 45–57.

Sawik T. (2000) Mixed integer programming for scheduling flexi- ble flow lines with limited intermediate buffers. *Mathematical and Computer Modelling* 31(13): 39–52.

Shanker, K. and Srinavasulu, A. (1989) Some solution methodologies for a loading problem in a flexible manufacturing system. *Int. J. Prod. Res.* 27:6, pp. 1019–103.

Shanker, K. and Tzen, YJ (1985) A loading and dispatching problem in a random flexible manufacturing system. *Int. J. Prod. Res.* 23:3, pp. 579–595.

Shantikumar, JJ and Yao, DD (1987) Optimal server allocation in a system of multi-server stations. *Manage. Sci.* 33, pp. 1173–1182.

Shantikumar, JJ and Yao, DD (1988) On server allocation in a multiple center manufacturing system. *Oper. Res.* 36, pp. 333–342.

Singhal, K., Fine, CH , Meredith, JR and Suri, R. (1987) Research and models for automated manufacturing. *Interfaces* 17, pp. 5–14.

Sridhar, J., & Rajendran, C. (1993). Scheduling in cellular manufacturing system: A simulated annealing approach. *International Journal of Production Research*, 31(12), 2927–2945.

Stecke, KE (1983) Formulation and solution of non-linear integer production planning problem for flexible manufacturing system. Manage. Sci. 29:3, pp. 273–288.

Stecke, KE (1985) A hierarchical approach to solve grouping and loading problems of flexible manufacturing systems. *Eur. J. Oper. Res.* 24:3, pp. 369–378.

Stecke, KE and Morin, TL (1985) The optimality of balancing workload in certain types of flexible manufacturing systems. *Eur. J. Oper. Res.* 20, pp. 68–82.

Stecke, K.E. and Kim, I. (1988) A study of FMS part type selection approaches for short-term production planning. *International Journal of Flexible Manufacturing Systems*, 1, 7–29.

Suri, R., (1998) *Quick Response Manufacturing*, (Productivity Press: Portland, OR).

Swarnkar, R., & Tiwari, M. K. (2004).Modeling machine loading problem of FMSs and its solution methodology using a hybrid tabu search and simulated annealing-based heuristic approach. *Robotics and Computer- Integrated Manufacturing*, 20, 199–209.

Taillard, E., 1990, Some efficient heuristic methods for flow shop sequencing problems. *European Journal of Operational Research*, 47, 65–74.

Tiwari, MK , Hazarika, B. , Jaggi, P. , Vidyarthi, NK and Mukhopadhyay, SK (1997) A heuristic solution approach to the machine loading problem of an FMS and its petrinet model. *Int. J. Prod. Res.* 35:8, pp. 2269–2284.

Van Looveren, AI , Gelders, JLF and Van Wassenhove, LN Kusiak, A. (ed) (1986) A review of FMS planning models. *Modeling and Design of Flexible Manufacturing Systems* pp. 3–31. Elsevier, Amsterdam — edited by.

Zhang, F., Zhang, Y.F., and Nee, A. Y. C. (1997) Using genetic algorithm in process planning for job shop machining. *IEEE transactions of evolutionary computation*, 1(4), 278–288.

Zolfaghari S, Liang M. Jointly solving the group scheduling and machine speed selection problems: a hybrid tabu simulated and simulated annealing approach. *Int J Prod Res* 1999;37(10): 2377–97.

8

Integrating AGVs and Production Planning with Memetic Particle Swarm Optimization

Sri Krishna Kumar[1], M.K. Tiwari[2] and J. Harding[1]

[1]*Wolfson School of Mechanical and Manufacturing Engineering, Loughborough University UK*
[2]*Department of Industrial Engineering and Management, IIT Kharagpur, India*

Abstract

Automated guided vehicles (AGVs) play an important role in advanced manufacturing system where jobs are being transferred from one machine to another using AGVs on guided paths. The focus of this research is on the development of an integrated production planning and AGV control system in the real time environment. This problem consists of simultaneous assignment, scheduling and conflict free routing of the AGVs with bi-directional movement along with machine and job scheduling. A variant of particle swarm optimization, Enhanced Memetic Particle Swarm Optimization (EMPSO), has been utilized for optimizing the integrated production planning and AGVs system. EMPSO Optimization scheme incorporates local search techniques in the standard Particle Swarm Optimization (PSO) algorithm, resulting in an efficient and effective optimization method. Results obtained justify the superiority of the Memetic approach.

Keywords: Production planning, AGV, PSO, EMPSO

8.1 Introduction

An automated guided vehicles (AGVs) system is a set of driverless vehicles used to transfer materials and goods between the two

Manoj Tiwari and Jenny A. Harding (eds.) Evolutionary Computing in Advanced Manufacturing, (151–168) © Scrivener Publishing LLC

points. AGVs systems have been successfully adopted in a large variety of industries such as aerospace, automotive, chemical, electronics, food and beverage, plastic, and container ports. As defined by the Material Handling Industry of America (MHIA, 1999), an AGVs system provides better production planning and control, safety, cost reduction, and flexibility. In the paradigm of advanced manufacturing, with frequent changes in production planning and scheduling, AGVs systems have been proved to be efficient material handling equipment (Asef-Vaziri and Goetschalckx, 2008). They are used to transport goods and materials between workstations (machines) and storehouses of an automated manufacturing system through pre-determined guide paths, and their routings can be altered arbitrarily according to transportation requests. Thus, an AGV system enables an advanced manufacturing system to be flexible in material routing and dispatching between machines and is well suited to environments where product specification and production priorities vary continuously.

The advanced manufacturing system is composed of various working stations, each with a specific function such as milling, drilling, polishing, washing, or assembly. These work stations are connected with the guide path network by pick-up/delivery (P/D) points where AGVs transfer the unfinished material (as shown in the Figure 8.1). For efficient performance of the advanced manufacturing system, optimal design and control of the AGVs system should be adopted (Maza and Castagna , 2005).

The design and control processes of AGVs involve many issues and can be divided in three levels: 1. Strategic 2. Tactical and 3. Operational. Strategic level decisions mostly involve guide-path design and have a strong impact on the decisions of other levels.

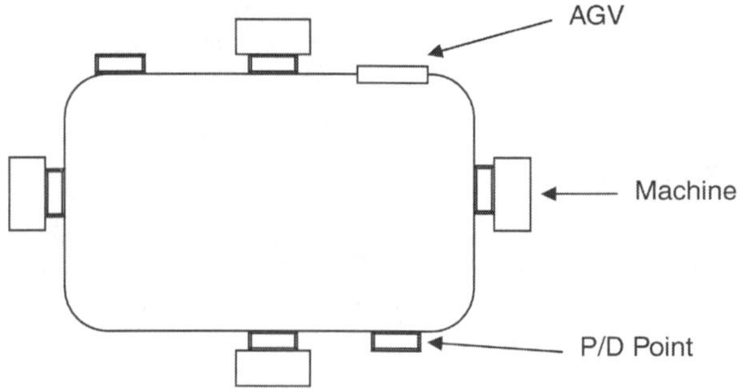

Figure 8.1 AGV network.

Depending on the manufacturing requirements, different guide paths are selected such as, unidirectional, bi-directional and tandem paths, for detailed descriptions see Le-Anh and Koster (2006). In a tandem guide path, work stations are divided into zones where an AGV serves each zone to avoid any collisions. However, tandem path design is efficient where the production sequences and specification do not vary. A bi-directional guide path provides higher productivity than a uni-directional design, but it still requires efficient control to avoid conflicts, collisions and backlogging.

Issues at the tactical level for an AGVs system include estimation of the number of vehicles, management of the power requirement for the AGVs, etc. Finally, vehicle routing, collision avoidance and bottleneck prevention problems are dealt with at the operational level.

As discussed above, various problems such as conflicts, deadlocks, collisions, blockings, etc. can arise in the AGV system. If two AGVs serve different work stations, but follow the same route, this can cause conflict, AGVs moving in opposite directions in the same route can cause a collision and if stopped, blockage and deadlock situations will arise. Therefore, these vehicle management problems must be carefully considered in the design and control of an AGV system at strategic, tactical and operational levels.

In advanced manufacturing contexts, in most cases, strategic and tactical level issues such as guide path and number of AGVs are predefined and only the operational level decisions need to be taken in accordance with the production planning and scheduling. However, at this level, production scheduling and AGVs scheduling and routing problems are dealt with separately, which generates a suboptimal solution for the overall planning (Seo and Egbelu, 1999). In fact these two problems are tightly connected as production scheduling generates the transport request, which in turn generates the AGVs scheduling and routing request. Similarly, AGVs scheduling and routing determines the start and finish of a job. In order to obtain a proficient manufacturing system an integrated production scheduling and AGVs scheduling routing is needed as these are inter-related aspects of the manufacturing system. The description of production scheduling and AGVs routing is as follows:

8.1.1 Production and AGVs Scheduling

The aim of production scheduling is to allocate the machine (work station) for each job and to determine the sequence of operations in accordance with precedence relations, whereas, AGVs scheduling assigns the vehicles for the transfer of jobs between the work stations. These scheduling activities are carried out under certain constraints such as deadlines, limited resources etc. with the objective of minimizing the makespan, cost etc.

8.1.2 AGVs Routing

After the scheduling, the purposed of AGVs routing is to find a suitable route for the transfer of the job. The function of AGVs routing is therefore to determine an optimal route such as, shortest time path, shortest distance path, shortest waiting time path etc. These decisions are made taking into account the constraints of congestion, conflict and deadlock free path.

Practical manufacturing scheduling problems are very complex in nature, not only because of their combinatorial issues, but also because of constraints imposed by different production environments. From the perspective of production scheduling, these constraints include limited workstations, time limits, precedence relations etc. From the point of view of the AGVs system, the number of AGVs, route etc., are important. Unfortunately, classical scheduling approaches handle these issues independently because production scheduling and AGVs scheduling and routing are easier to handle in this way, however this simplification results in sub-optimal solutions for the overall scheduling problem, featuring resource constraints for both machines and vehicles. This chapter presents an integrated scheduling and routing problem for the production and AGVs system. The integrated approach is formulated as a mathematical programming model and is solved using an Enhanced Memetic Particle Swarm Optimization (EMPSO) technique.

8.2 Literature Review

In the manufacturing domain there has been a plethora of research dealing with production and AGVs planning separately. However, only a few researchers have integrated them in overall manufacturing planning. Bilge and Ulusoy (1995) formulated the integrated production and AGVs scheduling problem as a nonlinear problem with the objective of minimizing the makespan. He solved the problem using a time-window heuristic. Later on Ulusoy *et al.* (1997) studied the same problem and solved it using a genetic algorithm (GA) approach. Results obtained using the GA technique, were better than the "time-window" heuristic. Lee and DiCesare (1994) used the shortest path routes between the machines for integrating the production and AGVs scheduling. They used a Petri-net method for solving the problem with the objective of minimizing the makespan. Sabuncuoglu and Karabuk (1998) presented an integrated scheduling problem assuming limited capacity of the machines; however they ignored the conflict avoidance issue. Scheduling of AGVs along with loading and unloading issues was considered by Smith *et al.* (1999). Anwar and Nagi (1998) studied the integrated scheduling problem with precedence constraint of the jobs. A more detailed literature review can be found in Gen *et al.* (2009).

8.3 Mathematical Model

8.3.1 Problem Statement

This problem considers that a machine layout and AGV numbers and guide path already exist. A production and AGVs planning and sequencing is required for a set of jobs, each consisting of different operations and precedence relations, in order to minimize the makespan. The detailed description of the model is given below. This mathematical model includes the precedence constraints between the job and their operations. For AGVs, this model considers the collision and congestion constraints and provides the scheduling and routing plans.

8.3.2 Mathematical Programming Model

A production order contains a set of jobs I = {1,2,3,.... I} and the number of operations for each job O = $\{O_1, O_2, \ldots, O_I\}$ on a set of machines K = {1,2,3,....., K}.. A set of operations for each job, $\varphi_i = \{j', j''\}$ defines the precedence relation. More notations used to describe the problem are as follows:

L_{ij} = a set of machine capable of performing the ith job's k^{th} operation

ν = index for AGV; ν = {1,2,3,......, V}.

P_{ij}^{k} = processing time for the ith job's j[th] operation on the kth machine, $\forall i \in I, j \in J, k \in L_{ij}$.

$A_{kk'}$ = set of paths connecting the machine k and k'

$T_{kk'}^{a}$ = travelling time of AGVs between the machines k and k' using the path a, $a \in A_{kk}$.

$M_{kk'}^{a}$ = set of machines lying in the path joining the machines k and k' using path a,

$$a \in A_{kk'}, k, k' \notin M_{kk'}^{a}$$

R_{ij}^{m} = ready time for the machine for performing the ith job's k[th] operation.

R_{ij}^{ep} = ready time for the AGV to go for the empty trip for the transfer of the ith job's k[th] operation

R_{ij}^{lp} = ready time for the AGV to go for the loaded trip for the transfer of the ith job's k[th] operation

S_{ij} = starting time for the ith job's k[th] operation

8.3.2.1 *Decision Variables*

$\eta_{jj'}^{i}$ = 1 if j[th] operation is carried out after the j'[th] operation for the ith job, else 0

x_{ij}^{k} = 1 if machine k performs the ith job's k[th] operation, else 0

$y_{iji'j'}^{k}$ = 1 if machine k performs the ith job's j[th] operation after the i'[th] job's j'[th] operation, else 0

eu_{ij}^{va} = 1 if the vth AGV uses the ath path for the transportation of the ith job's k[th] operation in the empty trip, else 0

lu_{ij}^{va} = 1 if the vth AGV uses the ath path for the transportation of the ith job's k[th] operation in the loaded trip, else 0

$w_{iji'j'}^{va}$ = 1 if the vth AGV performs the transportation of the ith job's j[th] operation after the i'[th] job's j[th] operation, else 0.

S_{ij} = starting time of the ith job's j[th] operation.

Γ = makespan (finishing time of all the jobs)

Now, the objective function
Objective function:

$$\text{Min } \Gamma \tag{8.1}$$

Subject to following constraint

$$\Gamma = \max\left(S_{ij} + P_{ij}^{k} x_{ij}^{k}\right) \tag{8.2}$$

The makespan time (Γ) should be equal to the end time of all the jobs and is called the finishing all jobs constraint.

For starting a job the following conditions need to be fulfilled:-

8.3.2.2 Previous Job Should Have Finished

$$S_{ij} \geq S_{ij'} + \sum_{k} \sum_{j',j' \neq j} P_{ij'}^{k} \eta_{jj'}^{i} \tag{8.3}$$

Constraint (8.3) ensures that the starting time of any job should be greater or equal to the end time of its preceding job.

8.3.2.3 Machine Should be Ready to Take the Job

$$R_{ij}^{m} \geq S_{i'j'} + \sum_{k} \sum_{i',i' \neq i} \sum_{j',j' \neq j} y_{iji'j'}^{k} P_{i'j'}^{k} \tag{8.4}$$

Constraint (8.4) ensures that the starting time of any job on the machine should be greater or equal to the finishing time of the previously assigned job on the machine.

8.2.3.4 Transfer of Job Should Have Finished

To transfer the assigned job, the AGV needs to complete an empty trip (to reach the machine where the job has finished) and a loaded trip (to transfer the job to the next assigned machine). For this, the AGV may need to wait in both its empty trip and its loaded trip. For calculation of these, we define two dependent variables $\alpha_{kk'}^{aij} = 1$ if the AGV needs to travel between k and k′, to transfer the job using the path a in the empty trip between machine k and k′, $\alpha_{kk'}^{aij}$ takes the value 1 when eu_{ij}^{va}, $w_{iji'j'}^{va}$, x_{ij}^{k}, $x_{i'j'}^{k'}$ simultaneously take values of 1.

Another dependent variable $\beta_{kk'}^{aij} = 1$ if the AGV needs to travel between k and k′, to transfer the job using the path a in the loaded trip between machine k and k′, $\beta_{kk'}^{aij}$ takes the value 1 when lu_{ij}^{va}, η_{jj}^{i}, x_{ij}^{k}, $x_{i'j'}^{k'}$ simultaneously take values of 1.

Starting time for empty trip

$$R_{ij}^{ep} = \max\left\{\left(\sum_{i',i'\neq i}\sum_{j',j'\neq j} S_{i'j'} w_{iji'j'}^{va}\right), \left(\sum_{k}\sum_{k'}\sum_{i'',i''\neq i,i'}\sum_{j'',j\neq j,j'} S_{i''j''}\alpha_{kk'}^{ai''j''}\right)\right\} \tag{8.5}$$

The first part in the max function calculates the finishing time of the AGV's previously assigned transfer job. In this model we have neglected the transfer time of the load from the AGV to the machine, so the finishing time of the AGV's loaded trip will be the starting time of the transferred job. The second part in the max function ensures that the AGV cannot move in the assigned path unless it is cleared,

so, the finish time for the empty trip

$$(FET_{ij}) = R_{ij}^{ep} + \sum_{k}\sum_{k'}\sum_{a} T_{kk'}^{a}\ eu_{ij}^{va} \tag{8.6}$$

Similarly, the starting time for the loaded trip can be given by the equation

$$R_{ij}^{lp} = \max\left\{\left(\sum_{i',i'\neq i}\sum_{j',j'\neq j} S_{i'j'} y_{iji'j'}^{k}\right), \left(\sum_{k}\sum_{k'}\sum_{i'',i''\neq i,i'}\sum_{j'',j\neq j,j'} S_{i''j''}\beta_{kk'}^{ai''j''}\right), FET_{ij}\right\} \tag{8.7}$$

So, the finish time for the loaded trip (FLT_{ij}) =

$$(FLT_{ij}) = R_{ij}^{lp} + \sum_{k}\sum_{k'}\sum_{a} T_{kk'}^{a} lu_{ij}^{va} \tag{8.8}$$

So the starting time of a job can be given by:

$$S_{ij} = \max(R_{ij}^{m}, FLT_{ij}) \tag{8.9}$$

8.2.3.5 *Machine Assignment Constraint*

$$\sum_{k} x_{ij}^{k} = 1 \tag{8.10}$$

Constraint (8.10) ensures that an operation is assigned to one machine only.

8.2.3.6 *Vehicle Assignment Constraint*

$$\sum_{a} eu_{ij}^{va} = \sum_{a} lu_{ij}^{va} \tag{8.11}$$

$$\sum_{v}\sum_{a} eu_{ij}^{va} = \sum_{v}\sum_{a} lu_{ij}^{va} = 1 \tag{8.12}$$

Constraint (8.11) ensures that transportation of a job is carried out by the same vehicle in both the empty and loaded trips, while constraint (8.12) ensures that a job should be assigned to a single AGV using the single path.

8.2.3.7 *Machine Work Order Determination*

$$\sum_{i,i\neq i'}\sum_{j.j\neq j'} y_{iji'j'}^{k} \leq 1 \tag{8.13}$$

8.2.3.8 *Job's Operation Order Determination*

$$\eta_{jj'}^{i} = 1, \text{if}\,(j, j') \in \varphi_i \tag{8.14}$$

$$\eta_{jj'}^{i} + \eta_{j'j}^{i} = 1 \text{ if}\,(j, j') \notin \varphi_i \tag{8.15}$$

Scheduling and Routing of an AGV are considered to be a variation of the Vehicle Routing Problem (VRP), which is NP-hard. In this model we have integrated the production planning, so the system is far more complex than normal scheduling and routing of AGVs. Various artificial intelligence (AI) techniques have been successfully implemented to solve such combinatorial problems. In this chapter we will demonstrate how an enhanced memetic particle swarm optimization (EMPSO) technique can be used for obtaining near optimal solutions for the above model. The next section describes the basics of particle swarm optimisation (PSO) and EMPSO.

8.4 PSO and EMPSO

The PSO methodology is used for finding optimal regions of complex search spaces through the interaction of individuals in a population of particles. This type of algorithm imitates the metaphor of social interaction and searches a space by adjusting the movement of individual vectors, called 'particles', as they are conceptualised as moving points in multi-dimensional space. The individual particles are drawn randomly toward the positions of their own previous best performance and the best previous performance of their neighbours (Kumar *et. al.*, 2010).

While implementing PSO in any problem, a population of particles is initialised with the random positions $\vec{x_t}$, and velocities $\vec{v_{it}}$, for which the fitness function *f*, is evaluated. At each iteration, the position vector and the velocity vector for every particle is adjusted and the fitness function evaluated with the new position vector. When a particle's position vector $\vec{v_{it}}$, gives a better solution than the previous selected one, it stores it as a particle best position vector $\vec{p_i}$. Similarly, the global best position (best in the whole population) vector $\vec{p_g}$ is also stored. Particles adjust their velocities according to cognitive and social learning. Equation (8.16) states the way particles update their velocities.

$$\vec{v}_{it+1} = \vec{v}_{it} + \varphi_1(\vec{p}_i - \vec{x}_{it}) + \varphi_2(\vec{p}_g - \vec{x}_{it}) \tag{8.16}$$

The first part of the equation is the velocity component in the previous iteration. The second part is called the cognitive component as it interacts between the particle's current position and its best position achieved so far. This part is called the social learning as it interacts between the particle's current position with the particles best position. φ_1 and φ_2 are the constants. The particle's new position will be updated according to the equation (8.17).

$$\vec{x}_{it+1} = \vec{x}_{it} + \vec{v}_{it+1} \tag{8.17}$$

The performance of any algorithm depends on its ability of exploration (global search) and exploitation (local search). In PSO, particles are attracted towards the global best position and prematurely converge towards the specific points, i.e. exploration takes over the exploitation. Evolutionary algorithms have proved to be very useful in many applications. It is necessary to improve local search abilities in optimization (Angeline 1998). This chapter has integrated the memetic algorithm (MA), which incorporates the better local search ability, with PSO and has called it the enhanced memetic particle swarm optimization (EMPSO). In order to improve the local search ability, EMPSO combines the solutions (position of the particle) in a similar manner to the use of crossover in GAs (ElMekkawy and Liu, 2009). Recombination is carried out for a randomly selected $\rho\%$ of the population. If the recombined solutions show better fitness values, then the original solutions will be replaced by the recombined ones. As the recombined solutions, are generated closer to the original solutions, they perform the local search and bring the balance between exploration and exploitation. A flow chart of the EMPSO is shown in the Figure 8.2.

8.5 Example

In order to demonstrate how to use EMPSO this chapter considers the simple example (Gen *et al.* 2009) as shown in the Figure 8.3 and the machine and AGVs layout as shown in Figure 8.4. A product has to be manufactured from the material using eight operations. The product to be manufactured is shown in Figure 8.3 and the rectangular block of material (labelled "Stock") is transformed into the required product by eight milling and drilling operations shown in Figure 8.3 as v1 to v8. The operation types and the times required for each of the operations v1 to v8 on different machines has been given in the Table 8.1. Figure 8.5 depicts the precedence relations between the jobs.

In order to generate the initial solution we randomly generate the solution strings (particles) in accordance with precedence relation and machine constraints (machine on which operation can be performed). Figure 8.6 shows the solution string generated for the job sequence and the machine number on which it will be operated.

The sequence of operations has been shown in red boxes and the machine number selected for each operation is shown in yellow, i.e. at first operation 1 will be carried out on machine 1 then operation 4 to be carried out on machine 3. In the next step, an AGV will be assigned for each transfer of the job. This example considers a single job, so a single machine will be operated at any time, and there will not be any conflict for AGV movement and therefore, for this problem, we can neglect the route determination for AGVs and assume that they will take the shortest path. However, in a multiple job environment, with multiple operations, a solution string should consider the empty and loaded travel routes. Now, this problem only needs to assign the AGV for each transfer and this has been shown in the green boxes (Figure 8.7) i.e. AGV 3 has been assigned

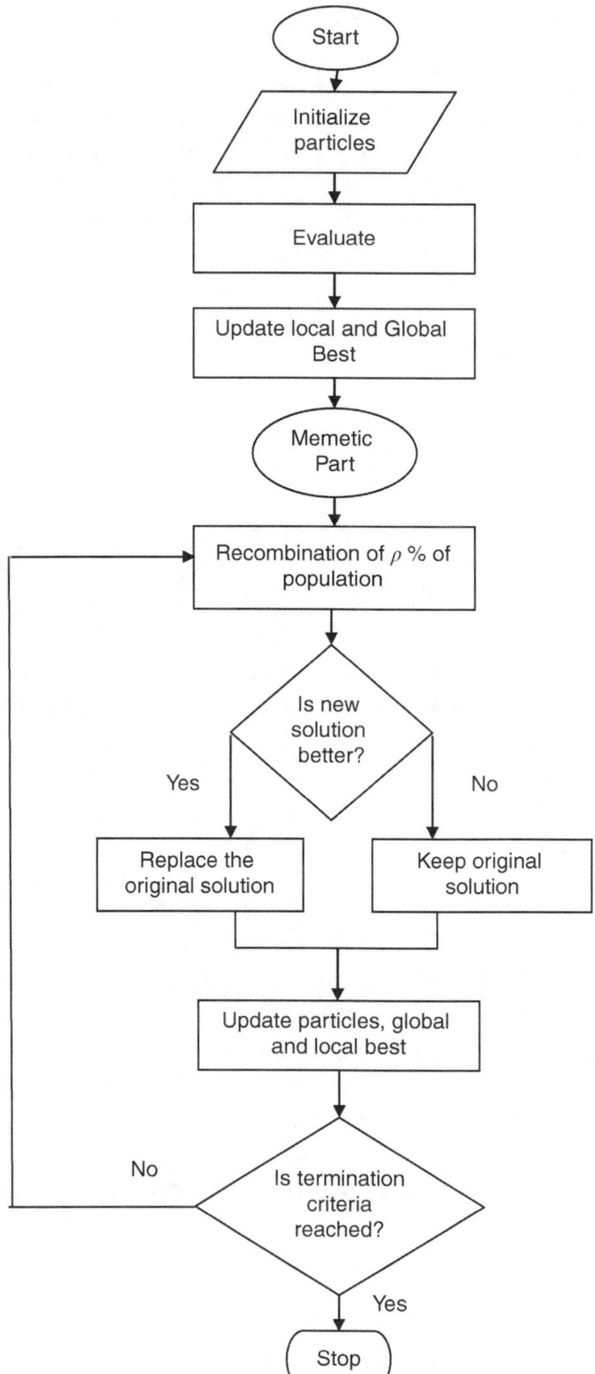

Figure 8.2 Flow chart of EMPSO.

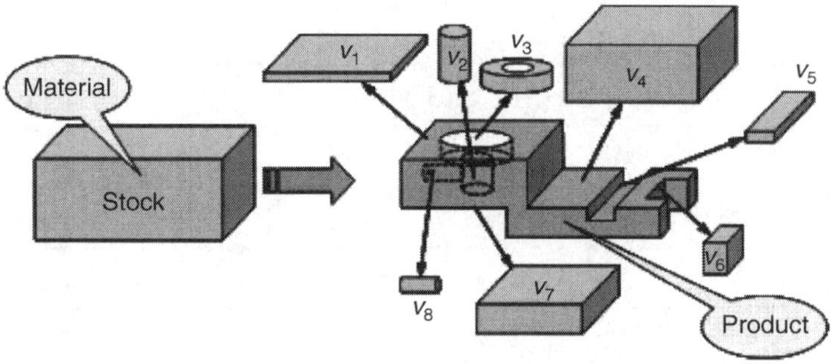

Figure 8.3 Job Requirement.

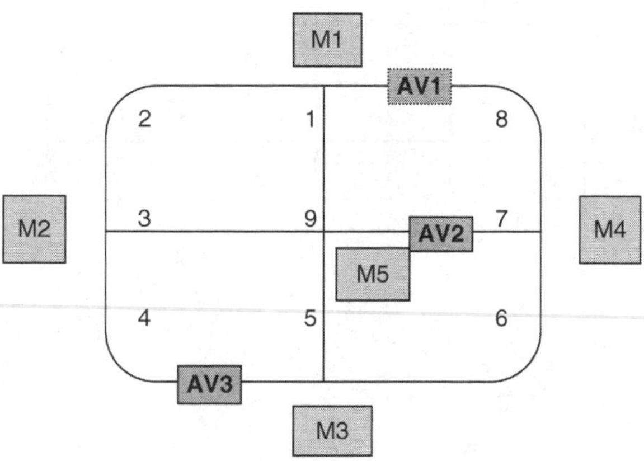

Figure 8.4 Machine and AGVs layout.

to transfer the job after finishing operation 1 on machine 1 for the execution of operation 4 on machine 3.

Assuming a fixed travelling time between the adjacent machines (lets say 5 minutes), we can easily calculate the overall finishing time (makespan) for each generated solution.

8.6 Recombination (Local Search)

As described earlier, the recombination operation in EMPSO is similar to the crossover of GA. The recombination operation randomly selects the two solution strings (particles) and swaps the solution from a randomly chosen bit point (as shown in Figure 8.8).

The best two solutions are selected from the four and the particles position vector is updated accordingly.

Table 8.1 Processing time.

Job	Machine and Time					Type of Operation
	M1	M2	M3	M4	M5	
v1	20	x	25	X	22	Milling
v2	x	25	X	30	x	Drilling
v3	x	30	X	20	x	Drilling
v4	25	x	20	X	30	Milling
v5	40	x	25	X	35	Milling
v6	15	x	20	X	20	Milling
v7	20	x	25	X	15	Milling
v8	x	20	X	25	x	Drilling

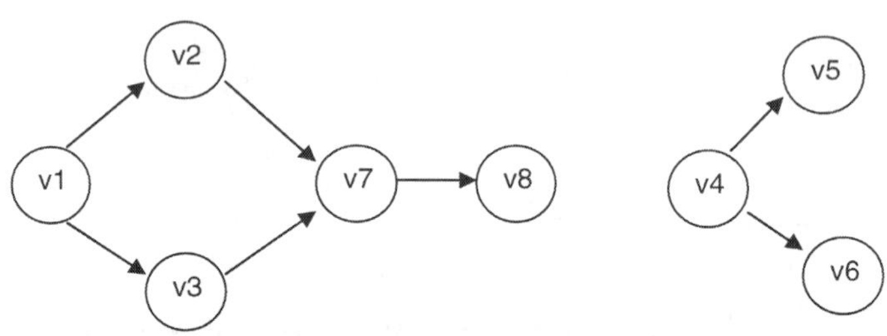

Figure 8.5 Precedence relation of the job.

Figure 8.6 Solution String.

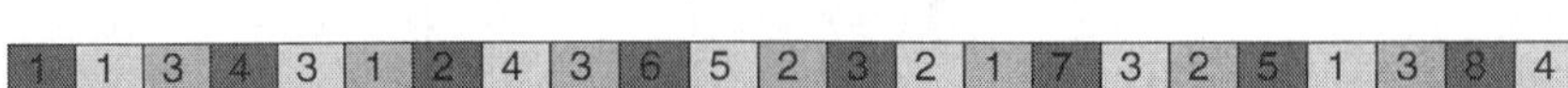

Figure 8.7 Solution string with AGV.

In order to demonstrate the effectiveness of EMPSO over PSO, discussed above, we have taken a complex production and AGV planning and scheduling problem to evaluate the computational efficiency of EMPSO. Assuming machine and AGV layout as shown in Figure 8.4, we have used the following (Table 8.2) normal distribution for data generation:

Table 8.2 Data generation.

Data	Jobs	No of Operation	Processing Tine
Normal Dist.	N(1,10)	N(1,10)	N(5,50)

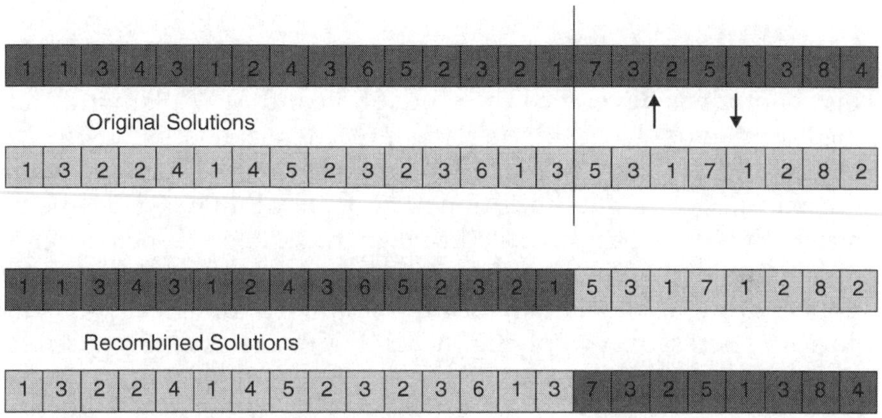

Figure 8.8 Recombination of the strings.

For travelling time we have considered that to reach the adjacent machines, an AGV takes 5 minutes. In the experiment we have considered population size (number of particles) of 100 and constants φ_1 and φ_2 as 0.5. 20% of the population are set for recombination. Figure 8.9 shows that EMPSO outperforms the PSO due to its better local search ability.

Although we have performed our experiment with these parameters readers should notice that parameter tuning cannot be generalized and it varies from problem to problem making this step of paramount importance for successful implementation of the algorithm. Exhaustive experiments are required to finalize the parameters and are called parameter tuning.

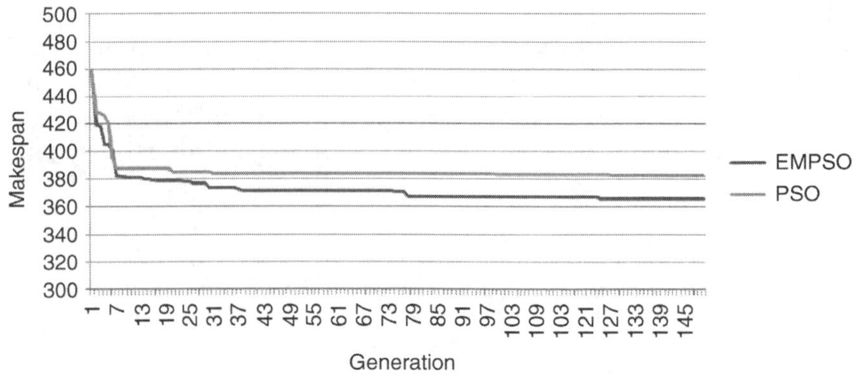

Figure 8.9 Convergence graph.

8.7 Summary

This chapter has developed the production and AGV planning and scheduling model considering the precedence relation between the jobs and avoiding collision between the AGVs. Fundamentals of Particle Swarm Optimization (PSO) and Enhanced Memetic Particle Swarm Optimization (EMPSO) have been discussed and implemented on a general problem. This should help readers in understanding and mathematically mapping a practical problem and with successfully applying the EMPSO to get the best possible solution.

References

Angeline, P. J. (1998). Evolutionary optimization versus particle swarm optimization: philosophy and performance differences. In V. W. Porto, N. Saravanan, D. Waagen & A. E. Eiben (Eds.), *Evolutionary programming* (Vol. VII, pp. 601–610). Berlin: Springer.

Anwar,M.F., Nagi, R., 1998. Integrated Scheduling of material handling and manufacturing activities for just in time production of complex assemblies. *International Journal of Production Research* 36 (3), 653–681.

Asef-Vaziri, A. and Goetschalckx, M., Dual track and segmented single track bidirectional loop guide path layout for AGV systems, *European Journal of Operation Research*, Vol.- 186, Pages – 972–989, 2008.

Bilge, U., Ulusoy, G., 1995. A time window approach to simultaneous scheduling of machines and material handling system in an FMS. *Operations Research* 43 (6), 1058–1070.

ElMekkawy, T.Y. and Liu, S., A new memetic algorithm for optimizing the partitioning problem of tandem AGV systems, *Int. J. Production Economics* 118 (2009) 508–520.

Gen, M.,Lin, L. and Zhang, H., Evolutionary techniques for optimization problems in integrated manufacturing system: State-of-the-art-survey, *Computers & Industrial Engineering* 56 (2009) 779–808.

Kumar, S.K., Tiwari, M.K. and Babiceanu, R.F., Minimisation of supply chain cost with embedded risk using computational intelligence approaches, *International Journal of Production Research* Vol. 48, No. 13, 1 July 2010, 3717–3739.

Le-Anh, T. and De Koster, M.B.M., A review of design and control of automated guided vehicle systems, *European Journal of Operational Research,* Vol.- 171, Pages- 1–23, 2006.

Lee, D.Y., Di Cesare, F., 1994. Integrated scheduling of flexible manufacturing systems employing automated guided vehicles. *IEEE Transactions on Industrial Electronics* 41 (6), 602–610.

Maza, S. and Castagna P., A performance-based structural policy for conflict-free routing of bi-directional automated guided vehicles, *Computers in Industry,* Vol.- 56, Pages – 719–733, 2005.

Sabuncuoglu I., S. Karabuk, "A Beam Search based Algorithm and Evaluation of Scheduling Approaches", *IIE Transactions,* Vol: 30, pp: 179–191, 1998.

Seo, Y. and Egbelu, P.J. , Integrated manufacturing planning for an AGV-based FMS, *International Journal of Production Economics,* Vol.- 60, Pages- 473–478, 1999.

Smith, J., Peters, B., Srinivasan, A., 1999. Job shop scheduling considering material handling. *International Journal of Production Research* 37 (7), 1541–1560.

Ulusoy, G., Funda, S.-S., Bilge, U., 1997. A genetic algorithm approach to the simultaneous scheduling of machines and automated guided vehicles. *Computers and Operations Research* 24 (4), 335–351.

9

Simulation-Based Aircraft Assembly Planning Using a Self-Guided Ant Colony Algorithm

Sai Srinivas Nageshwaraniyer[1], Nurcin Celik[1], Young-Jun Son[1] and Roberto Lu[2]

[1]*Systems and Industrial Engineering, The University of Arizona, Tucson, AZ, 85721, USA*
[2]*Technical Fellow, The Boeing Company, Seattle, WA, 98124, USA*

Abstract

The aerospace industry plays a key role in today's U.S. economy with its total sales comprising 1.4% of the U.S. gross domestic product. The increasing competition in the aerospace industry between domestic and foreign competitors is motivating aerospace companies to seek for new and innovative ways to decrease the costs and lead-time for new product development and production by implementing flexible and integrated assembly systems. In this work, we present a simulation-based optimization framework for aircraft assembly planning using a self-guided ant colony algorithm (SGAC), where the objective is to achieve line balancing in the assembly of aircraft. More specifically, the goal is to minimize the number of workstations and maximize average utilization of resources in the stages of assembly. Based on the current state of the facility, the framework devises an optimal assembly plan. This plan is executed until the next time interval when the framework is again invoked to deliver the optimal assembly plan. The generality and validity of the proposed approach have been successfully tested for aircraft assembly planning under various conditions via a designed experiment.

Keywords: Aircraft assembly, shop floor control, ant colony algorithm, dynamic planning

Manoj Tiwari and Jenny A. Harding (eds.) Evolutionary Computing in Advanced Manufacturing, (169–196) © Scrivener Publishing LLC

9.1 Introduction

Aerospace manufacturing (including manufacturing of aircraft and spacecraft) contributes significantly to the U.S. economy with its total sales comprising 1.4% of the U.S. gross domestic product in 2008 (Aerospace Facts and Figures, 2008; Bureau of Labor Statistics, 2008; Tinseth, 2008). In particular, aircraft manufacturing comprises 2.8% of the nation's manufacturing workforce with over 500,000 Americans in high skilled and high-wage jobs (Platzer, 2009). Today, the aerospace industry is experiencing increased competition from domestic and foreign competitors with labor representing one of the highest costs as the labor skills required by the aircraft industry are highly transferable and sought after by other industries. This competition is forcing aerospace companies to seek for innovative ways to decrease the costs and the lead-time for new product development and production. Similarly, several premier manufacturers of commercial jetliners (e.g., Boeing and Airbus) and leaders in defense and space systems are searching for ways to improve quality, decrease costs, and reduce development time by implementing flexible assembly systems.

Entering into the new century, aircraft manufacturing and assembly is facing many new challenges and requirements such as high quality, high speed and small batch production. Many aircraft manufacturing enterprises still apply old shop floor control mechanisms, which cannot fulfill these new market requirements (Celano and Fichera, 1999). Currently, the development of new assembly process technologies effectively increases the efficiency of aircraft-related assemblies; however, the progress of shop management technologies for such assemblies is lagging behind. (Liu, 2006; Caridi and Sianesi, 2000). In this work, we propose simulation-based planning for aircraft assembly and this planning is performed frequently at predetermined time intervals. Based on the current state of the facility, the proposed framework first devises an optimal assembly plan. This plan is executed until the next time interval, when the framework is again invoked to deliver the optimal assembly plan. Figure 9.1 shows various feeder lines associated with the aircraft assembly process in general, and Figure 9.2 shows different types of sub assemblies that generally take place in the aircraft assembly.

Towards this goal, we develop a self-guided ant colony algorithm (SGAC), an algorithm for finding optimal paths based on

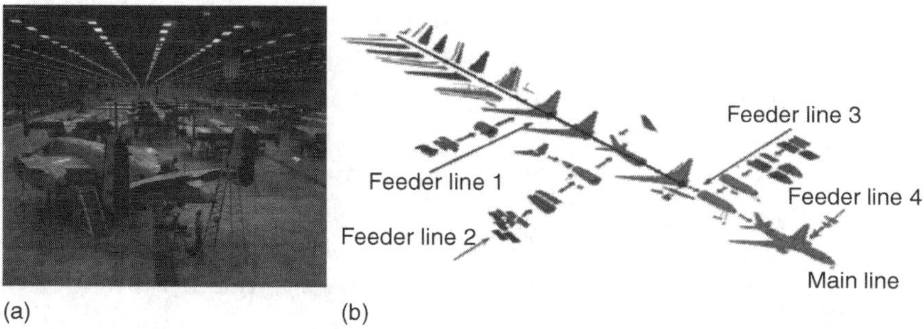

Figure 9.1 (a) Stationary aircraft assembly lines in shop floors (Library of Congress); (b) feeder lines in aircraft assembly (Betancourt, 2007).

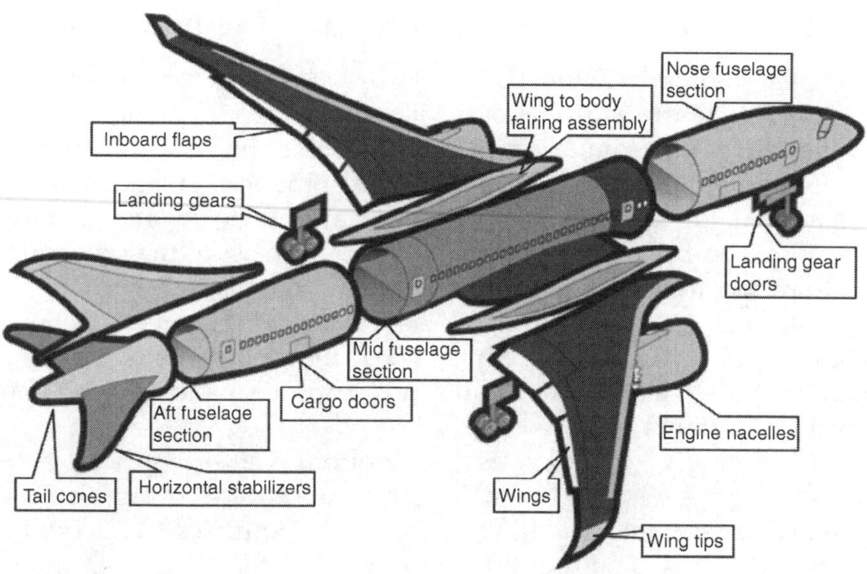

Figure 9.2 Cross sections of aircraft assembly (Boeing 787).

the behavior of self-guided ants searching for food, to resolve the assembly line balancing problem discussed above, which is very complex (it is proven to be NP-hard) and has a large decision space. The working procedure of SGAC is based on the search behavior of an ant species called 'T.Albipennis' for a new nest after abandoning its old nest. This activity of the 'T.Albipennis' ant is called nest hunting. These ants show remarkable decision making while nest hunting under adverse environmental conditions (Marshall and Franks, 2007). These ants while nest hunting look for all possible nests (solutions) before they arrive at the optimal nest. However, they may find an incorrect optimal solution due to judgment error under time constraints. If this judgment error can be avoided, it is possible to identify the optimal solution. Since our problem is characterized by the involvement of numerous and multi faceted options, it is analogous to the problem faced by the ants. Therefore, we utilize the nest hunting behavior of these ants to arrive at the optimal solution to our problem while avoiding the judgment error mentioned above.

9.2 Background and Literature Survey

9.2.1 Assembly Planning in Aircraft Manufacturing

Engineer-to-order (ETO) (Lu *et al.*, 2009) is a manufacturing philosophy where customers are heavily involved throughout the entire design and manufacturing process for the end products (http://www.bigmachines.com/eto.php). The end-products in manufacturing environments employing that philosophy are built to unique customer specifications based on their order. Therefore most of the time, each end-product requires a separate set of bill of materials, item numbers and routings. Hicks and Braiden (2000) explain that proper assembly planning in an ETO environment (e.g., the aircraft manufacturing industry) has the ability to significantly improve manufacturing performance significantly.

An assembly operation is the motion that brings two sub-assemblies together to form another sub-assembly or the final assembly. Assembly planning, in general, comprises the planning of the activities including but not limited to facilities design, line balancing, ergonomics, process design, material handling and throughput analysis (PLM Software, 2009). In general, there are two goals of assembly planning in aircraft manufacturing. The first goal is to generate plans to achieve the performance objectives of the shop floor operations (e.g., minimizing the lead-time). The minimization of product development time is critical for aircraft manufacturers to stay competitive in the market. However, it is difficult to formulate an efficient process plan for aircraft assembly due to the complex processes and various customer specifications (e.g., orders including various types of wings required for different aircrafts may come at the same time). Arnold *et al.* (2004) investigate the effectiveness of assembly simulation as a process plan development aid. They study the process plans from a case study of an existing aircraft manufacturing company with the help of process planners. The process planners evaluate different characteristics of assembly simulation of each process plan. The plans approved by the planner are tested and observed to result in reduction of time taken to develop a process plan. Through their analysis, they conclude that assembly simulation is an effective tool as a process plan development aid and also that the experience of a process planner positively affects the magnitude of process development time saved. Another study by Zhang *et al.* (2008) presents a new method for the generation of work breakdown structure in aircraft assembly scheduling using assembly orders. This structure aims to effectively improve fine control ability of an assembly workshop by utilizing the assembly process information. The application of the presented method in the field of aircraft assembly demonstrates the functionality of the method.

The second general goal of assembly planning in aircraft manufacturing is to identify and reduce the existing inefficiencies in production. In aircraft assembly, due to its complexity and the large number of manual interventions, non-value added activities consume a significant amount of time. Ramamoorthy (2007) studied the non-value added activities in the upper main door assembly process of a business jet. He implemented a Lean Six-sigma integration based approach to address them and achieved a reduction of 16 days in the lead-time for the above mentioned assembly process. Jin *et al.* (2008) proposed a metric called non-value added ratio (NVAR), which is the ratio between the time consumed by all non-value added activities and the total assembly time. This metric, assigned to each considered assembly procedure, serves as an indicator of the utility of the shop floor design. They devised a closed-loop systematic approach for the implementation of this metric for continuous productivity improvement and demonstrate their approach on a case study involving an aerospace manufacturing company.

The decisions in assembly planning essentially involve fixing the spatial location of a component and the sequence in which that component has to be assembled with respect to other components within the same assembly. Aircraft assembly planning is complex and may have long lead times due to various factors including the large number of sub-assemblies (e.g., up to 500 parts and 800.000

facets) and disassemblies and the physical constraints in each assembly (Jones and Wilson, 1998). Therefore, assembly operations also account for a large percentage of aircraft manufacturing cost (see Figure 9.3 for multiple tool control systems deployed in an aircraft manufacturing shop floor to perform the sub-assemblies simultaneously). Hence, decisions in assembly planning are critical to the cost and lead-time of shop-floor operations in the aircraft manufacturing industry. In literature, the assembly planning problem is broken into two major problems, which are namely 1) assembly sequence planning problem, and 2) assembly line balancing problem, and these are resolved separately.

An assembly sequence is a sequence of operations that constructs a product from its constituent parts. The assembly sequence planning is constrained by geometry and mechanics of the rigid parts of the product, without considering the constraints on the resources that execute the assembly motions. The resources here can be humans or robots; and are assumed capable of performing the final assembly sequence without laying any constraint on the assembly sequence planning. The assembling sequence planning usually proceeds using the CAD model of the product. The sequence of the sub-assemblies of the product affects the efficiency of any assembly process. Moreover, developing the optimal assembly sequence is difficult due to numerous

Figure 9.3 Multiple tool control systems deployed to do the sub-assemblies simultaneously (Snap-on Industrial).

geometric constraints present in the assembly operations of the product. Wilson and Albuquerque (1995) propose a dual approach for dealing with constraints in assembly sequence planning. In the first part of their approach a set of assembly sequences are validated using CAD models of the products. Then the user is asked to enter constraints into the computer system that were not possible to be addressed in the validation of the assembly sequences via the CAD model. The computer system then generates and validates a new assembly sequence based on the constraints provided by users. Christiand and Yoon (2007) proposed a path and sequence planning algorithm based on a genetic algorithm for a virtual environment that they considered. As an effort to improve the assembly process, Buckingham (2007) proposed and demonstrated snake-arm robots for replacing manual interventions on an existing wing assembly. The motivations were that the future wing designs would not provide adequate access for manual assembly methods, and the manual interventions in wing assembly have health and safety concerns along with the low efficiency. A discussion on the future prospects of the snake-arm robots for assembly in forthcoming wing designs was also given in that paper.

Assembly line balancing is another important category of assembly planning and had been a popular manufacturing research topic for many years. Until 1955, the only way to design assembly lines was by trial and error. Salveson (1955) published the first article on order of a line. Line balancing is the process of allocating sets of tasks to stations on an assembly line, where the loads in the stations are well balanced. A task in the assembly line is a smallest possible or indivisible job on the line, and the time taken to complete each task is referred to as task time. The time taken to complete all the tasks assigned to a station is called station time. The cycle time of the line is the greatest station time of the line (i.e. that of the bottleneck station). In this paper, we differentiate this cycle time of the line from total lead-time of the aircraft, which is the time taken to produce the final aircraft starting with its sub assemblies on the shop floor. In a balanced line, all stations have similar station times or the difference between highest station time and lowest station time is low. The fundamental aspect of assembly line balancing is that the precedence requirements are met while searching for ideal combinations of task sequences. Traditional assembly lines had all the workstations in the form of a straight line. Later, Japanese companies (e.g., Toyota) employed U-type and S-type assembly lines, where multi-skilled work force can be allocated at different stations to perform various tasks during the same cycle of production. This is not possible in straight assembly lines due to precedence constraints in them. Therefore, the topology of the assembly line removes certain constraints on the mode of resource allocation (i.e., whether they can be allocated to multiple stations or not). While there have been advances in general assembly planning in literature, the aircraft manufacturing shop floors have stationary assembly lines because of the large size of the work pieces (Figure 9.1).

9.2.2 Self-Guided Ant Colony Algorithm

The assembly line balancing problem is an NP-hard problem under the class of combinatorial optimization. This means that all possible ways of allocating tasks to work stations have to be analyzed to find the optimal solution to problem. Besides, an increase in the number of tasks significantly increases the computational time required to arrive at the solution to this problem. In this study, we first map our assembly line balancing problem into a traveling salesman problem (TSP), which is a widely studied combinatorial optimization problem. As the TSP is very challenging to solve using deterministic methods, various heuristics and their variants have been proposed such as tabu search (Fiechter, 1994), genetic algorithm (Snyder and Daskin, 2006), ant colony algorithm (ACO) (Dorigo and Gambardella, 1997; Dorigo, 1992), evolutionary optimization (Fogel, 1988), elastic net method (Durbin and Willshaw, 1987) and particle swarm optimization (Onwubolu and Clerc, 2004). In this study a variant of ACO, self-guided ant colony algorithm is developed to solve our assembly line balancing problem.

The ACOs are said to belong to the ACO set. Ant systems were the first algorithms proposed in ACO set. They were proposed by Dorigo (1992) as an algorithm to determine the optimal path in a graph based on behavior of ants seeking to find the shortest path between their colony and a source of food which is similar to TSP. As the concept of optimization through ants foraging for food was beginning to be applied to solve problem types other than the TSP, this set expanded to comprise other ACOs including rank-based ant system, elitist ant system, max-min ant system, ant colony system and best-worst ant system (Dorigo, 2007). The self-guided ant colony algorithm is an ACO, which was originally proposed by Tripathi *et al.* (2009) to solve a disassembly modeling and sequencing problem. In this work, we have customized the self-guided ant colony algorithm to find the optimal solution to our assembly line balancing problem (see Section 9.4 for details).

9.3 Specifications of the Considered Aircraft Assembly

In this work, we consider a hypothetical aircraft final assembly shop floor as our case study. The considered shop floor performs the final assembly operations via its necessitated materials and resources. Here, materials include aircraft sub-assemblies including wing, fuselage and tail which are comprised of composites in large scale such as fasteners, sealants, etc. These materials are procured via ships, air or road and stored on the shop floor.

The resources are further divided into two categories including the human resources, and machine resources. First, the human resources are categorized into three different types based on their expertise, including engineers, technicians, and computer experts. While not included in this study, air carriers, finance experts, computer experts and even other aircraft manufacturers can be additionally considered as part of the human resources class depending on the goal of a study. These human resources are assigned to specific operations involving assembly, testing, rework and maintenance. Second, the machine resources are divided into three groups, which are computational, tooling and transport equipment. Computers are used as computational resources for designing process schedules and simulating the whole assembly process. Tooling equipments such as robots are used for shimming, drilling, fastener installation, and sealing. The transport equipments are mostly cranes, automatic guided vehicles (AGV), and automatic carriers. In the considered facility, in order to move the aircraft components from their storage area to their process area, an advanced storage and retrieval system is first employed. This system handles items of a small size right up to components of four meters in length. Kits of parts are then carried to the particular workstation using driverless, automatic guided vehicles (AGV). Once the components are retrieved from their stored area for assembly, the process or subprocess of assembling takes place.

The considered shop floor focuses on four major operations, which are namely assembly, testing, rework and maintenance while there are other minor operations. Aircraft assembly is complex and relies on high-tech design and fabrications. Assembly of a larger-heavier aircraft is much more difficult than a smaller-lighter craft. The numerically controlled (NC) machines enable every detail of the complicated sub-assemblies to be automatically programmed. In the facility considered in this work, many components or sub-assemblies of the aircraft are outsourced to other manufacturers both in the U.S. and increasingly around the world (e.g., inboard flap from Japan and tail cone from China). This may be either for cost savings in production, specialist development or as an incentive for that country to buy other company products. As mentioned earlier, since the components that are assembled to form a final aircraft are huge, the assembly line is a stationary assembly line. The shop floor is comprised of multiple stationary assembly lines, where each line may contain aircrafts in various stages of production (see Figure 9.1). Here, one entire line may be dedicated to a single customer order, and the machines and human resources working on each line may have to be shared with other lines from time to time.

There are two major types of assembly operations performed as part of the production in the considered shop floor. These are pre-final assembly (prefal) and final assembly (fal). Prefal focuses on the assembly of the mid section of the fuselage. The mid section is divided into two halves. The two halves are brought together, and the floor plate is inserted in between these two halves. The plate is riveted to the halves. Then splices are inserted into the inner section of this incomplete fuselage, reamed, and riveted. This process is done at a station with anthropomorphic robots. The sequence of assembly processes that takes place in fal on the aircraft shop floor is as follows: fuselage sections assembly, forewing assembly, tail assembly and engine assembly onto the aircraft. There exist various sizes and weights of the wings, fuselages, and tails, which have to be accommodated into various types of aircrafts. The fuselage (involved in the first assembly process) is divided into the mid section, nose section and aft section (see Figure 9.2). These three components of the fuselage are brought together via automatic carriages. At sections where they are joined together, they are reamed and sealed. Temp fastening is done at the joined sections. Then these sections are riveted. Developing technology again has enabled the forewing skin panels to be machined from solid billets, which can be over

72 ft in length, depending on the aircraft type for which it is designed. The forewings are then brought via cranes while there are supports to hold onto them when they are riveted onto the fuselage. Tails are brought on by cranes and assembled onto the rest of the aircraft. After the tail assembly, the wings are fitted with engines. The aircraft is then ready for painting in a closed chamber.

Testing, rework and maintenance are performed from time to time during the above mentioned assembly processes. Final tests on the aircraft are performed after the final aircraft is assembled and ready. The final aircraft is taken for trial runs outside the shop floor for various tests including flutter and cross wind tests. After testing, some aircrafts are again worked upon for repair. After each assembled aircraft completes its test runs successfully, it is shipped or flown to its customers.

9.4 Proposed Simulation-Based Assembly Planning Framework

9.4.1 Overview of the Proposed Framework

The aircraft assembly process on the shop floor described in Section 3 is constrained by time, resources and customer specifications. The sub assemblies within the aircraft assembly, including fuselage, wing and tail sub assemblies, have to be completed before these assemblies can be joined to form the whole aircraft. In general, the final aircraft assembly has to be completed within a six-month time window. As mentioned earlier, the resources that are required for aircraft assembly can be divided into two types, including the machines and human resources. While it can be assumed that there is no limitation on the machines availability, and they can be allocated any time; the human resources who are highly skilled, and limited in number, have to be shared among various assembly lines working in the shop floor. Therefore, not only is the time window a constraint but the allocation of human resources to the assembly has to be planned in order to distribute the resources efficiently among other sub assemblies. In addition, it should be ensured that the quality of the final aircraft falls within the customer specifications. If there is a low tolerance for the design to deviate from the customer specifications, a very minor production error may lead to the order being rejected resulting in significant losses to the manufacturer. Furthermore, given the time and resource constraints, one or more of the design flaws may not be attended to during the course of the time window. Due to these reasons, to aid the manufacturer in formulating an efficient assembly plan, we propose a simulation-based assembly planning framework, where the self-guided ant colony algorithm (SGAC) is used to solve the line-balancing problem in the aircraft assembly.

The proposed simulation-based planning framework is composed of two main modules: the SGAC algorithm and the simulation model (see Figure 9.4). At the start of the predetermined time window, the SGAC algorithm is first invoked in order to solve the line-balancing problem. The SGAC algorithm is used to filter the set of near optimal solutions from the large decision space of the line-balancing problem that minimizes the total number of workstations. The algorithm filters the assembly plans whose total number of workstations is above

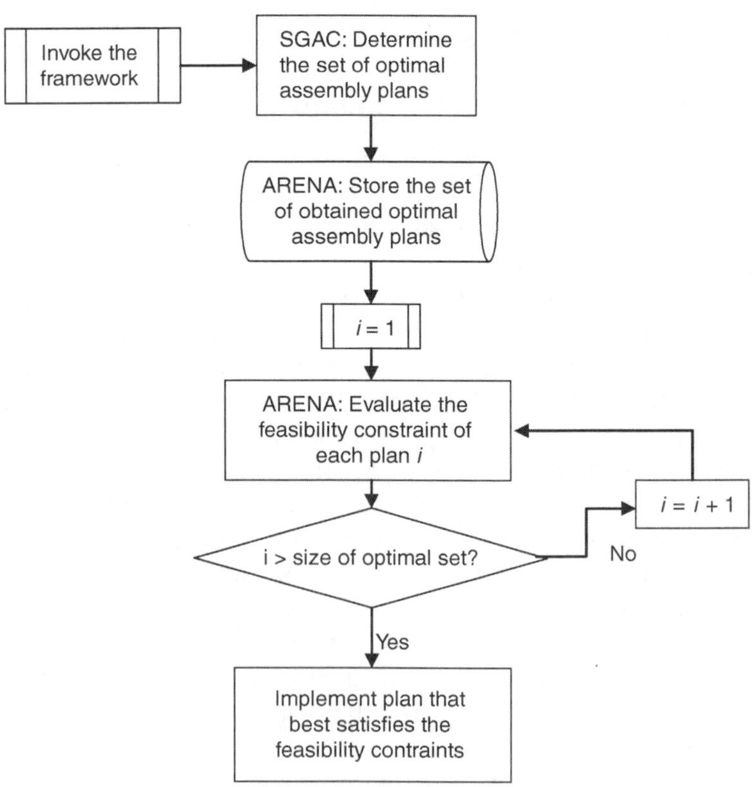

Figure 9.4 Proposed simulation-based assembly planning framework.

a set maximum number of workstations that a plan is allowed over the rest of the time window. Those plans that remain unfiltered are said to belong to the optimal set of assembly plans.

The simulation model then obtains and evaluates this optimal set of assembly plan from the SGAC algorithm and runs each plan in this set to determine the plan that best satisfies the feasibility constraints on the shop floor. The plan that satisfies the feasibility constraint the best (maximizes the average workstation utilization and minimizes the variance in work station utilization) is the optimal assembly plan. In order to ensure implementation of the optimal assembly plan at any point during the time window, the framework is invoked periodically after every time interval equal to the planning interval from the start of the time window. The planning interval is fixed at the start of the time window. Each time the algorithm is invoked, it is fed with the inputs that collectively define the state of the shop floor and re-determine the current optimal plan by comparing the objectives of the current optimal plan with those of the currently operated assembly plan. This is due to the fact that the progress of the assembly operations on the shop floor may not be as predicted by the currently operated assembly plan. At the end of this comparison, if the currently operated assembly plan is not the same as the current optimal plan the latter is implemented.

In this study, experiments are performed on a simulation model of a shop floor performing aircraft assembly. The whole assembly operation can be divided into a number of tasks. There is a particular sequence that has to be followed while going from one task to another. In addition, for a particular task to be completed, some tasks must have already been completed. A precedence graph indicates the precedence constraints in the sequence of tasks. The precedence graph defining the aircraft assembly operations that we consider in this work is provided in Figure 9.5. Prefal 1 and Prefal 2 represent pre-final assemblies mentioned in Section 9.3. They are performed prior to the final assembly of aircraft on the given shop floor. Tasks 1, 9, 11, 2, 10, and 18 represent fuselage mid-section 1 sub-assembly or pre-final assembly 1; tasks 7, 15, 23, 8, 16, and 24 represent fuselage mid-section 2 sub-assembly or pre-final assembly 2; tasks 3, 11, 19, 4, 12, and 20 represent nose section sub-assembly; tasks 5, 13, 21, 6, 14, and 22 represent tail section sub-assembly; tasks 25, 29, and 33 represent joining fuselage mid-section 1 and wings; tasks 26, 30, and 34 represent preparing nose section for further assembly; tasks 27, 31, and 35 represent joining tail and rudder; tasks 28, 32, and 26

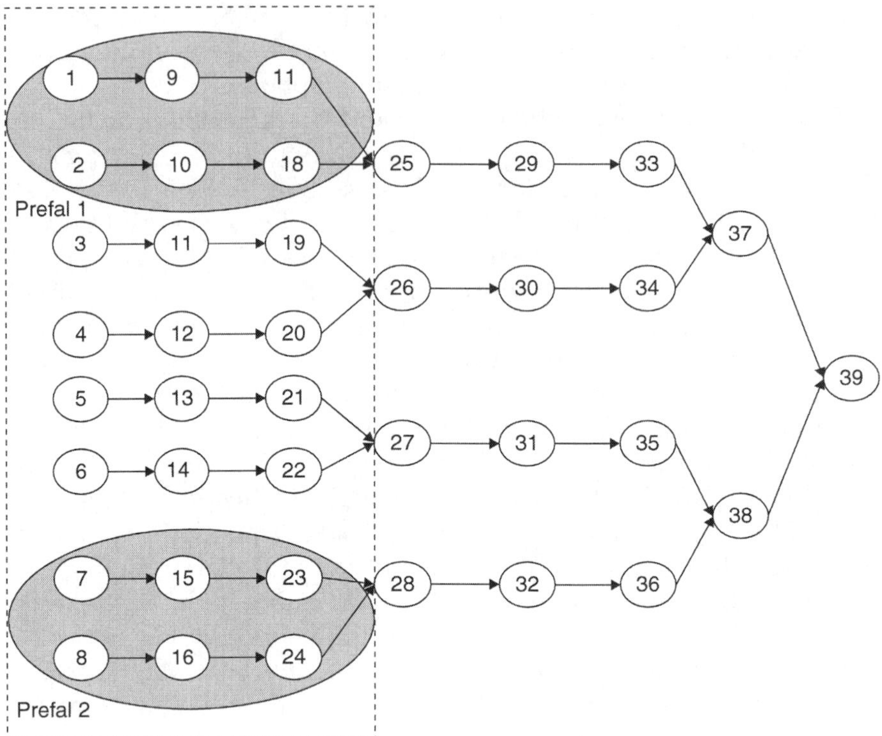

Legend

Tasks [1, 9, 11, 2, 10, 18]	- Fuselage mid-section 1 sub-assembly or Pre-final assembly 1
Tasks [7, 15, 23, 8, 16, 24]	- Fuselage mid-section 2 sub-assembly or Pre-final assembly 2
Tasks [3, 11, 19, 4, 12, 20]	- Nose section sub-assembly
Tasks [5, 13, 21, 6, 14, 22]	- Tail section sub-assembly
Tasks [25, 29, 33]	- Joining Fuselage mid-section 1 and wings
Tasks [26, 30, 34]	- Preparing Nose section for further assembly
Tasks [27, 31, 35]	- Joining Tail and rudder
Tasks [28, 32, 26]	- Preparing Fuselage mid-section 2 for further assembly
Task [37]	- Joining Fuselage mid-section 1 and Nose section
Task [38]	- Joining Fuselage mid-section 2 and Tail
Task [39]	- Engine Installation and Paint

Figure 9.5 Overview of the aircraft assembly processes with precedence constraints (precedence graph for an instance of 39 tasks).

represent preparing fuselage mid-section 2 for further assembly; task 37 represents joining fuselage mid-section 1 and nose section; task 38 represents joining fuselage mid-section 2 and Tail and lastly, task 39 represents engine installation and paint.

9.4.2 Mathematical Formulation

As mentioned earlier, our goal is to minimize the total number of work stations on the aerospace shop floor. Equation (9.1) specifies the objective function that minimizes the number of workstations used for the whole aircraft assembly process. The constraint given in Equation (9.2) limits the values of x_{ij} and y_j; x_{ij} is 1 if task i is assigned to workstation j, otherwise it is 0; y_j is 1 if workstation j is used, otherwise it is 0. The constraint in Equation (9.3) limits the number of tasks that can be assigned to a workstation using a fixed maximum cycle time value and depending on whether the workstation is used. The constraint in Equation (9.4) makes sure that workstation $j + 1$ is used only if j is already used. The constraint in Equation (9.5) makes sure that only one workstation is assigned to a task i. The constraint in Equation (9.6) presents the precedence constraints based on product sequencing requirements: if $x_{ij} = 1$, then the RHS of Equation (9.6) must assume a positive value over each immediate predecessor h of i.

9.4.2.1 *Indices*

- i, h: task
- j, k: workstation
- I: Set of all tasks
- J: Set of all workstations
- x_{ij}: if task i is assigned to workstation j
- y_j: if workstation j is used

9.4.2.2 *Parameters*

- N: total number of tasks $i = (1,\ldots\ldots, N), i \in I$
- M: total number of workstations $j = (1,\ldots\ldots, M), j \in J$
- CT_{max}: upper bound on the cycle time of a work station
- t_i: task time of task i
- $P(i)$: Set of immediate predecessors of task i

9.4.2.3 *Objective Function*

$$\sum_{j=1}^{M} y_j \tag{9.1}$$

9.4.2.4 Constraints

$$x_{ij} \text{ and } y_j = 0 \text{ or } 1 \tag{9.2}$$

$$\sum_{i=1}^{N} t_i x_{ij} \le CT_{\max} y_j, \forall j \in J \tag{9.3}$$

$$y_{j+1} \le y_j, \forall j, (j+1) \in J \tag{9.4}$$

$$\sum_{j=1}^{M} x_{ij} = 1, \forall i \in I \tag{9.5}$$

$$x_{ik} \le \sum_{j=1}^{k} x_{hj}, \forall k \in J, \forall i \in I, \forall h \in P(i) \tag{9.6}$$

9.4.3 Details of Self-Guided Ant Colony Algorithm (SGAC)

In this section, we explain the details of the working procedure (see Figure 9.6) of SGAC by first providing the definitions of some of the terms used in SGAC for convenience. The mathematical structure of SGAC is as given in the following sub sections:

9.4.3.1 The Initialization Criteria

As mentioned earlier, the line-balancing problem in this paper is represented in the form of TSP graph, which contains nodes and arcs. Initially, the pheromone count on each arc is assumed to be equal. The initial pheromone concentration $\tau_0(n_0, s)$, on each arc is calculated by Equation (9.7); $l(n_o, s)$ denotes the length of the arc between the nodes n_o and s; Q_0 is the amount of pheromone that is distributed uniformly over each arc. A set of ants is generated and each ant is positioned on a randomly chosen node.

$$\tau_0(n_o, s) = \frac{Q_0}{l(n_o, s)} \tag{9.7}$$

There is a tabu-list that keeps track of the nodes that have been visited more than the limit known as the visitation limit. The value

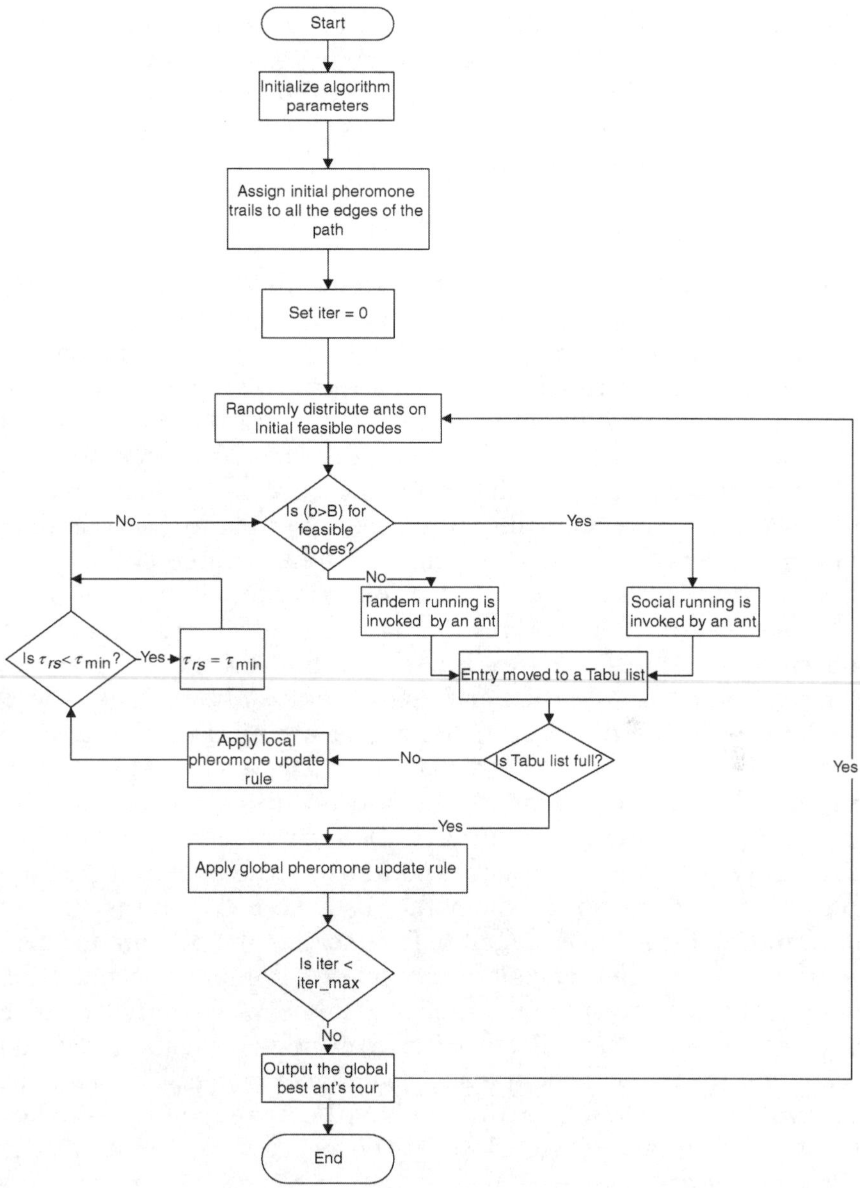

Figure 9.6 Working procedure of SGAC (based on ASGA by Tripathi *et al.* (2009)).

of the visitation limit is fixed at the time the SGAC algorithm is invoked. The nodes that belong to the tabu-list are known as visited nodes and the nodes that do not belong to the tabu-list are known as feasible nodes. The list is used to direct an ant to another node if the node it is on is a visited node.

9.4.3.2 *SGAC: State Transition Rules*

A set of computational agents is used to search for the optimal solution to our line-balancing problem. The search behaviors of these agents are similar to that of T.albipennis ants' nest hunting behavior and thus, are termed as artificial ants. The artificial ants can make use of three different rules for calculating the probability of transitioning to a particular feasible node. They are social carrying rule, forward tandem running rule, and reverse tandem running rule. The social carrying rule is preferred over the forward tandem running and reverse tandem running rules, and when the social carrying rule is not applicable, the forward tandem running rule. The concepts of quorum threshold (Franks, 2003) and pheromone concentration threshold are utilized for altering the probabilities of transitioning to a particular feasible node. A trade off in speed versus accuracy of the convergence of the algorithm is made when a particular rule is chosen. If there is an urgent need for speedy convergence (harsh environment), the quorum threshold is low; otherwise it is high (benign environment). It can be noticed that while making a transition to a feasible node using the above mentioned rules, the ant does not consider the distance between the current node and that feasible node but only considers the pheromone concentration on it. This concept of utilizing pheromone concentration when making a transition is unique to only SGAC compared to all other ACOs set, where distance between nodes is also considered as a factor for making the transitions between nodes. This way the ants will be prone to take paths that have been frequently visited by other ants and thus will get closer to the local optimal solution after every iteration.

9.4.3.2.1 Social Carrying Rule

We also define the quorum quotient (Equation (9.8)) as the ratio number of ants that have passed through a given feasible node from the current node to the total number of ants in the colony.

$$b(n_o, v) = \frac{N_{n_o v}}{N} \tag{9.8}$$

where $N_{n_o v}$ is the number of ants that have passed through the arc connecting n_o with feasible node v in the last iteration and N is the total number of ants in the colony. The quorum quotient is calculated for each path connecting the current node n_o and the feasible node v.

Let an ant be at a node n_o, and let B_{n_o} be the set of feasible nodes around this node, whose quorum quotient with respect to n_o is greater than quorum threshold ($q_{threshold}$). That is v is an element of B_{n_o} if $b(n_o, v) > q_{threshold}$. The social carrying rule is applied if there exists at least one node v in B_{n_o}. In such a case the transition probability is given in Equation (9.9).

$$p(n_o, v) = \begin{cases} \dfrac{b(n_o, v)}{\sum_{s \in B_{n_o}} b(n_o, s)} \\ 0, \quad \text{otherwise} \end{cases} \tag{9.9}$$

9.4.3.2.2 Forward Tandem Running Rule

Let T_{n_o} be the set of feasible nodes surrounding n_o such that the pheromone concentration in the arc joining n_o and each node in T_{n_o} is greater than the pheromone concentration threshold T. Given an ant at a node n_o this rule is applied if there exists no element v in B_{n_o} and if the pheromone concentration in atleast one of the arcs connecting n_o to a surrounding feasible node is greater than pheromone concentration threshold T. The transition probability corresponding to this rule is mathematically modeled as shown in Equation (9.10).

$$p(n_o, v) = \begin{cases} \dfrac{\tau(n_o, v)}{\sum_{s \in T_{n_o}} \tau(n_o, s)} \\ 0, \quad \text{otherwise} \end{cases} \tag{9.10}$$

9.4.3.2.3 Reverse Tandem Running Rule

In cases where set B_{n_o} and T_{n_o} are empty, given a node n_o, the reverse tandem running rule is applied. Therefore, this rule is applied when both the forward tandem running rule and the social carrying rule cannot be applied. The transition probability is modeled as

$$p(n_o, v) = \begin{cases} \dfrac{\tau(n_o, v)}{\sum_s \tau(n_o, s)}, & s \text{ is a feasible node} \\ 0, & \text{otherwise} \end{cases} \tag{9.11}$$

9.4.3.3 *The Pheromone Update Rules*

Two pheromone update rules are utilized to update the pheromone concentration lying between the paths of the nodes.

9.4.3.3.1 Dynamic Pheromone Variation Rule

The amount of pheromone on an arc between two nodes varies due to two events in the computational process. The first event corresponds to the evaporation of pheromone (Equation (9.12)). The second event corresponds to the addition of pheromone on a particular arc as an artificial ant passes over it (Equation (9.13)).

$$\tau^{new}(n_o, s) = (1 - \rho(n_o, s))\tau^{old}(n_o, s) \tag{9.12}$$

$$\tau^{new}(n_o, s) = \tau^{old}(n_o, s) + \rho(n_o, s)\left(\frac{Q(n_o, s)}{l(n_o, s)}\right) \tag{9.13}$$

where $\rho(n_0, s)$ is the local pheromone evaporation parameter on the arc connecting n_o with s and $Q(n_0, s)$ is the amount of pheromone on the arc connecting n_o with s.

9.4.3.3.2 Global Pheromone Update Rule

After the end of each iteration, the pheromone content of the best tour traveled by each of the N ants is updated according to Equation (9.14) and Equation (9.15).

$$\tau^{new}(n_o, s) = \tau^{old}(n_o, s) + \rho(n_o, s)\Delta\tau(n_o, s) \tag{9.14}$$

$$\Delta\tau(n_o,s) = \sum_{k=1}^{N} \Delta\tau^k(n_o,s)$$

$$\Delta\tau^k(n_o,s) = \begin{cases} Q(n_o,s)\left(\dfrac{1}{\dfrac{L_{best}}{L_k} * D * l(n_o,s)}\right), & if\ (n_o,s) \in k^{th} \text{ ant's tour} \\ 0, \text{ otherwise} \end{cases} \tag{9.15}$$

L_{best} and L_k are the tour lengths of the global best ant and the kth ant respectively. D is a dynamic quantity between 0 and 1 and its value changes after every iteration. The above equation is designed such that $\Delta\tau^k\ (n_{0,}\ s)$ obtains a non-zero value only for paths that are traversed by the global best ant. After a global pheromone update, all the ants are reinitialized and their tabu-lists are emptied. The iterations are continued until the count of the completed iterations reaches the preset maximum number. The optimal set of assembly plans (see Section 4.1) that is obtained at the end of the termination of the SGAC is evaluated by the simulation model.

9.5 Experiment and Results

In this section, the experiments conducted to test the performance of the simulation-based optimization framework proposed in this study have been summarized. After SGAC determines the optimal set of assembly plans by solving the line balancing problem, these plans are evaluated by the simulation model of the given shop floor (which was built using the Arena simulation package) to test their ability and satisfy the feasibility constraints. The assembly plan that best satisfies the feasibility constraints is selected as the optimal assembly plan. In this work, the SGAC algorithm was implemented using MATLAB programming language. The experiments were run by setting the parameter ρ of the algorithm to 0.65. The number of ants is equal to the number of tasks in our precedence graph (39 in our test case). In addition, the maximum number of iterations in SGAC was set to 1000. Exemplary screenshots showing the operations of the developed algorithm to find the optimal tour path that minimizes the number of workstations and iterative course of SGAC are shown in Figures 9.7 and 9.8, respectively.

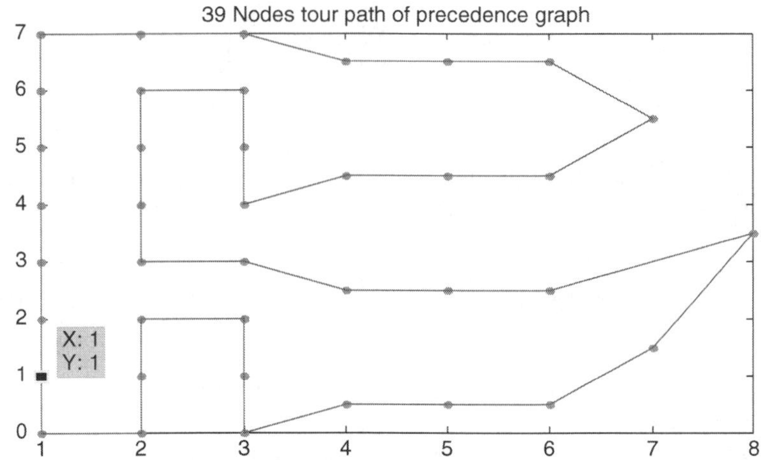

Figure 9.7 Optimal tour path that minimize the number of workstations.

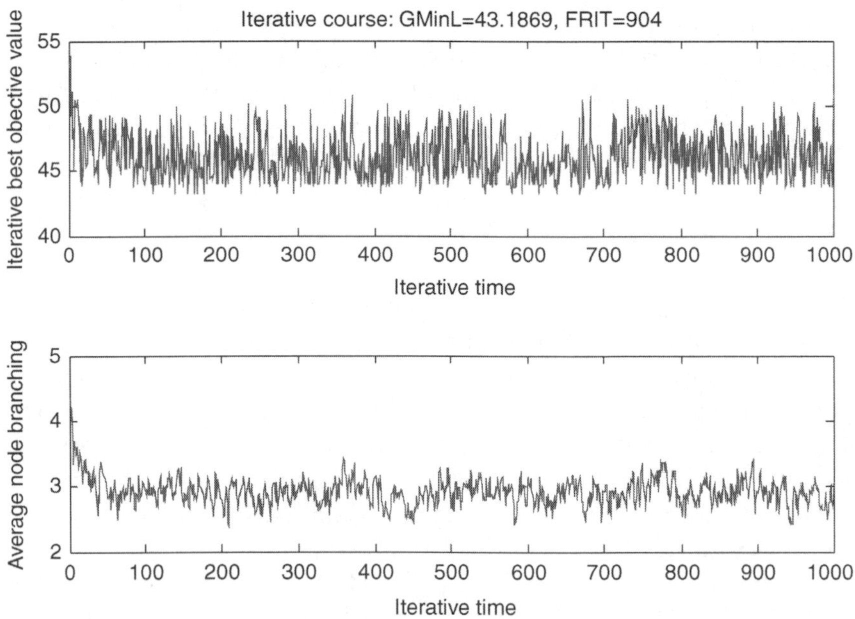

Figure 9.8 Iterative course of SGAC.

Figure 9.7 gives the optimal tour path that minimizes the number of workstations on the precedence graph in Figure 9.5 determined by the SGAC. Figure 9.8 gives the iterative global optimal solution determined by SGAC. The details of the experiments have been explained in the following sub-sections.

9.5.1 Effect of Rework on the Total Lead Time

In Section 9.3 where the case study considered in this work was explained, we noted how resources and time on the shop floor are spent on testing, rework, and maintenance. Here, after every test, it is determined whether rework has to be performed. Therefore, at any point in time during the assembly operations, the resources may have to be allocated to perform rework. In order to represent the effect of rework on the total lead time, we performed experiments and simulated rework during the course of the assembly operations. One significant reason causing rework is the inefficiency of a newly trained workforce. The need for rework decreases as time progresses, therefore we simulated rework only in the initial days after the start of assembly operations. As soon as the need for rework appeared, the status of the shop floor was fed into the simulation-based optimization framework and new optimal assembly plan was devised for the existing number of workstations. Table 9.1 depicts the effect that rework has on the lead-time of the current order on the shop floor. From Table 9.1, it can be observed that the increase in lead time due to rework may vary depending on the number of rework operations. The manager can avoid the increase in lead time either by increasing the capacity of the work stations or by creating extra work stations. However, the balance between the costs of doing the same versus the estimated increase in lead time should be maintained. If meeting the delivery schedule of the aircraft becomes the top priority, the number of workstations can be increased to meet the demand.

Table 9.1 Effect of rework on total lead-time.

S. #	Day of Assembly Operation	# of Reworks	Precedence Graph Size	Predicted Lead Time	Total Lead Time	Increase in Lead Time
1	0	0	39	43.18	43.18	0
2	4	1	36	39.18	39.87	0.69
3	4	2	37	39.18	41.18	2
4	6	2	35	37.18	39.18	2
5	6	3	36	37.18	40.19	3.01

9.5.2 Effect of Size of the Order on the Average Utilization of Workstations

The sequence and size of the order that the aircraft manufacturing company receives differs from customer to customer. The differences in these orders mainly occurs due to differences in the customer specifications in the designs of the customized aircraft. The order size affects the distribution and utilization of machine and human resources within the shop floor. A small order size may result in under utilization of the machine and human resources whereas a large order size coupled with a short and disproportionate time window may result in over utilization of the same resources leading to unfavorable wear and fatigue in machine resources and exhaustion in human resources. With the help of the framework developed in this study, we analyze the average utilization for varying size of orders.

Table 9.2 depicts the effect of order size on average utilization of work stations on the shop floor. From Table 9.2 it can be noted that as the order size decreases, the average utilizations drop for the same amount of resources (for the cycle times 4 and 5). Larger order sizes (i.e., of value 39) lead to utilization of 0.83 for workstations with cycle time 4 and therefore can lead to wear and fatigue. It can also be concluded that for order sizes less than or equal to 39, lower values of cycle times are optimal for the workstations, whereas for larger order sizes cycle times of 5 or greater than 5 may be optimal.

Table 9.2 Effect of order size on average utilization of workstations.

S. #	Order Size (Number of Tasks)	Average Utilization	Cycle Time of Workstations
1	39	0.72	5
2	39	0.83	4
3	30	0.58	5
4	30	0.68	4
5	20	0.55	5
6	20	0.63	4

9.6 Conclusion and Future Work

In this paper, we proposed a simulation-based planning framework to determine the optimal assembly plan for an aircraft shop floor.

The process of assembly of an aircraft on a shop floor is constrained by time, the shop floor's resources and the demand in accuracy of the aircrafts' design as desired by the customer and conforming to specifications laid down by the regulating agencies. The demand in accuracy of the final aircraft's design leads to frequent and unpredictable increases in the number of tasks to be completed on the shop floor as the assembly operations progress. Thus, it is not be optimal to stay with the initially obtained assembly plan. Therefore, this framework has been designed to help obtain the optimal assembly plan at any point during the time window of the assembly operations.

A self-guided ant colony algorithm has been developed and embedded in the proposed planning framework to solve the line balancing problem of the shop floor and devise an optimal set of assembly plans. Then the simulation model is invoked to evaluate those candidate plans and determine the optimal assembly plan that has the most favorable feasibility constraints (maximum average work station utilization and least variance of work station utilization) on the shop floor. The feasibility constraints were maximum average utilization and minimum variance in utilization of workstations. Preliminary experiments have been conducted to show the effect of rework on total lead-time of an aircraft, and to show the effect of order size on average utilization of resources on the shop floor. It was shown that rework in general increased the total lead-time of an aircraft. It was also shown that lesser number of work stations were optimal for the small order sizes that were tested. Future venues of this study include conducting more exhaustive experiments by running detailed simulations of the optimal set of assembly plans to be able to observe their performances based on different metrics across time on the shop floor and filter the best assembly plan.

References

Aerospace Facts and Figures. (2008), Aerospace Industries Association, 56th ed.,14.

Arnold, J., Ramulu, M. and Rao, P.M. (2004) Importance of assembly simulation as an aid for process planning for an aircraft assembly operation: Perspective from experience, *International Journal of Manufacturing Technology and Management*, 6(5), 434–456.

Betancourt, L.C. (2007) ASALBP: The alternative subgraphs assembly line balancing problem, formalization and resolution procedures, *Doctoral Thesis submitted to Institute of Industrial and Control Engineering.*

Buckingham, R. (2007) Snake-arm robots: A new approach to aircraft assembly. AeroTech Congress and Exhibition, Los Angeles, CA, USA, Session: Robotics and Component Assy (Part B).

Bureau of Labor Statistics, Aerospace product and parts manufacturing, Career Guide to Industries, 2008–09 Edition, available at http://www.bls.gov/oco/cg/home.htm.

Caridi, M. and Sianesi, A. (2000) Multi-agent systems in production planning and control: An application to the scheduling of mixed-model assembly lines, *International Journal Production Economics*, 68, 12–14.

Celano, G. and Fichera, S. (1999) An evolutionary approach to multi-objective scheduling of mixed model assembly lines, *Computers and Industrial Engineering*, 37, 69–73.

Christiand, N., and Yoon, J. (2007) Optimal assembly path planning algorithm for aircraft part maintenance, *International Conference on Control, Automation and Systems*, COEX, Seoul, Korea, 2190–2194.

Dorigo, M. (1992) Optimization, Learning and Natural Algorithms, PhD thesis, Politecnico di Milano, Italie.

Dorigo, M. (2007) Ant Colony Optimization. *Scholarpedia.*

Dorigo, M. and Gambardella, L.M. (1997) Ant colonies for the travelling salesman problem, *Biosystems*, 43(2), 73–81.

Durbin, R. and Willshaw, D. (1987) An analogue approach to the travelling salesman problem using an elastic net method, *Nature*, 326, 689–691.

Fiechter, C.-N. (1994) A parallel tabu search algorithm for large traveling salesman problems, *Discrete Applied Mathematics*, 51(3), 243–267.

Fogel, D.B. (1988) An evolutionary approach to the traveling salesman problem, *Biological Cybernetics*, 60(2), 139–144.

Ford, T. (1998) Continuing wing production, *Aircraft Engineering and Aerospace Technology*, 70(1), 9–14.

Franks, N.R., Dornhaus, A., Fitzsimmons, J.P. and Stevens, M. (2003) Speed versus accuracy in collective decision making, *Proc. R. Soc. Lond. B*, 270(1532), 2457–2463.

Hicks, C. and Braiden, P.M. (2000) Computer-aided production management issues in the engineer-to-order production of complex capital goods explored using a simulation approach, *International Journal of Production Research*, 38 (18), 4783–4810.

Jones, R.E. and Wilson, R.H. (1998) On constraints in assembly planning, *IEEE Transactions of Robotics and Automation*, 16(6), 849–863.

Jin, Y., Curran, R., Butterfield, J. and Burke, R. (2008) A quantitative metric for workstation design for aircraft assembly. Collaborative product and service life cycle management for a sustainable world, *Advanced Concurrent Engineering*, Springer-London, Part 13, 523–540.

Library of Congress. Assembling B-25 bombers at North American Aviation, Kansas City, Kan[sas], available at http://hdl.loc.gov/loc.pnp/fsac.1a35291

Liu, S. (2006) Advanced aircraft assembly techniques and prospects, *Aviation Manufacturing Techniques*, 10, 38–41.

Lu, R.F., Peterson, T.D. and Stoch, R.L. (2009) Asynchronous stochastic learning effects in engineering-to-order customization process, *International Journal of Production Research*, 47(5), 1309–1329.

Marshall, J.A.R. and Franks, N.R. (2007) Computer Modeling in behavioral and evolutionary ecology: whys and wherefores, *Modeling biology: structures, behavior, evolution, The Vienna series in Theoretical biology*, 335–353.

Onwubolu, G. C. and Clerc, M. (2004) Optimal path for automated drilling operations by a new heuristic approach using particle swarm optimization, *International Journal of Production Research*, 42(3), 473–491.

Platzer, M.D. (2009) U.S. aerospace manufacturing: Industry overview and prospects, Congressional Research Service Report for Congress.

PLM Software (2009) Siemens, General assembly planning for automotive suppliers, 1–8.

Salveson, M.E. (1955) The assembly line balancing problem, *Journal of Industrial Engineering*, 6(3), 18–25.

Snap-on Industrial. Aerospace Industry Aircraft Manufacturing, available at http://www1.snapon.com/Aerospace/AircraftManufacturing.nws

Snyder, L.V. and Daskin, M.S. (2006) A random-key genetic algorithm for the generalized traveling salesman problem, *European Journal of Operational Research*, 174(1), 38–53.

Ramamoorthy. S. (2007) Lean Six Sigma Applications in Aircraft Assembly, Thesis, Department of Industrial and Manufacturing Engineering, Wichita State University.

Tinseth, T. (2008) Current Market Outlook 2009–2028 for Boeing, available at http://www.boeing.com/commercial/cmo/.

Tripathi, M., Agrawal, S., Pandey, M.K., Shankar. R., Tiwari, M.K. (2009) Real world disassembly modeling and sequencing problem: Optimization by algorithm of self-guided ants (ASGA), *Robotics and Computer Integrated Manufacturing*, 25(3), 483–496.

Wilson, R.H., Albuquerque, N.M. (1995) Minimizing user queries in interactive assembly planning, *IEEE Transactions on Robotics and Automation*, 11(2), 308–312.

Zhang J., Li, Y., Yu, J.F., and Zhang K.F. (2008) A new method for generating work breakdown structure on aircraft assembly scheduling.

10

Applications of Evolutionary Computing to Additive Manufacturing

Candice Majewski

Wolfson School of Mechanical and Manufacturing Engineering, Loughborough University, UK

Abstract

This chapter focuses on the current and potential applications of Evolutionary Computing within the field of Additive Manufacturing (AM). As a relatively new field of manufacturing, there are many areas in which AM can benefit through Evolutionary techniques, with substantial interest generated by the idea of combining these two areas of expertise.

The chapter begins with a brief introduction to Additive Manufacturing, highlighting the main advantages of these technologies, and the generic process chain covering most of these technologies. The remainder of the chapter then focuses on the main sections of the process chain in which Evolutionary Computing techniques provide (or have the potential to provide) benefits to the process, or to the resultant parts. Due to the large amount of research in this area it is not possible to include in detail every example; rather this chapter aims to introduce the reader to a wide range of applications within the field of AM, referring him to more detailed references where appropriate.

It is hoped that the information contained in this chapter will serve to identify some of the major areas in which Evolutionary techniques will benefit the AM industry, as well as identifying those areas in which further work is required in order to take full advantage of the benefits of using AM processes.

Keywords: Additive manufacturing, design for additive manufacturing, process planning, digital design, adaptive slicing

Manoj Tiwari and Jenny A. Harding (eds.) Evolutionary Computing in Advanced Manufacturing, (197–234) © Scrivener Publishing LLC

10.1 Introduction

Additive Manufacturing (AM)[1] can be defined as 'the use of a computer-aided design (CAD)-based automated manufacturing process to construct parts that are used directly as finished products or components (Hopkinson *et al.*, 2005)'. AM encompasses a range of technologies, all of which begin with the generation of a 3D CAD model. This model is normally converted to .stl format (a triangulated surface representation of the part), sliced into discrete layers (see Figure 10.1), and built up one layer at a time. Post-processing of parts may be required, depending on the specific AM process.

The major advantage of this approach is the absence of any requirement for tooling. This necessarily leads to lower costs and shorter lead times, particularly for low to medium volumes (Hopkinson and Dickens, 2001). The lack of tooling requirement also provides benefits through reductions in Design for Manufacture and Assembly (DFMA) restrictions, allowing the

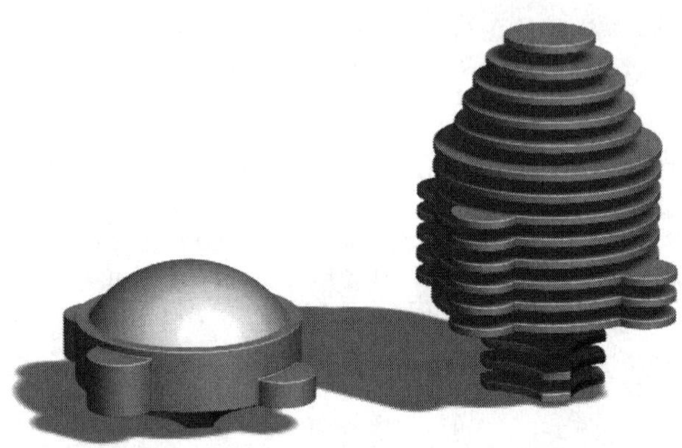

Figure 10.1 Original CAD model converted to discrete slices for manufacture. Image courtesy of James Brennan-Craddock, Additive Manufacturing Research Group, Loughborough University, UK.

[1] The reader should note that, while the term Additive Manufacturing is now the standard for the technologies discussed here, several other terms are still often used interchangeably. These include, but are not limited to, Rapid Prototyping or Manufacturing, Freeform Fabrication, Layer Manufacturing or Tool-less Manufacture.

manufacture of highly complex parts (see Figure 10.2) and the manufacture of fully working assemblies as single units. Figure 10.3 shows a Jaguar Cars door handle assembly manufactured as a single unit, where the number of individual components was reduced from 11 to one.

The generic, or most common, process chain for part manufacture is as shown in Figure 10.4, although it should be noted that not all stages are required for every AM process.

Examples of evolutionary computing can be found in almost all stages of the AM process, from the original design of a part or component, through to planning the optimum manner in which to produce it. Whilst it is not practical to include here every possible occurrence of this, this chapter aims to introduce the main areas of use, with a view to stimulating further developments and improvements in these areas.

Figure 10.2 Complex AM sculpture. Design by, and courtesy of, Dr Guy Bingham, Design School, Loughborough University.

Figure 10.3 Door handle assembly produced as a single unit.

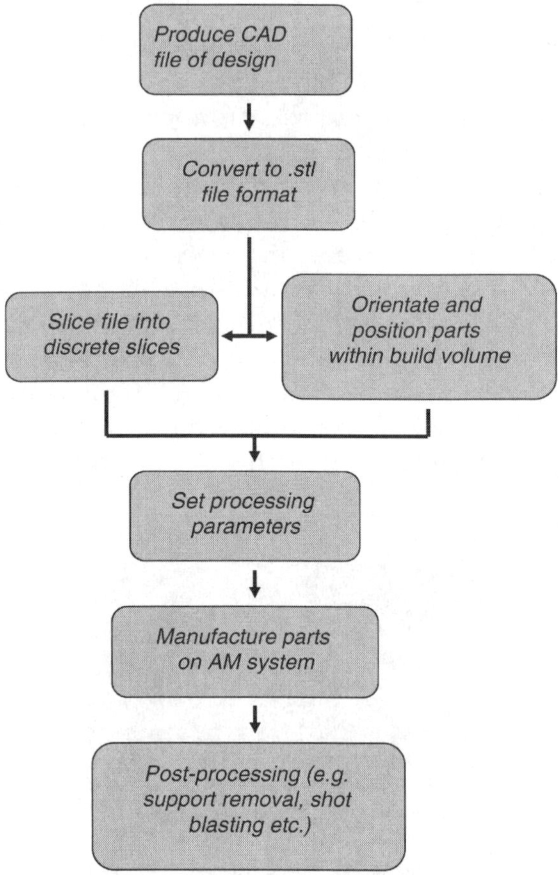

Figure 10.4 Generic process chain for AM.

10.2 Design for Additive Manufacturing

By far the largest amount of interest in evolutionary computing within the AM community is for the design of parts and components. The following sections discuss some of the major areas of significance for the use of evolutionary computing in design for AM.

10.2.1 Structural Design

In recent years there has been an increasing amount of interest in the ability of AM processes to reduce the amount of waste generated compared with traditional manufacturing techniques, and the fact that the geometric complexity achievable enables more optimised designs for specific applications.

For traditional manufacturing processes a large amount of material is wasted, both in generation of swarf, and through an inability to recycle or reuse this waste. The aerospace industry is particularly susceptible to large amounts of waste, where it is not uncommon to encounter 'buy-to-fly' ratios as high as 22:1 (Blose *et al.*, 2006).

In addition, for most traditionally manufactured products, the large number of manufacturing constraints precludes the production of an 'optimised' design. For example, a complex lattice structure may produce the optimal distribution of material to reduce the weight of a component whilst still withstanding the working loads encountered, but could be unfeasible (or in some cases impossible) to manufacture using traditional techniques. The use of AM techniques allows (within the constraints of each individual process) a greater capacity for producing geometrically complex parts, and therefore is a step closer to an optimised design.

Research at Loughborough University (ATKINS project, http://www.atkins-project.com) is investigating the effects of this in terms of reducing the carbon footprint of part manufacture. Initial work is focussed on demonstrating the improvements in carbon emissions throughout the overall supply chain which can be gained through the use of AM, following which the key findings will be used to further improve the carbon footprint of AM techniques. One key aspect of the project is the investigation into the best optimisation techniques for use with AM, and the effect of more optimised designs on carbon emissions. Various packages of work are attempting to optimise component designs. One example of this, carried out with Delphi Diesel Systems, used evolutionary techniques to optimise a diesel pump front plate, in order to both minimise weight and improved flow characteristics. Figure 10.5 shows the original part design (grey), and the same part designed for manufacture using AM (blue), with a ~37% saving in part volume.

Cellular structures (for example foams, lattices and honeycomb structures) have long been proposed as alternatives to the use of solid material for structural components, due to possessing high strength with relatively low mass. Of these the use of lattice structures has generated the most interest.

Previous research carried out by Georgia Institute of Technology has used Particle Swarm Optimisation techniques to carry out topological optimisation of lattice structures for structural integrity.

Figure 10.5 Diesel pump front plate. ATKINS project website, http://www.atkins-project.com.

A new method has now been proposed to combine structural optimisation with, for example, heat transfer or vibration absorption analysis (Rosen, 2007). The proposed methodology involves a two stage process, wherein the first stage is a topology optimisation to produce a design which not only meets the structural requirements of the product, but is also robust to adjustments of this topology. A further optimisation is then carried out using physics-based models, whereby the structure is modified (within set bounds) to meet the specified secondary objectives.

Elsewhere research has focussed on the use of lattices for deployable structures, stored in a compact manner, whereby a lattice sub-skin stores elastic energy and the final geometry is deployed upon application of internal air pressure (Maheshwaraa *et al.*, 2007). A dense grid of lattice elements is produced to fit the required overall geometry, following which the iSight (http://www.simulia.com/download/products/Fiper_Isight35_web.pdf) design exploration software iteratively adjusts the dimensions of the elements to maintain the required performance whilst minimising the deviation of the structure from the overall intended geometric profile. Penalisation schemes are included to encourage the use of thick (manufacturable) elements, and to remove elements which are too thin to reproduce physically.

It should be clear that, whilst AM offers greatly increased design complexity and flexibility, it is not necessarily the case that an optimised design will be capable of being manufactured, or that processing considerations will not render production of the part un-economical. Some attempts have therefore been made to incorporate processing constraints into design optimisation.

The Laser Engineered Net Shaping (LENS) process, where powder is fed directly into the path of a laser beam, has generated particular interest for the production of functionally graded structures, as it allows the possibility that feed powder composition could be varied throughout a layer to give different material properties where required (e.g. a hard-wearing outer surface with a thermally conductive interior). However, changes in composition cannot be made instantaneously and therefore incur a 'dwell time', adding time to the overall build process. An optimisation methodology has been proposed, based on a self-adaptive, real-valued evolutionary strategy, to optimise a multi-material structure for minimised weight, maximised stiffness and minimised processing time.

Elsewhere, Additive Manufacturing has been incorporated into the design process as a method of evaluating evolutionary product designs, rather than as the end product itself (using rapid prototypes for functional evaluation of evolutionary product designs). In this case the ease of producing 'one-off' components allows the production of the base design, and one or more 'optimised' designs, in order to test performance and validate the constraints and calculations included in the evolutionary algorithm. However, depending on the AM process selected, improvements in repeatability of mechanical properties and geometric tolerances will be required before truly accurate results can be obtained.

10.2.2 Functional Grading

Whilst there are some considerations which must always be taken into account, the specific issues or challenges associated with optimising for Additive Manufacturing are not standard across all processes. For many of the systems currently available, a binary system must be employed, whereby each individual element within the design space must be a '1' or a '0' (in other words, is either solid or a void). However, there is an increasing amount of interest in the idea of using variations in material properties throughout a part, known as functional grading.

Figure 10.6 shows a simple example of how such a structure could be produced for a cooling pipe, where the geometry ranges from a hard-wearing outer surface through to a thermally conductive interior.

In addition to the LENS process mentioned previously, another process with the capability of producing functionally graded parts

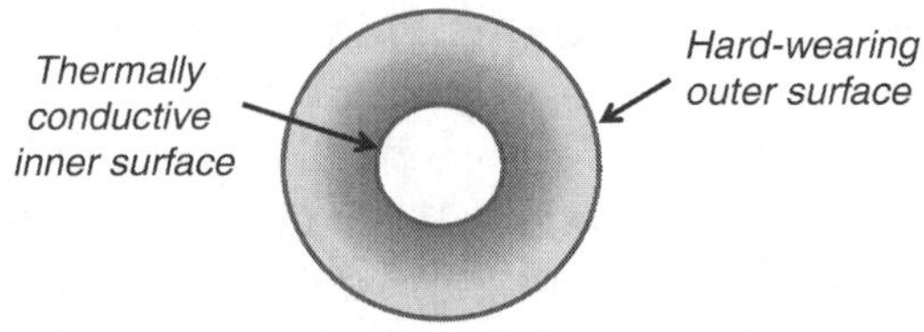

Figure 10.6 Potential example of functional grading.

is the Objet Connex (http://www.objet.com/3D-Printer/Connex_Family/), a process whereby raw materials are jetted in the required cross-section, and cured using ultra-violet light. Through the use of multiple print-heads, two different materials can be jetted simultaneously, pixel-by-pixel, and by varying the ratio of these materials within discrete sections of the design, a part can be produced which incorporates a variety of different mechanical properties (e.g. hardness) throughout.

However, despite the obvious advantages to be realised through changing material composition/performance throughout a part, relatively few AM processes currently allow true functional grading within parts.

Another challenge which must be overcome in order to fully realise the potential for functional grading is the limited availability of software required to efficiently model functionally graded structures. One example of software developed specifically for this type of application is the InnerSpace software from TNO (http://www.tno.nl/), but relatively few examples are currently available. In the Objet process, each different material composition must be applied to an individual shell, meaning a large part with varying composition throughout will require additional time spent at the design stage to assign the different properties.

As a result, some work is being carried out investigating the use of design to produce a form of functional grading through allowing individual elements to be comprised of different cell structures. In this way mechanical properties can be varied throughout the part, without the need to vary the base material composition.

A proposed method of achieving this is through the use of single material structures produced in varying volume fractions. These different volume fractions can be achieved by varying the parameters which describe them. Figure 10.7 demonstrates an example of this where the diameter of three hole features is varied, although different cell structures could also be used.

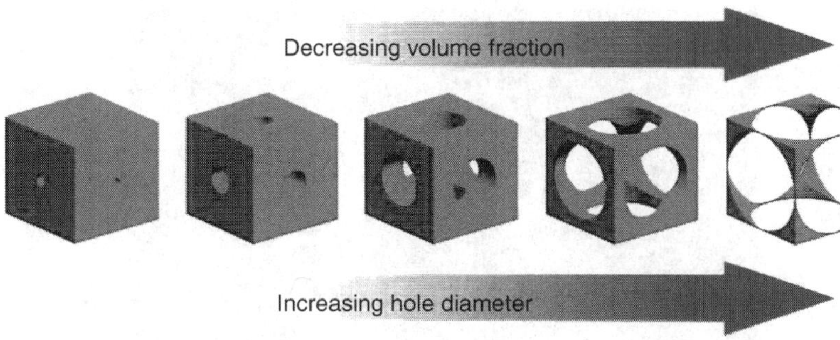

Figure 10.7 Varying cell volume fraction by alteration of a CAD model parameter, in this case hole diameter (Watts, 2008).

The proposed methodology in this case is the use of a multi-objective GA to perform a topological optimisation, either by using varied volume fractions of a single cell, or by using cells with the same volume fraction but different geometries.

10.2.3 Digital Design/Art

The requirements for evolutionary computing are somewhat different when considering art and design. In these cases the highest priority is the generation of aesthetically pleasing geometries, rather than the ability to withstand complex loading conditions. The following section highlights examples of this.

FutureFactories (http://www.futurefactories.com/) provides an example of the use of evolutionary computing for design. A variety of designs for consumer products have been produced which combine the geometric complexity achievable using AM with the use of evolutionary computing. This product is then built and finished using the relevant AM process. By careful control of the morphing process, and application of specific boundary conditions for the designs, individualised products are generated, whilst remaining instantly identifiable as part of the same brand.

One of the most widely-publicised of these designs is the 'Holy Ghost' chair collection. Designs were based on Philippe Starck's 'Louis Ghost' chair (http://www.starck.com/), combined with a Laser Sintered back (Figure 10.8).

Figure 10.9 details the process, scripted in 3DVIA Virtools (http://www.3ds.com/products/3dvia/3dvia-virtools/), used to generate the design of the seat back, beginning with creation and

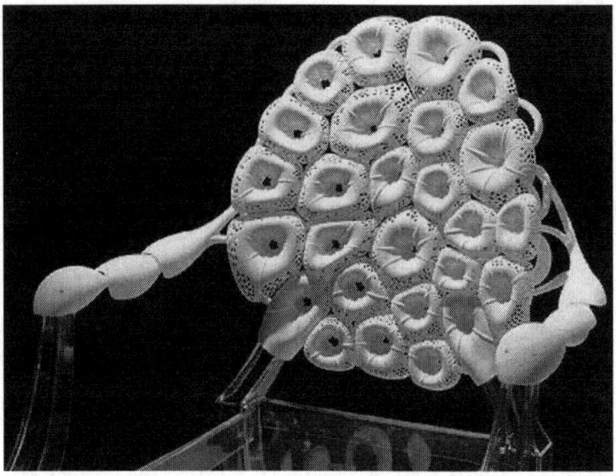

Figure 10.8 'Holy Ghost' design. Image courtesy of Lionel Dean, http://www.futurefactories.com.

placement of 'buttons' (the basic geometries from which the design is comprised), followed by uniform expansion of these buttons, and finally non-uniform expansion of these buttons to form a united geometry.

Whereas most evolutionary computing involves a large amount of skill and expertise, the *evoShape* (http://evoshape.co.uk/) product has been designed to allow designers and end-users to utilise the product with a minimum of training, in order to produce 3D CAD models of designs.

The process is controlled by a Genetic Algorithm which runs in combination with the geometric modeller within Siemens' NX software. The process begins with a population of 3D shapes, produced from randomly selected geometric primitives, which are combined, subtracted or intersected using the Genetic Algorithm. From this population, the user chooses one or more shapes he likes, which are then used to produce a new population, and the process is iterated until the user is satisfied with one or more of the designs generated. The NX geometric modeller can then be used to edit the shapes, add material properties, and so on. Figure 10.10 shows an example design created using the software.

Work is currently in progress to produce an improved touch-screen user interface for the software, which will then be used and tested by a group of collaborators for a range of applications, including jewellery, architecture and sculpture.

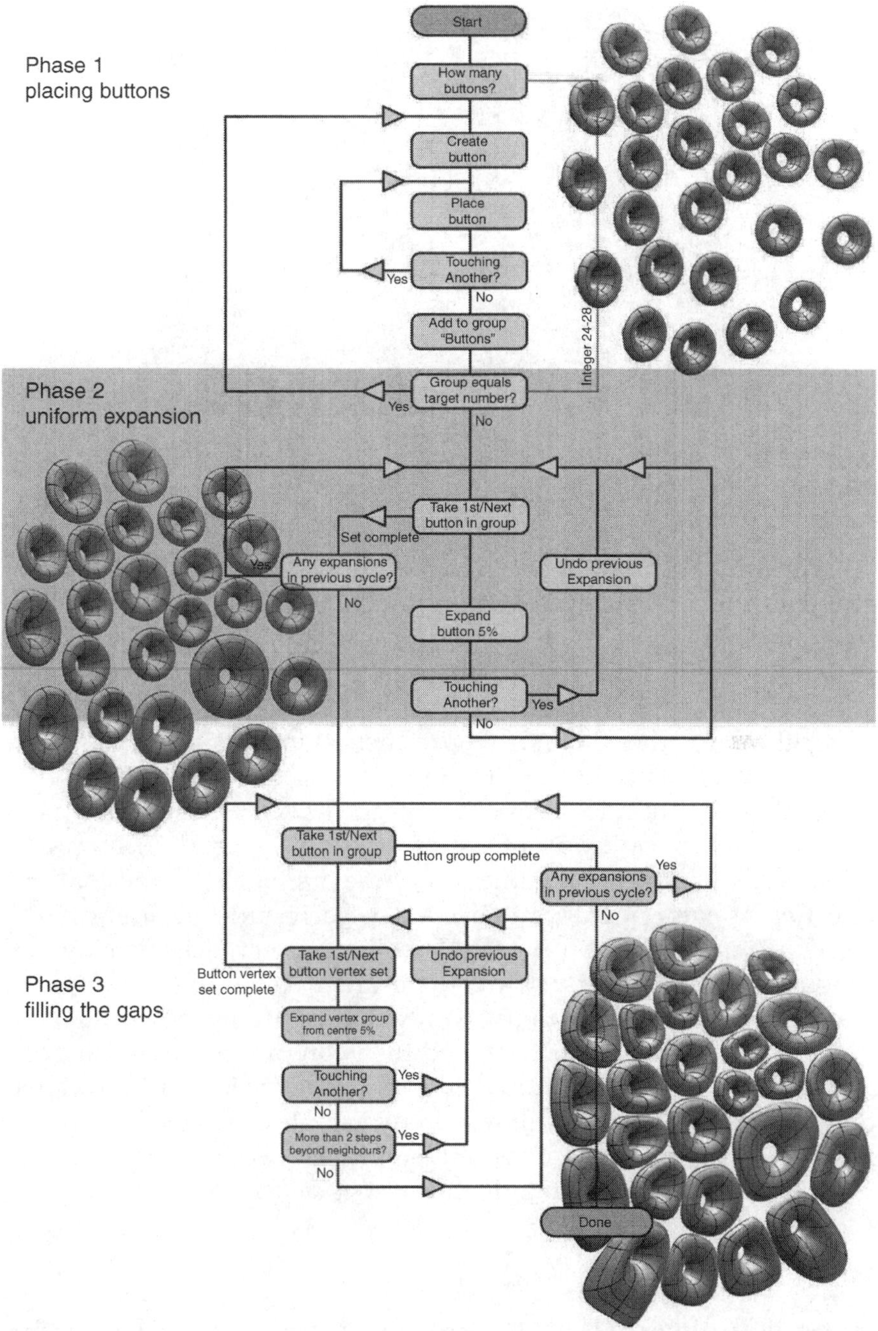

Figure 10.9 Diagrammatic representation of design generation process. Image courtesy of Lionel Dean, http://www.futurefactories.com.

Figure 10.10 Coffee machine design created using evoShape (http://evoshape.co.uk).

With the increased variety and complexity of products that can now be produced using AM techniques, it is important to ensure that parts are not 'over-designed', and that products are still aesthetically pleasing to the consumer.

Some research has been carried out into the use of Genetic Algorithms for the analysis of product styles (Hsiao *et al.*, 2010). In the example highlighted in this work, a consumer product is selected, and split into component parts. Each of these component parts is rated by a group of users, according to certain linguistic variables, and the importance of these variables. Combinations of different component parts are analysed in order to identify the 'best' design, based on the established user preferences. The overall process for a particular product therefore involves determining the main functions of a particular product, identifying potential solutions to achieve each function, and then combining these solutions and identifying the optimum solution. However, it should be noted that user intervention is still required at some stages of the process, in order to filter out proposed solutions which, for whatever reason, are actually unsuitable or impractical to manufacture.

10.2.4 Inspired by Nature

It is well-documented that Nature provides some of the best examples of optimised structures, and it has been proposed that these types of structure should form the basis of structural part design.

However, it is only with the advent of AM techniques that the ability to manufacture many of these types of structures has been realised.

A major area of interest has been for use in medical applications, and in particular the production of scaffolds for the growth and proliferation of tissue cells. The extremely complex nature of living tissue requires equally complex, and controlled, scaffolds in order to allow cell growth, and the geometric complexity achievable using AM renders it an ideal method for production of these structures, whilst also allowing a customisable external geometry. However, producing these structures with the use of conventional CAD software is an extremely time-consuming, manual, process, and several attempts are underway to develop a more automated manner in which to achieve the desired results.

Several authors have proposed simple methods of producing the design of such structures using repeating units of certain 3D shapes, or of selecting such shapes from a library, and mirroring them in the required planes to create the final geometry (Chua *et al.*, 2003; Liulan *et al.*, 2007).

At Loughborough University developments are underway to combine Object Oriented Programming methods with advanced CAD, in order to automatically generate the required geometrical elements, whilst ensuring that appropriate pore sizes, shapes, and interconnectivity are all maintained (Ramin and Harris, 2009). Due to the lack of detailed data regarding the exact parameters necessary to create suitable geometries for tissue engineering, the proposed methodology has been designed to allow flexibility in terms of additional variables and features as and when the need arises.

For the Fused Deposition Modelling process, a pre-processor has been developed to produce temporary biomechanical scaffolds which possess the correct porosity and interconnectivity (Masood *et al.*, 2005). Experimental work determined the effects of different processing parameters on the overall porosity of a test part, and a program was developed to select the optimum set of parameters to achieve a user-specified porosity.

Another approach for tissue engineering scaffolds starts with a non-porous block of material, and uses a topological algorithm to search for the optimal material distribution, taking into account porosity and mechanical specifications (Almeida and da Silva, 2010).

A somewhat different application which has been proposed for AM technologies is in the production of integrated robotic designs, where it has been suggested that a combination of electro-active polymers, Additive Manufacturing, and Genetic Algorithms could be used to develop robotic systems which closely replicate systems found in Nature (Melhuish *et al.*, 2001). Initially AM systems could be used as part of the optimisation process itself, where an evolutionary design approach would be used to select the 'best' designs from a set, and these designs would then be produced with the help of AM. These AM prototypes would then be tested under real-world conditions, and the results of this used to direct the next set of iterations. As developments in AM capabilities progress, it is then feasible that the entire robotic structure could be built as a single unit, either by pausing the process and inserting components (e.g. sensors and actuators), or by using combinations of materials within the AM process itself to generate the required performance characteristics of a section of the part.

10.2.5 Future Challenges

The preceding discussions have highlighted several areas of interest within the design stages of AM. The following section summarises the major areas where it is felt developments are most required in order to improve AM design even further.

10.2.5.1 Manufacturing Constraints

Whilst one of the major advantages of AM is the reduction in Design for Manufacture criteria associated with the processes, certain restrictions necessarily exist. This inevitably means that, even if it were possible to be certain that a design was the global optimum for a set of conditions, it is not necessarily possible to produce this exact structure.

Some of the constraints which must be considered are fairly obvious, and are relevant across all currently available systems – for example any design must take into consideration the minimum feature sizes which can be manufactured on each specific AM system, and every element within the design must be connected to at least one other, rather than allowing unconnected 'islands' within a part. Where a separate material is used to support overhanging areas within the build process, access must be allowed for removal of these supports, or for removal of powder or resin from interior features.

Less obvious factors to consider include variations in part properties throughout the build volume of a system, or difficulties in producing large, flat, cross-sections without warping of the part occurring.

If these types of constraints were to be incorporated into future algorithm developments it is envisaged that the likelihood of producing designs which are closer to the global optimum could be achieved.

10.2.5.2 Material Properties

It is obvious that any modelling or optimisation procedure can only produce results based on the input data it receives. However, at present there remains a lack of detailed, robust, data regarding the properties of AM parts. In particular the heterogeneity encountered in many processes, combined with effects of orientation, ensure that property data from other processes cannot be translated directly into AM designs. More detail on the properties of parts produced using specific materials, and on specific AM processes, will be required in order to ensure accurate modelling and assessment of future designs.

This highlights the fact that future progress will not depend simply on either experimental work or on algorithm developments, but on a combination of these areas, to ensure that results are as close to those intended as possible.

10.2.5.3 Computing Power

At present much optimisation is carried out on relatively small parts, or sections of parts, in order to demonstrate the efficacy of a particular approach. In addition, in many cases simplified assumptions are made within the optimisation process to reduce the large amount of analysis required, which can lead to over-simplification of the problem, and therefore to a sub-optimal design.

It is therefore clear, given the computational power required to optimise even the simplest of structures, that increasing levels of computing power, or indeed entirely new, less computationally-intensive optimisation methods, will be required if optimisation for AM is to become a feasible reality.

10.3 Data Handling

Whilst in the future it is likely that AM systems will slice and produce parts directly from a CAD file, at present most systems use an intermediate file format know as the .stl format. Originally developed for use in the Stereolithography process, .stl has become the de facto file format for use in AM, providing a standard file format which all systems can recognise. Almost all modern CAD packages provide an option to export a part or assembly in .stl format, allowing AM users to design in their preferred software.

The .stl format requires a 'watertight' CAD model (Figure 10.11 a), and approximates the surfaces of the model to an array of triangular facets. Depending on the resolution settings chosen, the accuracy of the resolution can be modified, although an increase in tolerances will lead to an increase in the overall file size. Figure 10.11 b and c demonstrate a low and high resolution .stl representation respectively.

Regardless of the resolution selected, some loss of information will occur when triangulating the original CAD model, and settings will therefore depend on the intended use of the part, and the resolution capabilities of the AM system itself. Additionally the .stl file cannot be manipulated beyond basic Boolean, cutting or extrusion operations, and does not contain any data regarding material properties.

Whilst the .stl format is acceptable for many applications, research is ongoing into methods of improving its functionality. Where high tolerances are used for complex parts, the file sizes generated can be difficult to manage, slowing even the simplest actions

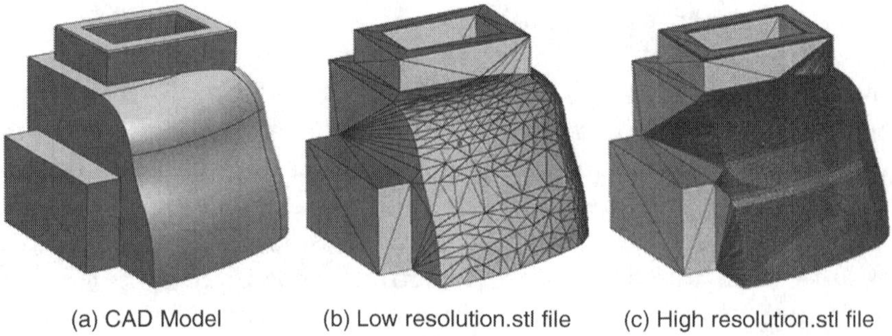

Figure 10.11 a) CAD model b) low resolution .stl file c) high resolution .stl file. Image courtesy of James Brennan-Craddock, Additive Manufacturing Research Group, Loughborough University, UK.

(e.g. rotating or translating parts), and in some cases the resultant file sizes are too large to be processed in the AM system itself.

Some research has been carried out to improve the .stl format by re-triangulating the generated mesh (Wang and Chang, 2008). Firstly the algorithm identifies and deletes any triangles which are completely surrounded by others, and where the surface normals are all similar (i.e. there is no significant curvature of the surface at this point). The voids which remain after deletion are then re-filled with an optimised layout of triangles, which are determined using a genetic algorithm which searches for the best configuration in terms of shape, smoothness and continuity of the mesh.

One major disadvantage of the .stl format is loss of accuracy due to the triangulation process, as unless a high tolerance is used, the .stl model (and therefore the physical part produced) can display obvious signs of faceting. Some research has focussed on smoothing the .stl file to provide a more accurate physical production of the finished part or component (Koc *et al.*, 1995). In this process the .stl file is sliced, and a Max-fit algorithm is used to fit smoothed curves to a series of selected points on a layer.

In the case of reverse engineering, whereby an existing part is scanned to a point cloud and converted to a 3D CAD model, large numbers of discrete points are captured in order to ensure an accurate representation of the part. However, in many cases the .stl file generated is extremely large, and the removal of certain triangles from the file can be carried out without causing a detrimental effect on the overall quality of a part produced using AM. One package of research (Chen and Wang, 1999) demonstrates a method for deleting unimportant triangles, and replacing them with a more optimal arrangement. Figure 10.12 shows an example of an 'as-produced' section of an .stl file (a), with all triangles removed from the interior (b) and replaced with an improved set of facets (c).

The first stage of the proposed methodology is the deletion of triangles, whereby the deviation of a triangle's surface normal from those of its neighbours is calculated, and those triangles with the lowest deviation (e.g. those comprising the 'flattest' areas, and therefore having the least influence on the part geometry) are selected for deletion. In this case the percentage of triangles to be deleted is defined by the user. Once the appropriate triangles have been deleted, each area is left with a number of individual vertices (as seen in Figure 10.12b), which must be re-connected to form new triangles.

During the re-triangulation process, a GA is applied in order to provide the best set of facets to meet specified conditions, namely that all triangles should be as close to equi-angular, smooth, and continuous as possible. In addition, all triangles created must have an outward-facing surface normal, and must share two vertices with each adjacent triangle. The creation of new vertices within the area to be filled is forbidden, and all edges of newly formed triangles must be within the area bounded by existing triangles (seen in Figure 10.12b).

Figure 10.13 shows examples of violations to these rules.

Whilst the standard method of representing a chromosome structure is in binary form, in this case the chromosome is comprised of a series of individual interior line segments, defined by the vertices they connect. Figure 10.14 shows a simple triangle set, along with its representation set (C) for purposes of the GA.

Whereas traditionally the initial population for a GA would be generated randomly, in this case an optimised initial population is produced, in order to ensure that no members are in violation of any of the rules introduced previously, and consequently

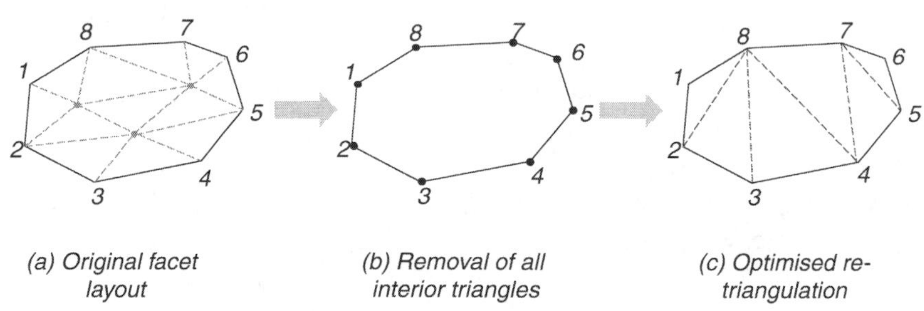

Figure 10.12 Re-triangulation process (adapted from Chen and Wang, 1999).

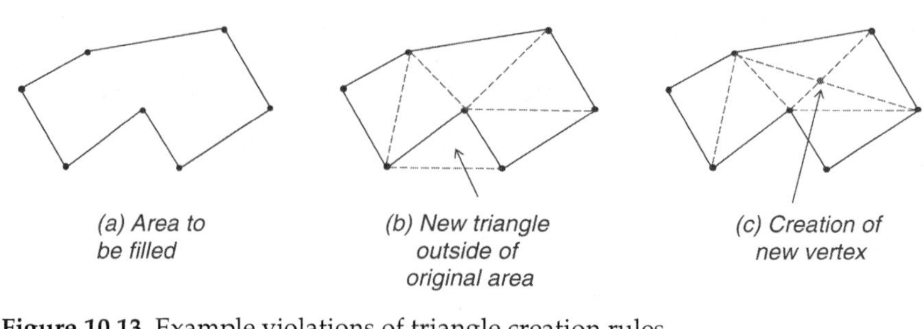

Figure 10.13 Example violations of triangle creation rules.

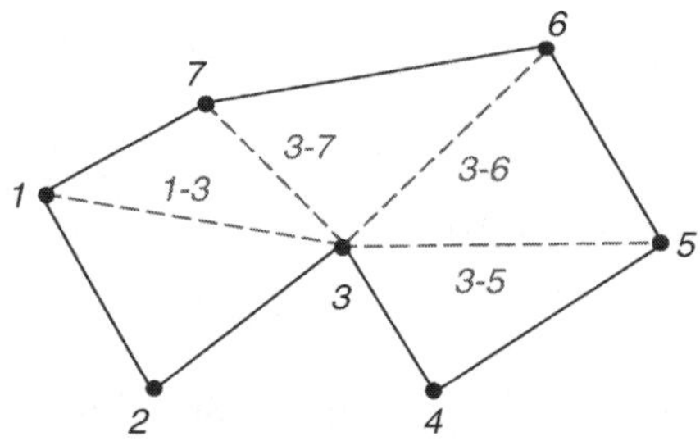

Figure 10.14 Example interior line segment representation.

to improve computational efficiency. A tournament selection process is employed here in order to maintain population diversity, whilst ensuring that the fittest chromosomes survive for inclusion in later generations. In this case pairs of chromosomes are selected randomly, and the fittest is selected as a parent for the following generation.

A single-point crossover is applied, whereby the back part of one parent is moved to the front of the other parent to create the offspring. In order to ensure the fulfilment of the criteria set out for the re-triangulation process, the crossover is constrained so that only offspring that contain no invalid segments are included in the next generation. A simple example of this would be the exclusion of any chromosomes containing duplicated line segments. A single point mutation is also applied, again constrained in the same manner.

The fitness of each chromosome is calculated based upon three criteria, F_1 – the equi-angularity of the triangles, F_2 – the smoothness of the triangular facets, and F_3 – the continuity of facets, as defined by the following equations.

$$F_1 = \sum_{i=1}^{n-2}\left[\frac{11}{9}\pi^2 - \sum_{j=1}^{\delta}\left(\alpha_j - \frac{\pi}{3}\right)^2\right] \tag{10.1}$$

Where α_j is an interior angle of a triangle, and n is the number of points to be re-triangulated

$$F_2 = \sum_{i=1}^{n-\delta}\left(\pi - \beta_i\right)^2 \tag{10.2}$$

$$F_\delta = \sum_{i=1}^{n} (\pi - \gamma_i)^2 \tag{10.3}$$

The overall fitness function is represented as:

$$F = w_1 F_1 + w_2 F_2 + w_\delta F_\delta \tag{10.4}$$

Where w_1, w_2 and w_3 are the weightings assigned to each factor. Termination of the routine is implemented when either most (user-defined percentage, in this case 95%) of the chromosomes in a generation have equal fitness values, or the pre-defined number of generations (200 for this work) is reached.

As discussed previously, a watertight 3D CAD model is a pre-requisite for the use of AM. Whilst in recent years most design work has been carried out using 3D CAD systems, there is a large amount of historical data held in the form of 2D engineering drawings. Some attempts have been made to automate the process of converting these drawings into 3D models, an example of which can be found in the Solidworks software (http://www.solidworks.com). Another example attempts to simulate the approach which would be used by an experienced designer, in order to transform a 2D drawing into a 3D CAD model (Chen and Feng, 2003). In this case a 2D drawing is scanned, and closed regions of black pixels are identified. Networks of these regions are then created, and a Genetic Algorithm is used to search for pathways which satisfy the requirements of all 2D views in the original drawing. Once the most appropriate connections have been established, a 3D model can be reconstructed.

10.4 Process Planning

10.4.1 Build Packing

One of the major advantages of some AM processes is the ability to nest parts within the build volume. For certain processes (e.g. Stereolithography) parts may only be positioned in a single level, whereas for others (for example polymer Laser Sintering) parts may be positioned above previous parts, throughout the entire build height.

This nesting capability allows the possibility for a single part to be duplicated many times within the build volume, thereby reducing the cost per part, or the production of a full set of components for a single product, allowing a more 'Just-in-Time' approach to product manufacture.

In order to take into account the pre- and post-processing times encountered during many AM processes, it is essential to fully utilise their nesting capabilities and fill as much of the build volume as is possible. For example, when Laser Sintering Nylon-12 materials, the build chamber is subjected to approximately a two hour pre-heat time, and, to prevent warpage, parts must be allowed to cool within the build chamber for several hours after the build is complete. These additional processing times render it essential to maximise utilisation of the build volume.

In most cases the major requirement is to minimise the build height (z direction), as in general a greater number of discrete layers will increase the build time. Previous research in Laser Sintering (Ruffo *et al.*, 2006) highlights the cost reductions associated with filling one full horizontal layer of a build, and again with completely filling the whole build volume, highlighting the necessity to pack the build as efficiently as possible (see Figure 10.15).

10.4.1.1 Bounding Box Approach

Most approaches to packing of parts within a build rely on a 'bounding box' approach. When using this approach, a geometry is approximated to a rectangular box fitting round the extremities of

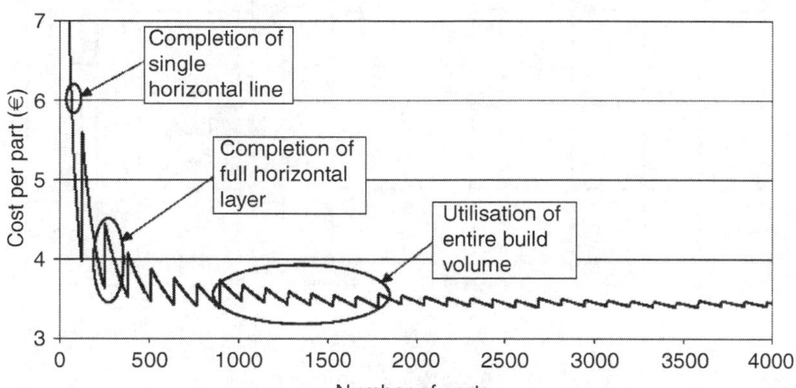

Figure 10.15 Cost vs. number of parts (adapted from Ruffo *et al.*, 2006).

the part dimensions. Utilising this technique for simple geometries (see Figure 10.16a) provides an accurate representation of the part for packing purposes. However, when considering a more complex geometry (Figure 10.16b), this technique necessarily leads to inaccurate part representation, whereby substantial amounts of free space are enclosed in the nominal part volume.

These inaccuracies lead to efficient packing for the simplest geometries, but for more complex parts the number of parts fitted to a particular area will be substantially lower than that achievable in practice. Figure 10.17 demonstrates that the bounding box method will allow maximisation of the number of simple geometries which can be fitted to the build area, whereas the space utilisation that would be achieved using this method for more complex geometries is substantially less than could be achieved using a different packing method.

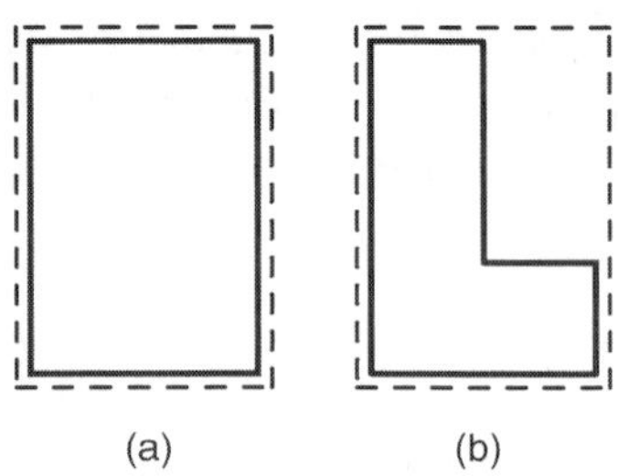

Figure 10.16 Bounding box for a) simple geometry b) complex geometry. Image courtesy of James Brennan-Craddock, Additive Manufacturing Research Group, Loughborough University, UK.

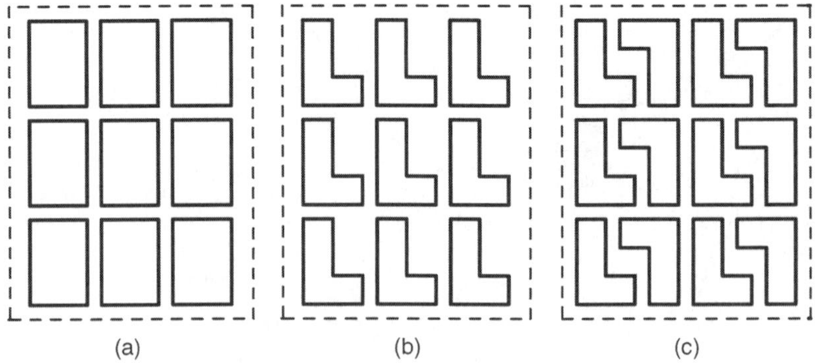

Figure 10.17 Build packing arrangement a) simple geometry (bounding box method) b) complex geometry (bounding box method) c) complex geometry (optimum packing).

10.4.1.2 Build Packing Solutions

Most build software for commercial AM systems includes some form of part placement algorithm as standard. However, as mentioned, most of these rely on the bounding box method, and are generally simplistic in their approach, allowing a minimum of user inputs. For example, this may include specifying the minimum distance between two parts, so that the next part imported into the software will automatically be placed at this spacing. However, most build software does not allow the optimisation of part placement within the build.

Many of the non-commercial systems which have been developed for various processes also make use of a bounding box methodology.

A pre-processing methodology for the Stereolithography process has been developed which employs a GA with a bottom-left placement rule to minimise unused space on the build platform (Canellidis *et al.*, 2006). Firstly a set of six candidate orientations are produced, taking into account the total number of layers to be fabricated, the downward facing areas (where support structures would be required), and the projected area which the part would occupy on the build platform. The user then selects his preferred orientation, and the GA is employed to place the maximum possible number of parts onto the platform so as to minimise the unused space.

Another, similar approach, also for the Stereolithography process, assumes that the user has orientated each part in the vertical direction to provide the best surface finish. A GA is then employed, using the bounding box of each part combined with a user-specified minimum part spacing, to pack the parts in order to maximise the total number of parts whilst also maximising the utilisation of the vat (Wodziak *et al.*, 1994). The algorithm allows the rotation of each part by +/–90° in the horizontal plane in order to achieve this.

Some attempts have been made to identify more accurate ways in which to achieve efficient packing, by moving away from the use of a bounding box methodology.

Commercially, the Smartspace module, available for the Magics AM software from Materialise, allows the automatic placement of parts, either based on the part geometry or on a bounding box method if required (Materialise Magics, http://www.materialise.com/Magics). The software optimises to minimise build height, and further improves on this by allowing small parts to be nested within larger parts, as long as the part can be removed afterwards. Rotation and position of specific parts can be restricted if required for specific reasons (e.g. surface finish requirements). Reductions in material usage of 30%, and build time of 15% have been quoted through use of the module. Figure 10.18 shows an example build packing solution with and without the use of Smartspace.

The module also includes automatic checks for collision and interlocking (where a small part is encased completely within a large, hollow part).

The GARP software, developed specifically for the packing of a Selective Laser Sintering build chamber, was designed to use a sequential GA to minimise the sum of distances of parts from the global origin, whilst ensuring there are no intersections between triangles of parts, or with the edge of the build chamber (Ikonen *et al.*, 1997). Due to the large number of triangles present in most .stl files, further developments have been made to use a distributed GA, making use of multiple CPUs to reduce the execution time (Lewis *et al.*, 2005).

Another approach is the use of a voxel (3D pixel) structure to represent the part, to avoid the inaccuracies incurred through use of a bounding box representation (Hur *et al.*, 2001). The methodology employs a modified bottom-left approach with a GA which aims to minimise build height whilst packing as many parts as possible in cases where all parts cannot physically fit into one build. The algorithm also aims to position parts as close to the centre of the build volume, where mechanical properties are generally higher, as possible. For one example, a reduction in build height of 50% is

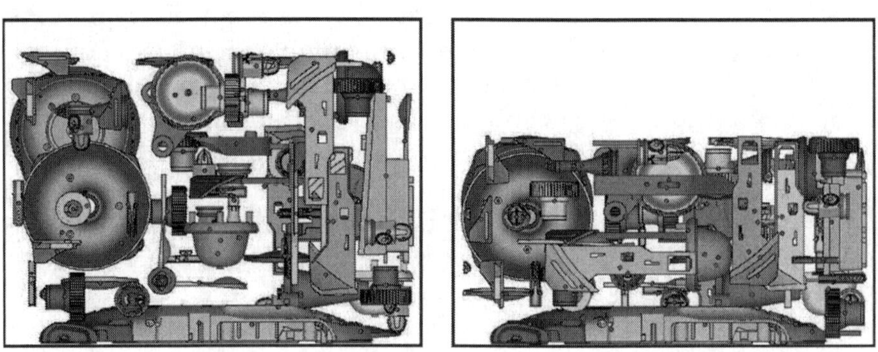

Figure 10.18 Example part nesting with (right) and without (left) Smartspace (http://www.materialise.com/materialise/view/en/92215-SmartSpace.html).

quoted when using the voxel-based approach, as compared with the bounding box approach.

Other researchers in the area have developed a system (Intellipacker) to obtain optimised build layout for Additive Manufacturing, based on part orientation and packing (Gogate and Pande, 2008). In the first of two stages, the user must rotate individual parts to select 'acceptable' part orientations, based on automated calculations of the overall roughness caused by stair-stepping at each orientation. Variations of these orientations are further modified by rotations around the Z axis, in increments of 45°, representing the orientations allowed for selection in the later GA-based packing stage.

An initial population of chromosomes is produced randomly, separated into three distinct components representing the packing sequence, orientation, and rotation. The recommended initial population size was between n and 2n for simple geometries (where n is the number of parts to be packed), with larger populations required in the case of particularly complex parts.

For the first of these components, each part to be packed within the build volume is assigned a real number, and the order of these within the string represents the order in which the parts will be packed. In the second stage, the orientations selected as being acceptable are each assigned a single digit number, which is then converted into a binary string. These individual strings are combined into a single string, in the packing order specified. Finally the rotation around the z axis is defined, based on another binary string of numbers relating to increments of 45° from the start position. Figure 10.19 demonstrates an example chromosome structure.

Crossover was carried out using a tournament-pairing scheme, whereby the chromosome with the highest fitness is paired with

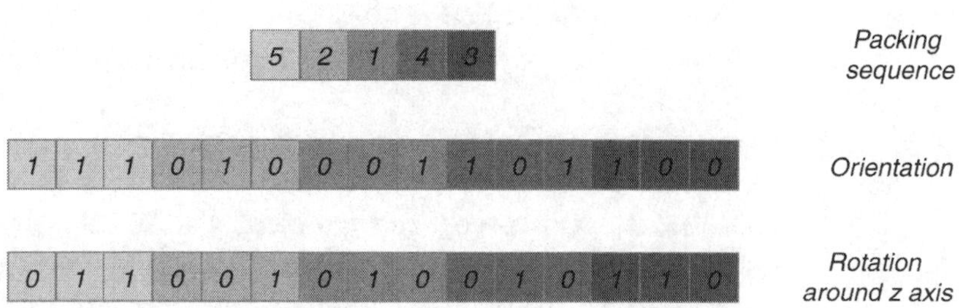

Figure 10.19 Chromosome structure (five parts, numbered consecutively from one).

the lowest fitness to undergo the crossover process. The packing sequence was subjected to a single-point, order-based, crossover, and the orientation and rotation strings were subjected to a simple crossover. It was identified that the probability of crossover should be greater than 0.75 to provide an efficient search.

Mutation was also implemented, by exchanging a 1 for a 0 in the orientation or rotation string, with a probability of ~0.2 providing the best results, and increasing this probability providing no advantage.

Also included in the algorithm is an elitism strategy, whereby the best chromosome in each new population is automatically included in the next generation to increase the speed of convergence. As this can also lead to reductions in genetic diversity, the elitism strategy is only implemented during the latter half of the generations specified. The number of generations suggested is between 1.5 and 2n.

In this case the first part specified in a particular chromosome is positioned at the farthest corner of the machine volume, and subsequent parts are positioned along the negative Z, Y and X axes, subject to collision and interference checks with existing parts and the extremities of the machine volume.

In this work four major factors are included, namely the total build height, staircase error factor (an indication of overall roughness of the part), and the volume and area of contact of support structures. Each of these factors is weighted by the user, and any that are irrelevant (e.g. support structures for Laser Sintering) can be ignored by assigning a weighting of zero. The fitness function is calculated as follows, with the requirement being to maximise this value:

$$Fitness = \frac{1}{(Wh\ x\ H)+(Wa\ x\ A)+(Wv\ x\ V)+(Ws\ x\ S)} \tag{10.5}$$

Where H = Actual build height/target height

V = Volume of supports/(volume of supports + volume of parts)

A = Surface area in contact with supports/total surface area of all parts

S = S (all parts) (Surface area of part x staircase error factor)/surface area of part and Wh, Wa, Wv and Ws are the user-specified weightings for each of the above respectively.

Once the best packing layout, orientation and rotation of each part has been identified for the AM system in question, a multi-part .stl file can be generated for export to the machine for manufacture.

10.4.1.3 Future Requirements

In addition to the developments described here, there are various other factors which must be considered when packing a build, many of which currently rely on the knowledge and experience of the build operator. For example, depending on the AM process selected, and in some cases the specific material used, the minimum separation between parts would differ. Other factors, such as placement of thin-walled sections, and avoidance of large, flat, cross-sections must also be taken into account.

In order to make systems more user-friendly, and accessible to those new to the field, these types of factors should ideally be included in any future algorithms developed.

10.4.2 Part Orientation

For every AM process, orientation of the parts within the build chamber can play a major part in the quality of the final product. This may be in terms of the surface finish (for example in the Stereolithography surface the best finish can be found on upward facing surfaces), the number of supports (and therefore the post-processing time) required for overhanging areas, or in the properties of the parts themselves (many processes produce anisotropic mechanical properties, whereby parts built parallel to one axis would be stronger than parts produced parallel with a different axis). Currently many of the parts produced using AM depend on operator skill or knowledge as to the best orientation in which to produce a part. Whilst this can be a relatively straightforward task for simple geometries, for more complex parts it can be difficult, or in some cases impossible, to select the optimum orientation manually, which has led to a substantial amount of research into automatically optimising this process.

Due to the 'stair-stepping' effects caused by the layer-by-layer nature of AM processes (see Section 10.4.3 for a more detailed description), the final manufactured part will necessarily vary from the original design to a certain extent, but careful orientation of the part within the build volume can aid in reducing this error. One method of quantifying the error associated with this is the calculation of the difference between the volume of material used in building the part and the volume specified by the CAD model, termed the 'volumetric error (Masood and Rattanawong, 2002). Research has been carried out at Swinburne University of Technology, whereby the volumetric error is calculated for each layer of a part at a series of different orientations, and combined to provide the total error for the entire part (Masood, 2003). Whilst this can be a useful tool, the process is limited in that it only assesses orientation around a maximum of two axes, and at user-specified angles of rotation.

Elsewhere a hierarchical particle swarm optimisation algorithm has been employed to optimise part orientation in terms of volumetric error, build cost, and operational efficiency (Ghorpade *et al.*, 2007). In this case the user provides weightings for the importance of each of these three factors, which are then used to determine the aggregate objective function for the orientation algorithm.

A further approach has constructed and trained neural networks, which are used in the evaluation function of a GA to optimise layer thickness and orientation in terms of volumetric accuracy and build time. In this case the accuracy is determined not just by the overall volumetric error but also by the deviation on each individual slice.

10.4.2.1 Support Structures

Some AM processes, for example when Laser Sintering polymeric materials, do not require additional support structures for overhanging features, as the un-sintered material acts to support these areas. However, many other processes (e.g. Stereolithography (SL) and Fused Deposition Modelling (FDM)) require additional, sacrificial supports to prevent warpage and allow the production of overhangs. Removal of these supports can be a time-consuming process, and in some cases leads to a poor surface finish on supported areas of the part. For this reason, minimising the amount of supports used in the building process can be a crucial part of the planning process for AM.

Some work assessing part orientation takes into account not only stair-stepping and build time effects, but also aims to reduce the number of supports generated during the build. One algorithm developed provides optimisation of orientation in terms of part accuracy, build time and quantity of supports for the Stereolithography process (Hur and Lee, 1998). The user can specify his order of preference for these three objectives, and the best orientation will be selected in order to optimise this. If several orientations meet the same minimum objective function, the orientation which best satisfies the requirements of the second objective is selected, and the process continues for the third objective if appropriate.

Other research also includes support structures in its approach, where a GA is used to orient a part according to user-defined ratings of the importance of minimising build time or maximising part quality (Byun and Lee, 2005). Supports are considered in achieving both of these objectives – firstly in terms of build time, as the fabrication of supports must add time to the overall process, and secondly in terms of part quality, whereby the total area of contact with supports will impact on the final surface finish.

Elsewhere a method has been developed for the reduction in processing time for the FDM process (Lam *et al.*, 1998). The time taken to deposit material for each layer of a solid part is a substantial proportion of the overall build time, and the use of hollow structures where the internal bulk material is not required for structural integrity could speed up the build process dramatically. However, in many cases a hollow structure could not be produced without the incorporation of internal support structures. To circumvent this difficulty, a method has been proposed whereby a thin shell is produced around the exterior of the part, with a user-specified offset from the overall geometry. A sub-boundary octree approach is used to produce interior reinforcements within this shell. For a particular component this approach has demonstrated a 43% reduction in processing time as compared to the production of the part as a full solid.

10.4.2.2 Post-Processing

Some of the developments discussed in the previous sections, for example minimising support structures, or reducing stair-stepping effects, will also have a direct secondary result on post-processing requirements for AM parts. However, in some cases the minimisation of post-processing has been addressed directly, as discussed here.

When considering the Stereolithography process, the surface finish following the removal of support structures is generally not acceptable, and further finishing (e.g. polishing) is required. In addition to being a time-consuming process, there is also a danger of causing damage to small or thin features. Whereas most developments have focussed on orienting parts to optimise the building process, some work has been carried out specifically to consider the post-processing stage as well as the build stage (Kim and Lee, 2005). Post-processing is split into sub-processes, some of which (e.g. post-curing) are assumed to be independent of orientation, and others (for example polishing or removal of support material) are assessed according to the area of the part that must be supported in each orientation. A steady-state GA is used to minimise both the build time and the post-processing time. Reductions in post-processing of over 50% have been demonstrated in some cases.

Other work has focussed on establishing which areas of a part would exceed a certain level of surface roughness and uses a steady-state GA to optimise for the best orientation, with the overall aim of reducing the regions of the part that will require post-machining (Ahn *et al.*, 2007).

10.4.3 Slicing

The layer-by-layer method of producing parts using AM leads to differences between the geometry described in the CAD model of a part, and the actual geometry produced, as each individual layer of the part can only be produced with vertical sides. Figure 10.20 demonstrates this effect using a cross-section of a simple geometry.

In some cases this stair-stepping effect is not important, whether due to the intended use of the part, or the fact that the geometry of the part is less susceptible to this effect (for example parts with straight, vertical, walls).

However, in many cases the poor finish caused by stair-stepping renders the use of AM inappropriate, or at the very least means

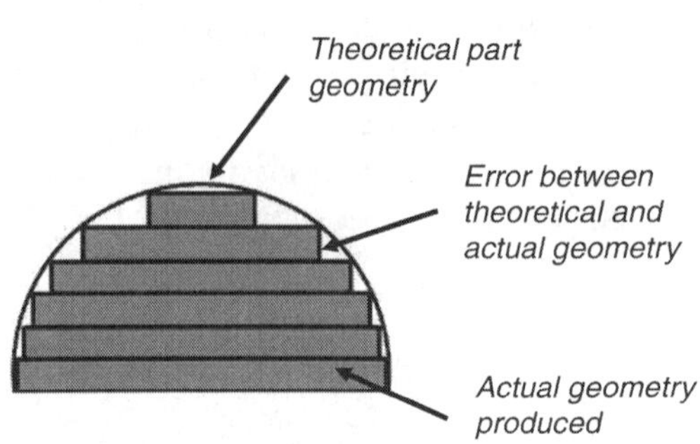

Figure 10.20 'Stair-stepping' effect in AM.

substantial time is spent post-processing the part to remove these steps. Various methods have been proposed to overcome these issues, as described here.

Within the field of AM, it is well-documented that a smaller layer thickness will provide a better surface finish through minimising the effects of stair-stepping. However, the smaller the layers the greater the number of individual layers in a part, which will increase the overall build time, and therefore the cost of the part.

One proposed solution is the use of different layer thicknesses throughout a part, depending on the geometry of each section, commonly termed 'adaptive slicing'. The easiest manner in which to achieve this is to assess the angle of a slope of a section or feature with respect to the horizontal plane, and use this to select the appropriate layer thickness. Figure 10.21 demonstrates this principle for slicing of a simple hemispheric geometry, where the use of constant slice thickness can be seen to provide a much less accurate reproduction of the part than is achieved using variable layer thickness.

A variety of methods and algorithms for generating the appropriate slice thicknesses have been developed, some of which are described here.

One accepted method of calculating the appropriate layer thickness throughout a part is to set an allowable value of 'cusp height' throughout the part. Figure 10.22 demonstrates what is meant by this term.

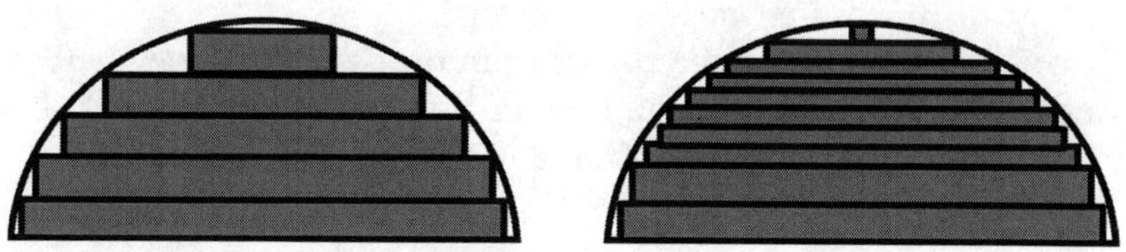

Figure 10.21 Hemisphere sliced with a) constant layer thickness b) variable layer thickness.

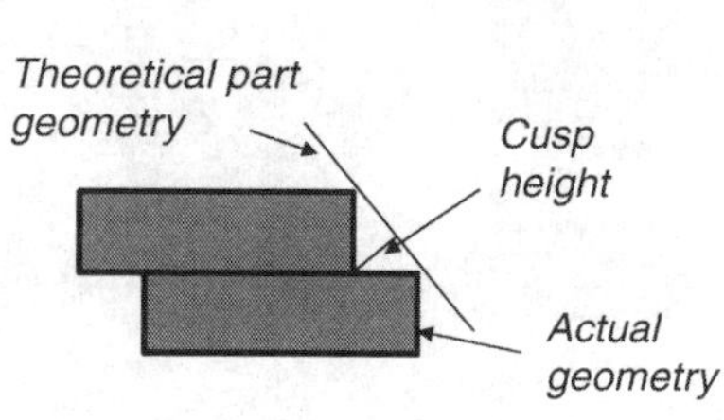

Figure 10.22 Definition of cusp height.

Some research (Xu *et al.*, 1997) has taken place regarding the development of a variable thickness slicer which is implemented in Pro-Engineer. The system employs a genetic algorithm which uses the curvature of the part to find the appropriate layer thickness at each z height in order to keep the cusp height within a specified tolerance throughout. The algorithm also ensures that slicing occurs at the beginning and end of any features, to further minimise loss of accuracy within the final part.

Other researchers have developed a method of slicing a model where the thickness of each layer is calculated in order to meet a user-specified maximum surface roughness (Byun *et al.*, 2006). This work also takes into account the fact that, while the edges of a part manufactured using AM are nominally vertical, in practice they possess some level of curvature at the edges. In this case the fillet and corner radii of each layer are incorporated into the roughness calculations.

Whereas conventional adaptive slicing will slice at a constant thickness across the entire cross-section, a further refinement uses the idea of local adaptive slicing, whereby the CAD model is divided into sub-sections, and each section is sliced according to the curvature of its surface. In this manner sections with straight (or near-straight) edges are sliced at the specified maximum thickness, allowing the minimisation of build time in these areas. Areas possessing higher degrees of curvature can be sliced using variable layer thickness, maintaining the maximum possible level of accuracy. Figure 10.23 illustrates this concept.

By utilising this method reductions in processing time of between 17 and 37% have been recorded, when compared with a part sliced using conventional adaptive slicing (Tyberg and Bhn, 1998).

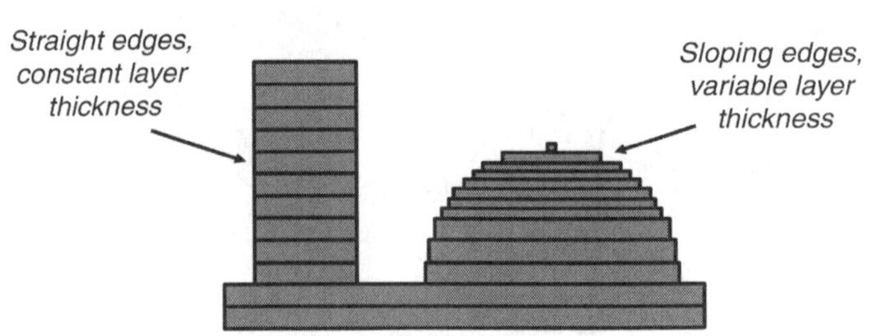

Figure 10.23 Local adaptive slicing (figure adapted from (Houtmann *et al.*, 2009)).

Elsewhere a system is employed whereby the exterior of a part is built using adaptive slicing techniques to achieve the required surface finish, whereas the interior is produced using thicker layers to minimise processing time (Sabourin *et al.*, 1997).

An alternative method proposed to reduce stair-stepping effects is the use of sloping edges to layers, whereby each individual layer is produced with an angled edge. This allows for the production of geometries which more accurately represent the intended design. Whilst very few technologies currently exist that can manufacture parts with anything other than vertical edges, developments in this area could provide a substantially greater degree of accuracy in the future.

10.4.4 Parameter Optimisation

For most AM processes, the selection of appropriate build parameters is crucial in achieving the required surface finish, geometric accuracy and mechanical properties. The type and number of parameters which must be considered is dependent on the AM process selected, and in some cases the specific material chosen. This necessarily means that any method of optimising processing parameters will be specific to an individual process.

One proposed method for optimising parameters in the Stereolithography process uses an adapted multi-objective optimisation method to achieve a balance of variables including orientation, layer thickness and four separate recoat variables, incorporating empirical data and analytical models to relate these variables to set objectives (West *et al.*, 2001). Objectives assessed are geometric tolerance, surface finish and build time, weighted according to user-specified preferences, and a compromise Decision Support Problem formulation is used to minimise the aggregate deviation from the specified objectives. Process specific constraints, for example support structures and large, flat sections, are also incorporated, using a penalty method.

Other work has focussed on hatch length and direction in AM processes. Within the Direct Metal Laser Sintering process, variations in hatch lengths have been shown to lead to heterogeneity of mechanical properties throughout the finished part, and a GA is used to optimise the hatch direction to provide a greater degree of homogeneity in the finished product (Nig *et al.*, 2005). Experimental work investigating the effects of material anisotropy and heterogeneity on part strength was used to identify the effect of hatch length on part quality. This data was then incorporated into an optimisation algorithm, whereby the Visual C++ software calls functions from the GAlib software developed by Massachusetts Institute of Technology and optimises hatch direction on each layer in order to minimise the amount of heterogeneity caused by variations in hatch length. The use of this technique has been shown to provide improvements both in terms of surface finish, and in reducing warpage of parts during the build process.

Elsewhere research has focussed on reducing the level of shrinkage encountered during the Selective Laser Sintering process (Wang *et al.*, 2009). Experimental investigations were carried out into the effects of different in-process parameters (e.g. layer thickness, laser power and scan speed) on part shrinkage, and fed into a back propagation neural network model to describe the process, which is then optimised using a GA. Beginning with a random population of different parameter sets, which are then assessed for predicted shrinkage, the best of these are selected and subjected to cross-over an/or mutation in order to identify the optimum parameters for minimal shrinkage.

One less process-specific approach takes the number of individual scanning vectors within a part as an indication of the expected build time, for example due to the need for deceleration of moving parts at the end of a single scan. A GA is used to minimise the total number of hatching segments, based on determining the optimum hatching direction (Zhao, 2006).

The common theme running through most, if not all, of the developments in parameter optimisation for AM is the need for robust experimental data to feed into the models and algorithms developed. Whilst much of the research carried out in this area is process-specific, it is certainly the case that many of the general approaches, combined with solid experimental results, could be relatively easily translated to other AM systems. The development of multi-objective optimisation algorithms/procedures, backed up with physical results, and allowing simultaneous optimisation of, for example, surface finish and specific mechanical properties, would provide major benefits in improving the stability of AM part production, and the confidence of the AM user in the processes.

10.4.5 Summary

It is hoped that this section has demonstrated the broad range of areas in which some form of evolutionary computing is, and could be, used to improve the process planning stages for AM.

However, some notable areas for further developments can be seen. Firstly, the mechanical properties of parts, and the repeatability of these properties, are crucial for increased uptake of AM technologies within industry. Whilst some attempts have been made to optimise certain processing parameters to meet specific objectives, more work is required to allow optimisation of properties for each AM process/material combination. Much of the work on modelling specific processes, and on optimisation of certain parameters (e.g. to minimise shrinkage), should be relatively easy to apply in a similar way in order to improve other objectives, and for a range of processes.

Due to the varying constraints seen in different AM processes, it is unlikely that it will ever be possible to develop a single meaningful algorithm which will be applicable across all AM technologies. In addition, whilst in some of the cases discussed attempts have been made to optimise more than one aspect of process planning (for example combining orientation and build-packing), due to the complex nature of the processes most developments have tended to focus on one single stage of the planning process. However, by definition all of the areas discussed in the section have an impact on each of the other stages of the process.

Arguably the most important goal for the future would be the development of a single software tool which would allow the user to select specific processes (and in some cases materials) for which to optimise, input, in order of importance, the objectives to be met, and output the 'best' orientation, placement and build parameters to achieve these objectives. Further developments could include AM processes and materials themselves in the algorithm, whereby the user could specify more than one process and/or material to include in the optimisation, and the algorithm would also identify the most appropriate of these to use for production.

As discussed in previous sections, in order to achieve this goal, rigorous and detailed experimental data will also be required to feed into the optimisation algorithms.

10.5 Concluding Remarks

It is clearly not possible, in such a short space, to describe in detail every example of Evolutionary Computing which can be found throughout the stages of Additive Manufacturing. However, it is hoped that this chapter has demonstrated the broad range of current, and potential, areas where evolutionary techniques could be implemented within or alongside AM processes. These areas range from the design process, whereby the geometric complexity achievable using AM can be complemented extremely effectively by the use of structural optimisation techniques, through the process planning stages.

It is hoped that the information presented in this chapter will go some way towards identifying some of the most relevant areas for Evolutionary Computing, as well as stimulating new developments where required for further improvements.

References

Ahn, D., Kim, H. and Lee, S., Fabrication direction optimisation to minimise post-machining in Layered Manufacturing, *International Journal of Machine Tools and Manufacture,* Volume 47, Issues 3–4, March 2007, Pages 593–606.

Almeida, H.A. and da Silva, B., Virtual topological optimisation of scaffolds for rapid prototyping, *Medical Engineering & Physics* 32 (2010) 775–782.

Blose, RE., Walker, B.H., Walker, R.M. and Froes, S.H., Depositing Titanium Alloy Additive Features to Forgings and Extrusions Using the Cold Spray Process, International thermal spray conference & exposition May 15–17 2006, Seattle, Washington, USA.

Byun, H.S., Determination of optimal build direction in rapid prototyping with variable slicing, *International Journal of Advanced Manufacturing Technology,* v 28, n 3–4, p 307–313, March 2006.

Byun, H.S. and Lee, K.H., Determination of the optimal part orientation in layered manufacturing using a genetic algorithm, *International Journal of Production Research,* v 43, n 13, p 2709–2724, July 1, 2005.

Canellidis, V., Dedoussis, V., Mantzouratos, N. and Sofianopoulou, S., Pre-processing methodology for optimising Stereolithography apparatus build performance, *Computers in Industry,* Volume 57, Issue 5, June 2006, Pages 424–436.

Chen, K. and Feng, X., Solid model reconstruction from engineering paper drawings using Genetic Algorithms, *Computer-Aided Design,* Volume 35, Issue 13, November 2003, Pages 1235–1248.

Chen, Y.H. and Wang, Y.Z., Genetic Algorithms for optimized re-triangulation in the context of reverse engineering, *Computer-Aided Design* 31 (1999) 261–271.

Chua, C.K., Leong, K.F., Cheah, C.M. and Chua, S.W., Development of a Tissue Engineering Scaffold Structure Library for Rapid Prototyping, Parts 1 & 2, *International Journal of Advanced Manufacturing Technology,* V21, 2003 pp 291–312.

Ghorpade, A., Karunakaran, K.P. and Tiwari, M.K., Selection of optimal part orientation in fused deposition modelling using swarm intelligence, *Proc. IMechE Vol. 221 Part B: J. Engineering Manufacture,* 2007, pp 1209–1219.

Gogate, A.S. and Pande, S.S., Intelligent layout planning for rapid prototyping, *International Journal of Production Research,* Vol. 46, No. 20, 15 October 2008, 5607–5631.

Hopkinson, N. and Dickens, P.M., Rapid Prototyping for direct manufacture, *Rapid Prototyping Journal,* V7, N4, 2001, pp 197–202.

Hopkinson, N., Hague, R.J.M. and Dickens, P.M., Rapid Manufacturing: an Industrial *Revolution for the Digital Age,* 2005 (John Wiley: Chichester, West Sussex).

Houtmann, Y., Delebecque, B. and Barlier, C., Adaptive local slicing in Stratoconception by using critical points, *Advances in Production Engineering & Management,* V4, 2009, pp 56–68.

Hsiao, S., Chiu, F. and Lu, S., Product-form design model based on Genetic Algorithms, *International Journal of Industrial Ergonomics,* Volume 40, Issue 3, May 2010, Pages 237–246.

Hur, J. and Lee, K., Development of a CAD environment to determine the preferred build-up direction for Layered Manufacturing, *International Journal of Advanced Manufacturing Technology,* V14, N4, 1998, pp 247–254.

Hur, S.M., Choi, K.H., Lee, S.H. and Chang, P.K., Determination of fabrication orientation and packing in SLS process, *J. Mater. Process. Technol.* 112 2–3 (2001), pp. 236–243.

Ikonen, I., Biles, W.E., Kumar,A., Ragade, R.K. and Wissel, J.C., A genetic algorithm for packing three-dimensional non-convex objects having cavities and holes. In: Proceedings of the Seventh International Conference on Genetic Algorithms. Michigan State University, July 19–23, 1997.

Kim, H.C. and Lee, S.H., Reduction of post-processing for stereolithography systems by fabrication-direction optimization, *Computer-Aided Design* 37 (2005), pp. 711–725.

Koc, B., Ma, Y. and Lee, Y., Smoothing STL files by Max-Fit biarc curves for rapid prototyping, *Rapid Prototyping Journal* V6, N3, 1995.

Lam, T.W., Yu, K.M., Cheung, K.M. and Li, C.L., Octree reinforced thin shell objects for Rapid Prototyping by Fused Deposition Modelling, *International Journal of Advanced Manufacturing Technology,* v 14, n 9, p 631–636, 1998.

Lewis, J.E., Ragade, R.K., Jumar, A. and Biles, W.E., A distributed chromosome genetic algorithm for bin-packing, *Robotics and Computer-Integrated Manufacturing,* Volume 21, Issues 4–5, August-October 2005, Pages 486–495.

Liulan, L., Qingxi, H., Xianxu, H. and Gaochun, X., Design and Fabrication of Bone Tissue Engineering Scaffolds via Rapid Prototyping and CAD, *Journal of Rare Earths,* Volume 25, Supplement 2, June 2007, Pages 379–383.

Maheshwaraa, U., Seepersad, C. and Bourell, D., Topology design and freeform fabrication of deployable structures with lattice skins, proceedings of the Solid Freeform Fabrication Symposium, Austin, Texas, 2007, pp 402–415.

Masood, M.H., A generic algorithm for a best part orientation system for complex parts in rapid prototyping, *Journal of Materials Processing Technology* 139 (2003) 110–1.

Masood, S.H. and Rattanawong, W., A Generic Part Orientation System Based on Volumetric Error in Rapid Prototyping, *International Journal of Advanced Manufacturing Technology,* V19, N3, pp 209–216

Masood, S.H., Singh, J.P. and Morsi, Y., The design and manufacturing of porous scaffolds for tissue engineering using rapid prototyping, *The International Journal of Advanced Manufacturing Technology* Volume 27, Numbers 3–4, 2005, 415–420.

Melhuish, C., Adamatzky, A. and Kennedy, B., Biologically inspired robots, *Proceedings of SPIE* Vol. 4329 (2001).

Nig, Y., Wong, Y.S. and Fuh, J.Y.H., Effect and control of hatch length on material properties in the DMLS process, *Proceedings of the Institution of Mechanical Engineers, Part B: Journal of Engineering Manufacture,* v 219, n 1, p 15–25, January 2005.

Ramin, E. and Harris, R.A., Advanced Computer Aided Design for Bone Tissue Engineering Scaffolds, *Proceedings of the Institution of Mechanical Engineers: Part H Journal of Engineering in Medicine,* Volume 223, Number 3, 2009, pp 289–301.

Rosen, D.W., Design for Additive Manufacturing: A method to explore unexplored regions of the design space, proceedings of the Solid Freeform Fabrication Symposium, Austin, Texas, 2007, pp 402–415.

Ruffo, M., Tuck, C. and Hague, R., Cost estimation for rapid manufacturing, *Proceedings of the Institution of Mechanical Engineers, Part B: Journal of Engineering Manufacture,* Volume 220, Number 9, 2006.

Sabourin, E., Houser, S.A. and Bhn, J.H., Accurate exterior, fast interior layered manufacturing, *Rapid Prototyping Journal,* v 3, n 2, p 44–52, 1997.

Tyberg, J. and Bhn, J.H., Local adaptive slicing, *Rapid Prototyping Journal,* v 4, n 3, p 118–127, 1998.

Wang, C. and Chang, T., Retriangulation in STL meshes for Rapid Prototyping and Manufacturing, *The International Journal of Advanced Manufacturing Technology,* Volume 37, Numbers 7–8, 2008, 770–781.

Wang, R., Li, X., Wu, Q. and Wang, L., Optimizing process parameters for selective laser sintering based on neural network and genetic algorithm, *International Journal of Advanced Manufacturing Technology,* v 42, n 11–12, p 1035–1042, June 2009 Database: Compendex.

Watts, D.M., A Genetic Algorithm based topology optimisation approach for exploiting Rapid Manufacturing's design freedom, PhD thesis, Loughborough University, 2008.

West, A.P., Sambu, S.P. and Rosen, D.W., A process planning method for improving build performance in Stereolithography, *Computer-Aided Design,* Volume 33, Issue 1, January 2001, Pages 65–79.

Wodziak, J.R., Fadel, G.M. and Kirschman, C., A genetic algorithm for optimizing multiple part placement to reduce build time, *Proceedings of the Fifth International Conference on Rapid Prototyping,* 1994, pp. 201–210.

Xu, F., Wong, Y.S., Loh, H.T., Fuh, J.Y.H. and Miyazawa, T., Optimal orientation with variable slicing in Stereolithography, *Rapid Prototyping Journal,* v 3, n 3, p 76–88, 1997.

Zhao, J., Research on the Optimization Method of Scanning Direction for Rapid Prototyping Based on GA, *Proceedings of the World Congress on Intelligent Control and Automation (WCICA),* v 2, p 7920–7923, 2006.

11

Multiple Fault Diagnosis Using Psycho-Clonal Algorithms

Nagesh Shukla and PKS Prakash

Warwick Manufacturing Group, University of Warwick, Coventry, UK

Abstract

Multiple Fault Diagnosis (MFD) is used as an effective way to tackle the problems of a real shop floor environment in order to reduce the total lifetime maintenance costs of the system. The major challanges associated with with MFD are: (i) large number of variables associated with response; (ii) low sample size; and (iii) ill-conditioned system caused due to system architecture. Further, the computational complexity associated with MFD increases exponentially with number of failure; thus, it warrants the application of heuristic techniques or artificial intelligence (AI) based optimization tools to diagnose the exact faults in real time. In this chapter, a methodology based on a Probabilistic Causal Model has been illustrated to resolve graph based multiple fault diagnosis problems. This methodology involves a new nature inspired algorithm know as the psycho-clonal algorithm for fault diagnosis. The proposed methodology collect the faults corresponding to each observed manifestation that can give the best possible result instead of finding all possible combinations of faults. Intensive computational experiments on well-known data sets witness the superiority of the proposed psycho-clonal algorithms existing state-of-art approaches proposed in the literature. Experimental results demonstrate the capability of proposed methodology in diagnosing the the exact fault in the minimum fault isolation time as compared to other approaches.

Keywords: Multiple fault diagnosis, probabilistic causal model (PCM), psycho-clonal algorithm

Manoj Tiwari and Jenny A. Harding (eds.) Evolutionary Computing in Advanced Manufacturing, (235–258) © Scrivener Publishing LLC

11.1 Introduction

The operational integrity of automated processes, namely, flexible manufacturing system (FMS), space shuttle, military aircraft, outer flight control system, *etc.*, consists of many complex elements paramount for both safety and successful completion of missions. In such systems, discrepancy in some part can propagate to subsequent stages and can lead to the development of complex and multiple faults. The maintenance processes of these types of system are tedious tasks, and are often attributed to the logistic requirements of design, equipment performance, repair and diagnosis processes. It is estimated that the lifetime maintenance cost of a complex system may increase from three to ten times the actual purchase cost (Pattipati, 2003). In addition, diagnosis of an individual machine or component may sometimes not coincide with the exact faults of the system. To accomplish diagnosis in the minimum isolation time, the manifestations of faults are determined and accordingly, faults are diagnosed. During the diagnosis process, it is advisable to avoid the guessing game of swapping components as much as possible because it increases both operational time and cost. A more global strategy of diagnosis is an exhaustive search algorithm that will give 100% guarantee of reliability but computational complexities increase exponentially with the increase in number of faults. Assume that for a given set of manifestations, there can be a maximum of '10' faults. For this case, there are 2^{10} (1024) possible combinations of faults. Thus, the aforementioned diagnosis processes are tedious in nature and converge slowly and therefore, they are difficult to implement in real life problems. The challenges with such problems can be removed by finding a proper combination between relative goodness of diagnosis and search strategy that will diagnose exact faults in the minimum fault isolation time.

In this chapter, we have employed a psycho-clonal based evolutionary algorithm to solve the multiple fault diagnosis problems. In literature, the Rollout algorithm combined with various heuristic algorithms was proposed for solving multiple fault diagnosis problems. Subsequently, such practices enhance the effectiveness of the underlying heuristic algorithm (Bertsekas *et al.*, 1997). Bertsekas and Tsitsiklis (1996) first proposed the Rollout strategy for the approximate solution of neuro-dynamic programming for policy iteration. Again, Bertsekas *et al.* (1997) applied the same for a sub-optimal solution method, which is inspired by policy iteration and dynamic programming. Castanon and Bertsekas (1999) also applied a Rollout strategy for a stochastic scheduling problem. Pattipati and Tu (2003) showed how Rollout strategies can be implemented efficiently in sequential fault diagnosis problems with a considerable saving in computational burden over the optimal algorithms. However, there is still a need to improve the computational efficiency of the fault diagnosis algorithms. Therefore, the psycho-clonal based evolutionary algorithm (Tiwari *et al.* 2005, Kumar *et al.* 2006 Akhilesh Kumar *et al.*, 2006; Sanjay Kumar Shukla, 2010; Singh *et al.*, 2006; Dashora *et al.*, 2008) is utilized in this chapter.

In this chapter, a graph based zero-time Multiple Fault Diagnosis (MFD) problem is taken into consideration. The psycho-clonal algorithm was applied to the MFD problem to diagnose the exact faults corresponding to the observed manifestations in minimum fault isolation time. The proposed psycho-clonal algorithm is implemented with a view to reducing the overall computational time for solving complex MFD problems. The proposed approach outperformed other approaches when applied on benchmark data sets available in literature as well as on simulated data sets.

The rest of the chapter is organized as follows. In section 11.2, the multiple fault diagnosis problem is discussed in detail and is mathematically formulated. Section 11.3 presents the psycho-clonal based solution approach and describes how it can be efficiently implemented. Benchmark test problems along with results and discussions are described in the Section 11.4. Finally, the chapter ends with concluding remarks and a note for future research.

11.2 Multiple Fault Diagnosis Problems

In recent years, various works on fault diagnosis models have been reported. For example, Fault Tree Analysis (FTA) is an efficient fault diagnosis method that can help maintenance personnel in finding the shortest path of fault diagnosis in a complex system (Tu and Pattipati, 2003). Similarly, Chang (Chang, 1991) advocated failure propagation tree knowledge to find the faults in the entire production system in terms of a hierarchical process plan. The model proposed by him globally depends upon the input and output of each machine by ignoring its functional details. Its inability to provide detailed diagnosis, such as a probabilistic inferencing function, limits its application to potential cases only. In order to diagnose the single fault in a graph-based zero-time and nonzero-time system,

Rao (1993a,b) proposed two single fault diagnosis algorithms (Forward and Backward). In single fault diagnosis, it is assumed that there exists only one fault at any time in the system. To ensure the aforementioned condition, frequent testing and maintenance is essential, which may lead to interpretation of uncertainties. However, single fault assumption is not relevant in many cases such as tolerant system and space based system where frequent testing and maintenance are not possible. Chessa and Santi (2001) proposed the D-fault algorithm for use while at most two failures occur in the system at a given time. They also introduced the concept of Sequential Fault Diagnosis (SFD) for an unknown number of faults. The major drawback associated with SFD is the online computation and dependency on source-alarm.

Kuo and Huang (1997) introduced a Petri-net model for multiple fault diagnosis in Flexible Manufacturing Systems (FMS). Their model is based on data measured from inspection machines and sensors. Fuzzy logic reasoning for fault diagnosis and its monitoring in FMS was developed by Chevalier *et al.* [1]. Qian (1990) employed fuzzy sets theory to describe fault propagation in a chemical process. Khoo *et al.* (1999) used fuzzy set theory and graph theory for developing an IDEF0 model-based intelligent fault diagnosis system in FMS that is able to solve real world problems upto a certain extent. However, it is not so popular because of inadequate diagnosis knowledge and other mandatory information.

Sequential fault diagnosis is another approach for both asymmetric (Biasizzo *et al.*, 2000) and symmetric (Pattipati and Alexandridis, 1990) problems. In symmetric cases, the fault free outcome is not necessarily a fault free component. It is suitable for cases where the costs of replacing components are high and one wants to only replace the definitely failed components. However, for a system where time of performance is the primary concern, it failed to make decisions in a timely manner. Peng and Reggia (1987) suggested a Probabilistic Causal Model (PCM) for cases where the structural relationship between the failure source and test are simple and the necessary probabilistic information are available. However, it is not applicable for those systems where time of performance is the primary concern.

In this chapter, graph based multiple fault diagnosis (GBMFD) problems are discussed. The GBMFD problem can be represented by a directed graph $G = (V, E)$, called a system graph, where $V = \{n_{ij}\}$ consists of a set of nodes, and E represents the digraph edges, specifying the functional information flow between the sets of failure sources and the set of tests in the system. The digraph model captures first order causal dependency effects such as n_i effects n_j and n_j effects n_k. The global dependency of n_k on n_i are inferred by the reachability analysis algorithm (Raghavan, 1996). Based on the fault propagation time, systems are classified as zero-time systems and nonzero-time systems. In a zero-time system, fault propagation appears instantaneously to an observer, and can be abstracted by assuming the fault propagation time to be zero. In contrast, in a non-zero-time system, the fault propagation time is slower than the response time of the observer.

The zero-time GBMFD problem consists of the following parameters.

1. Set of all possible faults in the system

$$D=\{d_1,d_2,d_3,d_4,......,d_k\} \tag{11.1}$$

2. Manifestations shown by the system corresponding to the fault set D.

$$M=\{m_1,m_2,m_3,m_4,......,m_n\} \tag{11.2}$$

3. A prior probability associated with each fault source.

$$P\ (d_i) = \{P(d_1),\ P(d_2),\ P(d_3),...,\ P(d_k)\} \quad \text{Where } (0 \le P(d_i) < 1) \tag{11.3}$$

4. A tendency matrix incorporates the causal probabilities of causal events. Fault d_i (where $d_i \in D$) and manifestation m_j (where $m_j \in M$) are known as causal events, when $m_j : d_i$ is true only if both d_i and m_j occurs together. Causal event reflects the concept of edge in an Information Graph (IG). Generally, $(m_j \wedge d_i) \supseteq (m_j : d_i)$ and is evidently a basic event. The causal probability of fault d_i with respect to manifestation m_j is represented by C_{ij} (Where $0 \le C_{ij} \le 1$). Mathematically, it represents how frequently disorder d_i causes manifestation m_j. Causal probability reflects a causal relation more accurately, when compared to the Bayesian method. Consider a case, where d_1 cannot cause m_i but d_2 can cause m_i. obviously, the causal relationship between d_1 to m_i is zero, if d_1 and d_2 occurred at the same time, while P $(m_j|d_1)$ would have a value greater than zero. Meanwhile, $P(m_j{:}d_1|d_2)=0$, since $m_j{:}d_1$ is always false.

5. M^+ is a subset of M that reflects the manifestations shown by the system.
6. M^- is a subset of M, which consists of manifestations that are not shown by the system corresponding to the fault.

$$M^- = M - M^+ \tag{11.4}$$

7. DI: Consists of exact disorders corresponding to manifestation M^+ in the system. Hence,

$$DI \subseteq D \tag{11.5}$$

In order to diagnose the exact faults corresponding to observed manifestations, Reggia, *et al.*, [18] proposed the Probabilistic Causal Model (PCM). This utilizes the relative likelihood function (L) to diagnose faults. Here, it is assumed that all faults are independent of each other and a manifestation does not exist without any fault. The relative likelihood value of fault set D_1 corresponding to the observed manifestation M^+ is given by:

$$L = L_1 . L_2 . L_3 \tag{11.6}$$

L_1 = Likelihood of fault set D_1 with respect to manifestation set M^+.
L_2 = Likelihood of fault set D_1 with respect to manifestations that are not shown by the system. Ideally, its value is closer to one.
L_3 = Likelihood of highly probable (very common) disorder d_j that contributes significantly in the overall relative likelihood value.

$$L_1 = P\left(M^+ \mid D_1\right) = \prod_{m_i \in M^+} P\left(m_i \mid D_1\right) \tag{11.7}$$

For a manifestation m_i (where $m_i \in M^+$)

$$\begin{aligned} L_1 &= L_{1,m_i} \\ L_{1,m_i} &= P\left(m_i : d_1 \wedge d_2 \ldots \wedge d_r \mid d_1 \wedge d_2 \ldots \wedge d_r\right) \\ &= 1 - \prod_{d_i \in D_1} \left(1 - C_{ij}\right) \end{aligned} \tag{11.8}$$

Hence, for all values of M^+

$$L_1 = \prod_{m_j \in M^+} \left(1 - \prod_{d_i \in D_1} \left(1 - C_{ij}\right)\right) \tag{11.9}$$

Similarly, L_2 is

$$\begin{aligned} L_2 &= \prod_{m_i \in M^-} P(m_i \mid D_1) \\ L_2 &= \prod_{d_1 \in D_1} \prod_{m_i \in M^-} (1 - C_{ij}) \end{aligned} \tag{11.10}$$

$$\text{and } L_3 \text{ is,} \quad L_3 = \prod_{d_j \in D_1} \frac{P_j}{(1 - P_j)} \tag{11.11}$$

Where P_j is a prior probability of disorder d_j.

Hence, the relative likelihood value of the fault set DI with respect to an observed manifestation M^+ is (following Eqn. 11.6):

$$L = \left(\prod_{m_i \in M^+} \left(1 - \prod_{d_j \in DI} \left(1 - c_{ij}\right)\right)\right) \times \left(\prod_{d_j \in DI} \prod_{M_i \in M^-} \left(1 - C_{ij}\right)\right) \times \left(\prod_{d_j \in DI} \frac{P_j}{\left(1 - P_j\right)}\right) \tag{11.12}$$

The desired set of faults corresponding to the observed manifestations M^+ is

$$DI = \max_{J \in DI_P} L\left(DI_J\right) \tag{11.13}$$

Where, DI_P includes all-possible fault sets observed in the system corresponding to the observed manifestation M^+, and L (DI_J) is the relative likelihood value corresponding to the fault set DI_J. Therefore, DI is the objective function that has to be maximized based on Eqn. 11.13.

Major hurdles encountered in multiple fault diagnosis processes are:

1. Computational complexity increases exponentially as the number of faults increases.
2. Measurement of relative goodness of faults.
3. Selection of a diagnostic strategy that can diagnose the exact fault of the system in the minimum fault isolation time.

Thus, the aforementioned problem can be viewed as a combinatorial non-linear optimization problem. That is, to find a disorder $DI_J \subseteq D$ among all 2^D possible combinations to maximize $L(DI_J)$.

The diagnosis of exact faults corresponding to the observed manifestations is an NP-hard problem (Maheshwari and Hakim, 1992). Hence, in order to run the manufacturing units smoothly, without any disturbances, an efficient fault diagnosis methodology that can diagnose the exact fault in minimum time is required. In recent years, various fault diagnosis methodologies have been suggested to overcome these complexities but due to the lack of some realistic assumptions, many of them are not applicable in shop floor situations. The objective of this study is therefore to develop a fault diagnosis approach from the implementation point of view. As mentioned in the earlier section, previous research has approached the diagnoses of the exact faults in the system, but the approaches taken have required long computational times. To overcome these difficulties, in this research, the psycho-clonal based evolutionary algorithm has been used to diagnose the exact faults of the system in the minimum fault isolation time. The following section details the background of the Psycho-Clonal algorithm and that is followed by the implementation section.

11.3 Background of Psychoclonal Algorithm

The proposed algorithm derives its traits from theories of motivation and the Artificial Immune System. These theories are briefly discussed in this section.

11.3.1 Artificial Immune System (AIS)

Nature has always been an interesting domain for human experts due to its complexity, sophistication and flexibility. In particular, the human body has constantly fascinated researchers, whether it is the mechanism of the brain or of the nervous system. Lot of optimization algorithms such as genetic algorithm (GA), artificial neural network (ANN) are inspired by the evolutionary mechanisms of living organisms, (Shukla, *et al.*, 2009). Recently, knowledge pertaining to immune systems has motivated researchers, not only from the biological point of view, but also from the computational perspective. As nervous systems inspired the development of artificial neural networks (ANN), in a similar manner, the immune system has inspired the emergence of artificial immune systems (AIS) as a novel computational intelligence paradigm (de Castro and Von Zuben, 2002). An AIS can be defined as an abstract or metamorphic computational system using ideas gleaned from the theories and components of immunology. An AIS aims to solve a wide range of tasks related to complex computational and engineering problems such as pattern recognition, machine learning and combinatorial optimisation.

The immune system recognizes any malfunctioning and disease causing elements, which are known as antigens 'Ag'. Antigens are heterologous proteins, which, when introduced, cause certain body cells to elaborate antibodies. Antigens are of two types, one being self and other non-self. Self-antigens belong to the body and are harmless, whereas non-self antigens are disease-causing elements. The distinction between self and non-self antigens depends on the main task performed by immune system. The most alluring characteristic of the immune system is the presence of receptor molecules, on the surface of immune cells, capable of recognising an almost limitless range of antigenic patterns. There are two major groups of immune cells; B-cells, and T-cells. These cells are similar in nature, but differ in the way they recognises antigens. Antigens free in solution (e.g. in the blood stream) can be recognised by B-cells, whereas T-cells recognize antigens present in other accessory cells. When an animal is exposed to an antigen, some sub population of its bone marrow derived from cells (B-lymphocyte) responds by producing antibodies (Ab). Antibodies are immune substances in the blood stream and on the surface of B-cells that recognize and bind to an Ag. Thus there is no distinction between a B-cell and its receptor antibody. Recognition of an antigen is the first necessity for the immune system to become active and perform a subsequent response. Recognition can only be activated, if the cell-receptor recognizes an antigen with an affinity greater than an affinity-threshold. If a B-cell encounters a non-self antigen with a suitable affinity-threshold, it proliferates and differentiates into memory and effector cells; a process called clonal selection.

The detailed procedure of clonal selection is described in the next sub-section.

11.3.2 Theory of Clonal Selection

Clone selection theory deals with the response of the immune system, when a non-self antigenic pattern is recognized by a B-cell (Ada and Nossal, 1987). The immune system responds to the antigenic attack by producing antibodies in large volume. The Ag stimulates the B-cell to proliferate and mature into terminal Ab (non-dividing) secreting cells, known as plasma cells. Proliferation in the case of immune cells is an asexual, amitotic process. The cells divide themselves (no crossover) to generate clones. During reproduction, the B-cells progenies undergo a hypermutation process that together with the strong selective pressure, results in B-cells with an antigenic receptor presenting higher affinities than with the selective antigen. This process is known as the maturation of the immune response.

Antibodies react specifically with the antigens that produced the antibody. The specificity is important as a protective measure against reinfection with the same antigen. This specificity forces B-cells to differentiate into long-lived B memory cells with a long-life span, in addition to proliferation. These memory cells are pre-eminent in future responses to the same antigenic pattern, or a similar one. The aforementioned, process of clonal, proliferation, and affinity maturation is schematically shown in Figure 11.1. The main features of clonal selection i.e. immune memory and affinity maturation are described in next sub-sections.

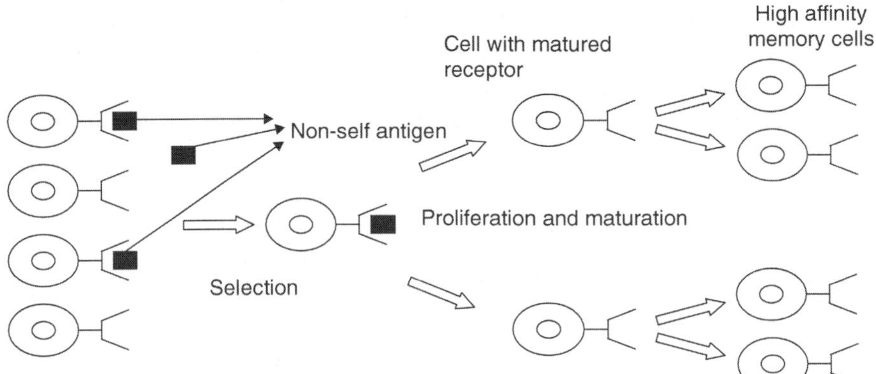

Figure 11.1 Process of Clonal selection, proliferation and affinity maturation.

11.3.2.1 *Immune Memory*

In immune system evolution, an entity is expected to come across a particular Ag repeatedly during its life span. At the initial stage of exposure to an Ag that stimulates, the adaptive response system is handled by a small number of low-affinity B-cells, each producing a different type of Ab. The presence of memory cells associated with the first infection, enhances the effectiveness of the immune response to a secondary encounter, as a high volume of Ab's are produced. In the case of clonal selection theory, instead of maintaining a large number of candidate solutions, it is proposed that a small set of the best individuals is kept. A clone will be created temporarily and progenies with low affinity will be discarded. Hence, a high quality solution is sought that exhibits parsimony. Such a strategy ensures both speed and accuracy of the immune system's response, which becomes successively higher after each interaction.

This is the most remarkable and attractive characteristic of the immune system from the engineering point of view, as the candidate solutions i.e. cells with higher affinity must somehow be preserved as high-quality candidate solutions and shall only be replaced by improved or matured clones, based on statistical evidence.

11.3.2.2 *Affinity Maturation*

Affinity maturation is carried out by two mechanisms, in which the repertoire of an Ag activated by a B-cell is diversified in a T-cell dependent immune response, and these are *hypermutation and receptor editing*. Abs present in a memory response have a higher affinity than those of the early primary responses. This phenomenon is referred as maturation of immune response and it is restricted to T-cell dependent responses. This maturation requires the Ag-binding sites of the Ab molecules to be structurally different from the ones present in the primary response.

Random changes are introduced into the genes as per antigen-antibody interaction. The hypermutation operator works in a similar fashion to mutation. The difference lies in the rate of modification, which depends upon antigenic affinity. Inferior antibodies are more prone to hypermutation. Generally, antibodies with low affinity are hypermutated at a higher rate whereas; antibodies with a higher affinity are hypermutated with a lesser rate. This phenomenon is known as receptor editing (i.e. the degree that an antibody resembles the optimum antibody) and governs the hypermutation. Thus, hypermutation guides the solution towards local optima while receptor editing helps in escaping from local-optima.

11.3.3 Maslow's Need Hierarchy Theory

Human Behavior is contingent in nature and is usually ruled by the strongest need. The basic concept behind such a theory is that people have certain fundamental needs and that people are motivated to engage in behavior that will lead to satisfaction of their needs. Psychologists claim that each need has a certain priority. As the more basic needs are satisfied, an entity seeks to satisfy higher needs. Maslow (1954), has given a framework that helps to explain the strength of certain needs, and this is known as the *Need Hierarchy Theory*. The theory hypothesizes that all people posses a set of five needs arranged in a hierarchy, from the most fundamental or basic survival need up to the most sophisticated needs of self-actualization. Figure 11.2 represents five needs arranged in a hierarchy; known as Maslow's need pyramid.

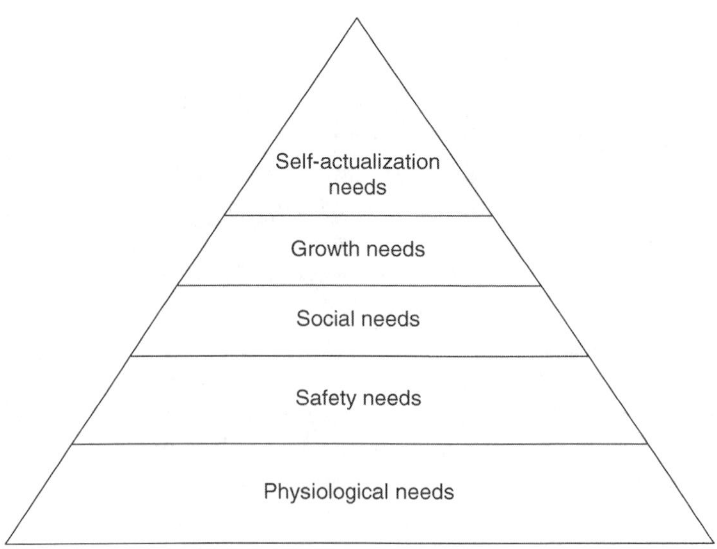

Figure 11.2 Maslow's Pyramid.

A. Physiological Needs Level

This refers to most indispensable survival needs for food, water and shelter from the environment to permit one's continued existence. These needs tend to have the highest strength until they are reasonably satisfied. In optimization, this refers to the generation of feasible solutions based upon the problem environment.

B. Safety Needs

Once the physiological needs are satisfied to a reasonable level, other needs prevail in the human mind. In the hierarchy, the need for safety comes next, and this is the need to be free of physical danger or self-preservation from external threats. External threats in the engineering milieu are constraints imposed on the problem. For continued existence, solutions are subjected to these threats or constraints and proper evaluation.

C. Social Needs

Next, social needs become dominant; a person will strive for meaningful relations with others. In the optimization, this refers to the selection of the candidate solution through interaction between candidates.

D. Growth Needs

Every entity desires to produce entities of its kind through reproduction. Here, a candidate solution diversifies to extend its search space. This is the basic mechanism of every evolutionary process.

E. Self-Actualization Needs

Self-actualization is the need to maximize one's potential and to realize one's capabilities. In the terms of Maslow's pyramid, "this need might be phrased as the desire to become more and more what one is, to become everything that one is capable of becoming". This is very true whether it is with respect to humans or to the solution of an optimization problem. Bearing in mind the above description, it has been concluded that the more the self-actualization needs are fulfilled, the stronger the individual become. This is the reason for carrying out a number of iterations to decide the near-optimal solution.

Based on the above needs and the clonal principal, the proposed psychoclonal methodology is conceived and it is described in the next section.

11.3.4 Pseudo Code for Psycho Clonal Algorithm

11.3.4.1 *Nomenclature*

Ab	:	Available set of antibodies.
Ab_d	:	Set of the new antibodies that will replace *d* amount of lower affinity Ab's from available set of antibodies.
$Ab_{k,n}$	:	Ab's from Ab with highest affinities.
Ag_m	:	Population of m antigens.
C_k	:	Population of N_c clones generated from $Ab_{k,n}$.
C_k	:	The set of hypermutated antibodies.
d	:	Receptor editing factor.
f_k	:	Vector containing values of objective function g(.) as the affinity of all Ab's.
f_k^*	:	Vector containing values of antigenic affinity for matured clones in relation to the antigen Ag_j.
j	:	a counter, varying from 1 to m.
k	:	a counter, varying from 1 to n.
N	:	The total number of antibodies
N_c	:	The total number of clones generated for each of the Ag's = $\sum_{i=1}^{n} R\,(\beta.\,N)$, i=1,2,..n.
R(.)	:	Step function operator.
β	:	Multiplying factor

The important steps of the proposed method are discussed in detail in the following section. These stages are as follow:

11.3.4.2 *Stage I*

Need level 1. *Physiological needs*: Defining a problem-specific objective function is a pre-requisite (for fault diagnosis its $L(DI_J)$). A randomly generated initial population of antibodies (a set of DI_J) is also required.

Need level 2. *Safety needs*: Here, the initial population is exposed to threats posed by antigens (constraints of the problem).

2.1 Randomly choose an antigen Ag_j from Ag_m and expose it to all Ab.

2.2 Based on the objective function, determine the vector f_k that contains the affinity of Ag_j to the entire N, Ab's in Ab.

Need level 3. *Social needs*: Here, interaction is carried out between antibodies with a view as to identifying their relations with each other.

3.1 Select n highest affinity Ab's from Ab to compose a new set $Ab_{k,n}$ of high affinity Ab's in the relation to Ag_j.

3.2 Then the selected Ab's will be cloned independently and proportionally to their antigenic affinities, generating a repertoire C_k of clones (higher the antigenic affinity, the higher the number of clones generated for each n selected Ab's).

Need level 4. *Growth needs*: Set C_k is submitted for hypermutation, inversely proportional to the antigenic affinity, generating a population C_k^* of matured clones (the higher the affinity, the smaller the mutation rate).

11.3.4.3 *Stage II*

After satisfaction of need level 4, it is necessary to check needs level 2, once again for these entities, as they are the new denizens of society. Thus, they must be exposed to the threats and properly evaluated as per the objective function.

Need level 2. *Safety needs*: Determine the affinity f_k^* of the matured clones C_k^* in relation to an antigen. Now, the whole of n Ab's are selected to compose the memory set.

Need level 5. *Self-actualisation needs*: finally replace the *d* lowest affinity Ab's from Ab_d and choose the best among them to fulfil the self-actualisation level. As mentioned above, this level becomes stronger and stronger after a number of generations. Thus, the process repeats till N = Ngen (maximum number of generation).

11.3.4.4 *Implementation of Psycho-Clonal Algorithm on Fault Diagnosis Problem*

As the psycho-clonal algorithm is used for fault diagnosis problems by following the aforementioned steps, every Antibody in the population represents a disorder hypothesis DI_J, i.e., a possible solution to the problem given M^+ and the population represents a set of possible solutions. Using the relative likelihood function $L(DI_J)$ as a fitness measure, the psyco-clonal algorithm is initiated to generate the next set of Antibodies. This process continues for a predefined number of iterations until the termination condition is satisfied. Then the best solution antibody is chosen to be the solution of the fault diagnosis problem.

11.4 Numerical Experiments

This section summarizes the numerical investigations performed to validate the effectiveness of the proposed methodology. The section is organized in two subsections, dedicated respectively to case studies and their numerical experiments (i.e., results and discussions).

11.4.1 Test Problems

Test problems have been conceived from different publications with an intention of analyzing the capability of the proposed algorithm. A dataset with increasing complexities has been presented as a benchmark to show the robustness and efficacy of the proposed approach. In addition, 10 more datasets are simulated using the proposed methodology. These test problems are outlined as follows.

- For the first case, a test problem has been conceived which maps the scenario described by Miller *et al.* (1993). In this case, the system consists of a maximum of 15 faults at any time and these faults can be observed by 10 different manifestations. i.e.

$$D = \{d_1, d_2, d_3, d_4, d_5, \ldots d_{15}\}$$
$$M = \{m_1, m_2, m_3, m_4, \ldots m_{10}\}$$

The prior probability $P(d_i)$ and tendency matrix depicted by the system corresponding to these faults are shown in Table 11.1.

- The second case study has been conceived with an increased complexity to determine the robustness of the proposed framework. It maps the scenario proposed by Bingline *et al.* (1993). In this case, the system

Table 11.1 Prior Probability and Tendency Matrix: 10×15.

	d_1	d_2	d_3	d_4	d_5	d_6	d_7	d_8	d_9	d_{10}	d_{11}	d_{12}	d_{13}	d_{14}	d_{15}
P_i	0.12	0.14	0.39	0.64	0.01	0.21	0.26	0.19	0.59	0.29	0.06	0.47	0.56	0.41	0.06
m_1	0.58	0.00	0.00	0.00	0.00	0.25	0.00	0.96	0.00	0.00	0.85	0.00	0.74	0.00	0.38
m_2	0.00	0.00	0.00	0.00	0.00	0.15	0.81	0.00	0.32	0.00	0.36	0.77	0.30	0.61	0.00
m_3	0.00	0.00	0.44	0.00	0.00	0.00	0.00	0.11	0.00	0.97	0.64	0.00	0.63	0.85	0.00
m_4	0.43	0.67	0.79	0.00	0.26	0.72	0.07	0.00	0.00	0.00	0.84	0.00	0.00	0.42	0.64
m_5	0.46	0.10	0.00	0.58	0.00	0.00	0.46	0.00	0.57	0.40	0.00	0.51	0.00	0.97	0.00
m_6	0.91	0.00	0.00	0.00	0.00	1.00	0.28	0.00	0.00	0.00	0.00	0.00	0.00	0.00	0.12
m_7	0.90	0.94	0.07	0.28	0.00	0.00	0.00	0.00	0.97	0.00	0.00	0.91	0.48	0.23	0.72
m_8	0.00	0.00	0.14	0.17	0.24	0.00	0.30	0.00	0.26	0.00	0.00	0.05	0.00	0.00	0.00
m_9	0.00	0.00	0.13	0.12	0.17	0.04	0.00	0.00	0.97	0.00	0.00	0.00	0.43	0.08	0.00
m_{10}	0.00	0.63	0.07	0.75	0.12	0.00	0.00	0.45	0.00	0.88	0.21	0.00	0.45	0.86	0.19

can consist of a maximum of 20 different faults and these faults can be identified by 20 different manifestations. i.e.

$$D = \{d_1, d_2, d_3, d_4, \ldots d_{20})$$
$$M = \{m_1, m_2, m_3, m_4, \ldots m_{20}\}$$

The prior probability ($P(d_i)$) and tendency matrix shown by the system corresponding to the aforementioned faults are shown in Table 11.2.

- In addition to these problems, ten more problems with increased complexity have been evaluated where the system can consist of a maximum of 10 to 50 different faults and these faults can be identified by 10 to 50 different manifestations.

11.4.2 Results and Discussions

From the research contributions in the area of Probabilistic Causal Model (PCM), it is found that a PCM requires a large amount of computational time to diagnose the exact faults in a system corresponding to observed manifestations. Therefore, it is not only desirable but also inevitable to develop an improved PCM that can solve a large size complex combinatorial optimization problem in less CPU time. Keeping in mind these primary requirements, we have proposed a solution methodology based on a psycho-clonal algorithm to diagnose the exact faults, corresponding to observed manifestations in minimum fault isolation time. The psycho-clonal algorithm based PCM incorporates the beauty of both the algorithms and significantly increases the computational efficiency of the PCM. The performance of the proposed methodology is superior, when compared with other existing methodologies on the test problems adopted from literature as illustrated in section 5.

For the first case exact faults are diagnosed. In this case corresponding to the observed set of manifestations (i.e. M^+)

$$M^+ = \{ m_1, m_2, m_4, m_5, m_7, m_8, m_9, m_{10}\}$$

Desired set of faults (i.e. DI) are

$$DI = \{d_3, d_4, d_9, d_{12}, d_{13}\}$$

and, its relative likelihood value (L) is $7.68528401831913e^{-02}$.

Table 11.2 Prior Probability and tendency Matrix: 20 × 20.

	d_1	d_2	d_3	d_4	d_5	d_6	d_7	d_8	d_9	d_{10}	d_{11}	d_{12}	d_{13}	d_{14}	d_{15}	d_{16}	d_{17}	d_{18}	d_{19}	d_{20}
p_i	0.17	0.07	0.03	0.12	0.135	0.18	0.075	0.03	0.14	0.05	0.026	0.014	0.054	0.06	0.003	0.023	0.048	0.079	0.098	0.027
m_1	0.00	0.53	0.00	0.00	0.00	0.00	0.00	0.29	0.39	0.00	0.00	0.00	0.00	0.00	0.00	0.00	0.00	0.00	0.00	0.00
m_2	0.06	0.00	0.54	0.74	0.00	0.00	0.00	0.34	0.00	0.74	0.00	0.00	0.00	0.00	0.00	0.00	0.00	0.00	0.00	0.00
m_3	0.00	0.81	0.00	0.00	0.72	0.09	0.22	0.00	0.00	0.00	0.00	0.00	0.00	0.00	0.00	0.00	0.00	0.00	0.00	0.00
m_4	0.68	0.09	0.00	0.00	0.00	0.00	0.46	0.00	0.20	0.00	0.00	0.00	0.00	0.00	0.00	0.00	0.00	0.00	0.00	0.00
m_5	0.00	0.85	0.45	0.52	0.00	0.66	0.21	0.00	0.90	0.00	0.00	0.00	0.00	0.00	0.00	0.00	0.00	0.00	0.00	0.00
m_6	0.10	0.00	0.90	0.00	0.00	0.00	0.76	0.00	0.48	0.00	0.00	0.00	0.00	0.00	0.00	0.00	0.00	0.00	0.00	0.00
m_7	0.51	0.00	0.59	0.65	0.00	0.00	0.00	0.00	0.38	0.00	0.00	0.00	0.00	0.00	0.00	0.00	0.00	0.00	0.00	0.00
m_8	0.00	0.13	0.00	0.00	0.49	0.00	0.00	0.25	0.00	0.27	0.00	0.00	0.00	0.00	0.00	0.00	0.00	0.00	0.00	0.00
m_9	0.00	0.34	0.00	0.32	0.00	0.00	0.00	0.00	0.00	0.00	0.00	0.00	0.00	0.00	0.00	0.00	0.00	0.00	0.00	0.00
m_{10}	0.00	0.85	0.29	0.00	0.00	0.44	0.43	0.00	0.00	0.00	0.00	0.00	0.00	0.00	0.00	0.00	0.00	0.00	0.00	0.00
m_{11}	0.00	0.00	0.00	0.00	0.00	0.00	0.00	0.00	0.00	0.00	0.00	0.00	0.00	0.00	0.72	0.00	0.05	0.00	0.00	0.00
m_{12}	0.00	0.00	0.00	0.00	0.00	0.00	0.00	0.00	0.00	0.00	0.31	0.00	0.29	0.00	0.92	0.00	0.80	0.04	0.12	0.00
m_{13}	0.00	0.00	0.00	0.00	0.00	0.00	0.00	0.00	0.00	0.00	0.00	0.00	0.00	0.00	0.00	0.00	0.00	0.00	0.00	0.00
m_{14}	0.00	0.00	0.00	0.00	0.00	0.00	0.00	0.00	0.00	0.00	0.85	0.00	0.64	0.88	0.07	0.00	0.00	0.26	0.00	0.43
m_{15}	0.00	0.00	0.00	0.00	0.00	0.00	0.00	0.00	0.00	0.00	0.00	0.00	0.15	0.00	0.47	0.00	0.73	0.00	0.00	0.18
m_{16}	0.00	0.00	0.00	0.00	0.00	0.00	0.00	0.00	0.00	0.00	0.00	0.00	0.11	0.00	0.00	0.32	0.00	0.23	0.00	0.11
m_{17}	0.00	0.00	0.00	0.00	0.00	0.00	0.00	0.00	0.00	0.00	0.03	0.50	0.00	0.00	0.00	0.26	0.00	0.69	0.95	0.00
m_{18}	0.00	0.00	0.00	0.00	0.00	0.00	0.00	0.00	0.00	0.00	0.00	0.00	0.72	0.27	0.00	0.00	0.58	0.00	0.00	0.00
m_{19}	0.00	0.00	0.00	0.00	0.00	0.00	0.00	0.00	0.00	0.00	0.00	0.00	0.62	0.00	0.73	0.00	0.00	0.00	0.00	0.00
m_{20}	0.00	0.00	0.00	0.00	0.00	0.00	0.00	0.00	0.00	0.00	0.00	0.00	0.00	0.00	0.96	0.00	0.00	0.51	0.67	0.00

Miller *et al.* (1993) showed that their results (using a GA along with local improvement operators) are better than the results obtained by Binglin *et al.* (1993) (using a GA).The result obtained using the proposed methodology is in tune with that was found by Miller *et al.* (1993). Extensive simulation shows that the result is an optimised one. In addition, the proposed methodology has better search efficiency than the methodology proposed by Miller *et al.* (1993). Hence, in real life when one wants to diagnose the fault in a minimum fault isolation time, the proposed methodology is a better option than the method proposed by Miller *et al.* (1993).

When the second test problem is solved by the proposed methodology, exact faults are diagnosed corresponding to different observed manifestation sets, as shown in Table 11.3. The desired set of faults and their relative likelihood values are in line with results obtained by Binglin *et al.* (1993). Moreover, the proposed methodology has better search efficiency, for instance, in the above diagnostic problem it requires at least 24 hours to compare all 2^{20} relative likelihoods to find the solution for a given set of observed manifestations (i.e. M^{+}). The GA proposed by Binglin *et al.* (1993) takes 10 minute to compute the desired set of faults, whilst the proposed methodology takes only a few seconds in the same environment. Hence, it can be applicable to real a world system where a large number of faults occur, and require quick and comprehensive diagnoses to eliminate those faults.

To demonstrate the consistent performance of the proposed algorithm, it has also been tested on ten simulated datasets with increasing complexity. Table 11.4 represents the results of these intensive computational experiments. The obtained results substantiate the fact that the proposed approach showed consistency in terms of fault isolation time.

11.5 Conclusion

In this Chapter, graph based multiple fault diagnosis problems for large-scale systems are considered. Moreover, the computational complexity of these types of problems is exponential in nature. Therefore, random search heuristics such as genetic algorithm, simulated annealing, Tabu search, *etc* are employed to resolve them. However, convergence of these methods is slow in nature. In order to alleviate the computational burden and to obtain better results, a

Table 11.3 Comparative results of psycho-clonal algorithm based approach and GA for Multiple Fault Diagnosis problem.

S.No.	Manifestation Observed(M^+)	Faults (DI)	Relative Likelihood Value	Comp. Time in Sec in GA Based Approach	Comp Time in Sec in Psycho-Clonal Algorithm Based Approach
1	$m_1, m_2, m_4, m_5, m_7, m_8, m_{13}, m_{15}$	$d_1, d_9, d_{10}, d_{14}, d_{17}$	$3.66878e^{-09}$	600	10.221
2	$m_7, m_8, m_9, m_{11}, m_{14}, m_{15}$	d_4, d_5, d_{17}, d_{20}	$1.32717e^{-10}$	600	10.233
3	$m_1, m_3, m_4, m_6, m_7, m_{11}, m_{13}, m_{15}, m_{14}$	$d_1, d_5, d_9, d_{14}, d_{16}, d_{17}$	$6.82107e^{-13}$	600	10.241
4	$m_1, m_2, m_3, m_7, m_8, m_{12}, m_{13}, m_{17}$	$d_4, d_5, d_8, d_{14}, d_{19}$	$2.49979e^{-09}$	600	10.239

Table 11.4 Results of Psycho-clonal algorithm based approach on ten standard test problems.

	Size of the Test Problem	Relative Likelihood Value	Comp Time in Sec
1	10 × 10	3.3858435e^{-03}	2.56352
2	15 × 15	5.6483456e^{-03}	3.25365
3	20 × 20	2.9034567e^{-05}	2.68465
4	25 × 25	3.3485643e^{-08}	3.35644
5	30 × 30	4.1434568e^{-10}	3.56445
6	35 × 35	2.6316578e^{-08}	4.35643
7	40 × 40	3.7578964e^{-09}	5.54323
8	45 × 45	6.3883435e^{-13}	6.56456
9	50 × 50	1.8983345e^{-12}	6.52432
10	50 × 50	4.1056047e^{-15}	7.34525

psycho-clonal algorithm with embedded features of a Probabilistic Causal Model has been proposed. It efficiently diagnoses the exact faults based on prior probability and tendency matrix knowledge. Furthermore, the obtained results substantiate the fact that the psycho-clonal algorithm with the embedded feature of a Probabilistic Causal Model is a better fault diagnosis methodology for complex systems than other random search techniques. The minimum fault isolation time can help in repairing procedures (both on-board and off-board), and efficiently reducing and managing the spare parts in supply chain management. However, the main drawback of the proposed algorithm is that it is only able to diagnose faults where prior knowledge exists, that is having data related to prior probability or tendency matrix, *etc.*

This research can be extended along several directions: dynamic fault diagnosis of tracking the evolution of system states, learning of parameters and models on line, adaptive inferencing, and agent–based architectures for distributed diagnostics and prognostics. In addition, the application of the proposed methodology on a complex real world system such as complex manufacturing system, aerospace, missiles and medical system are scope of future research.

References

Ada, G. L. and Nossal, G. J. V., 1987, *The Clonal Selection Theory*, Scientific American, 257(2), 50–57.

Akhilesh Kumar, Anuj Prakash, Ravi Shankar and M.K. Tiwari, "Psycho-Clonal algorithm based approach to solve continuous flow shop scheduling problem" *Expert Systems with Applications*, Volume 31, Issue 3, October 2006, Pages 504–514.

Bertsekas, D. P. and Castanon, D. A. Rollout algorithms for stochastic scheduling problems. *Journal of Heuristics* 1999; 5: 89–108.

Bertsekas, D. P. and Tsitsiklis, J. N., 1996, *Neuro-Dynamic programming*.Vol.1, 2, Athena Scientific, Belmont, (MA: Alhena).

Bertsekas, D. P., Tsitsiklis, J. N. and Wu, C., 1997, Rollout algorithms for combinational optimization. *Journal of Heuristics*, 3(3), 245–262.

Biasizzo, A., Zuzek, A. and Novak, F., 2000, Sequential diagnosis with asymmetrical test. *Comput. Journal*, 41(3), 191–197.

Binglin, Z., Tinghu, Y. and Ren, H., 1993, A genetic algorithm for diagnosis problem solving. *Proceedings of IEEE International Conference of Systems Man and Cybernetics*, 2, 404–408.

Chang, S., 1991, Failure propagation trees for diagnosis in manufacturing systems. *IEEE Transactions on Systems Man and Cybernetics*, 21(24), 185–194.

Chessa, S. and Santi, P., 2001, Operative diagnosis of graph-based systems in multiple faults. *IEEE Transactions on Systems Man and Cybernetics*, 31,112–119.

Chevalier, E., Martin, JA, Colomb, Blanch I, and Laserna, JML, 1997. "Hybrid manufacturing line supervision and diagnosis by means of fuzzy rules connected with a causal graph", *Proceedings of the Sixth IEEE International Conference on Fuzzy Systems*, 3, 1259–1264.

Dashora, Y., Tiwari, M. K., Karunakaran, K. P. "A psycho-clonal-algorithm-based approach to the solve operation sequencing problem in a CAPP environment" *International Journal of Computer Integrated Manufacturing*, Volume 21, Issue 5 (July 2008), Pages: 510–525.

de Castro, L. N.; Von Zuben, F. J. (2002). "Learning and Optimization Using the Clonal Selection Principle" *IEEE Transactions on Evolutionary Computation, Special Issue on Artificial Immune Systems (IEEE)* 6 (3): 239–251.

Khoo, L. P., Ang, C. L. and Zhang, J., 1999, An IDEFO model-based intelligent fault diagnosis system for manufacturing systems. *International Journal of Production Research*, 37, 35–48.

Kuo, C. H. and Huang, H. P., 1997, Colored timed Petri net based statistical process control and fault diagnosis to flexible manufacturing systems. *Proceedings IEEE International Conference on Robotics and Automation*, 4, 2741–2746.

Kumar V., PKS Prakash, Tiwari, M. K., Chan, F. T. S., 2006, Stochastic make-to-stock inventory deployment problem: an endosymbiotic psychoclonal algorithm based approach, *International Journal of Production Research*, Volume 44(11), 2245–2263.

Maheshwari, S.N. and Hakim, S.L., 1992 "On model for diagnosable systems and probabilistic fault diagnosis," IEEE *Trans. Comput.*, 41, 318–324.

Maslow, A. H., 1954, *Motivation and Personality*, New York: Harper & Bros.

Miller, A., Potter, W. D., Gandham, R.V. and Lapena, C. N., 1993, An evaluation of local improvement operator for genetic algorithms. *IEEE Transactions on Systems Man and Cybernetics*, 23, 1340–1351.

Pattipati, K. R. and Alexandridis, M, G., 1990, Application of heuristic search and information theory to sequential fault diagnosis. *IEEE Transactions on Systems Man and Cybernetics*, 20, 872–887.

Pattipati, K. R. and Tu, F., 2003, Rollout strategies for Sequential Fault diagnosis. *IEEE Transactions on Systems Man and Cybernetics*, 33(A), 86–99.

Pattipati, K.R, 2003, Combinatorial optimization algorithms for fault diagnosis in complex systems. *Proceeding of International Workshop on the IT-Enabled Manufacturing, Logistics and Supply Chain Management.* Indian Institute of Science, Bangalore 18–35.

Peng, Y. and Reggia, J. A., 1987, A Probabilistic Causal Model (PCM) for Diagnostic Problem Solving, Part 1: Integrating Symbolic Causal Inference With Numeric Probabilistic Inference. *IEEE Transactions on Systems Man and Cybernetics*, 17, 146–162.

Qian, D.Q., 1990, An improved method for fault location of chemical plants. *Computers and chemical engineering*, 14, 41–48.

Raghavan, V., 1996, Algorithms for Sequential Fault Diagnosis, Ph.D Thesis. Department of Electrical and Systems Engineering, University of Connecticut, Storrs, CT 06269–2157.

Rao, N. S. V., 1993a, Expected-value analysis of two single fault diagnosis algorithms. *IEEE Transactions on Computers*, 42, 272–280.

Rao, N. S. V., 1993b, Computational complexity issues in operative diagnosis of graph-based systems. *IEEE Transactions on Computers*, 42, 447–457.

Reggia, J. A., Nau, D. and Wang, P. Diagnostic Expert systems Based on a set Covering Model. *International Journal of Man-Machine Studies* 1993; 19 (5): 437–460.

Sanjay Kumar Shukla, M. K. Tiwari, Hung-Da Wan and Ravi Shankar. "Optimization of the supply chain network: Simulation, Taguchi, and Psychoclonal algorithm embedded approach " *Computers & Industrial Engineering*, Volume 58, Issue 1, February 2010, Pages 29–39.

Shukla, N., Tiwari, M K and Shankar, R. "Optimal Sensor Distribution for Multi-Station Assembly Process Using Chaos Embedded Fast Simulated Annealing", *International Journal of Production Research*, Volume 47, Issue 1, (2009), pages 187–211.

Singh, R. K., Prakash, Kumar, S. and Tiwari, M. K. "Psycho-clonal based approach to solve a TOC product mix decision problem" *The International Journal of Advanced Manufacturing Technology*, Volume 29, Numbers 11–12, 2006, 1194–1202.

Tiwari M .K., PKS Prakash, Kumar A., and Mileham, A., 2005, Determination of an optimal assembly sequence using the psychoclonal algorithm, Proceedings of the Institution of Mechanical Engineers, Part B: *Journal of Engineering Manufacture*, Volume 219(1), 137–149.

Tu, F. and Pattipati, K.R., 2003, Rollout Strategy for Sequential Fault Diagnosis, *IEEE Transactions on Systems, Man and Cybernetics, Part A*, 1, 86–99.

12

Platform Formation Under Stochastic Demand

D. Ben-Arieh and A.M. Choubey

Dept. of IMSE, Kansas State University, Manhattan, KS

Abstract

Daily global competition forces manufacturers to introduce an increasing number of products with shorter life span and at a lower cost. Moreover, this is achieved in an environment where demands are uncertain and with short lead times. Therefore, the focus of this chapter is a platform based production strategy that would ensure the economic production of a variety of products under demand uncertainty.

The chapter proposes a production strategy based on a platform architecture where every product variant in the family is assembled either directly from its components or from the platform. The problem is formulated as a two stage stochastic integer programming model with recourse. The chapter presents three solution approaches to the problem – an exact solution and two heuristic approaches. The results obtained from the exact method validate the model formulation and measure the contribution of the stochastic programming. An example provides insights into the stochastic model.

Keywords: Genetic algorithm, demand scenarios, probability distributions, mass customization

12.1 Introduction

In today's highly volatile market there is a growing concern for fulfilling the individual customer's wants and needs. "The customers now have plenty of choice ... they have become more aware...they select the product that most closely fulfills their opinion of being the best

Manoj Tiwari and Jenny A. Harding (eds.) Evolutionary Computing in Advanced Manufacturing, (259–288) © Scrivener Publishing LLC

value for the money..." (Hollins and Pugh, 1990). Therefore, "customers can no longer be lumped together in a huge homogeneous market..." (Pine, 1993) rather this competitive world of manufacturing requires the manufacturer to introduce an increasing number of products with shorter life span and at a lower cost. This requires the producer to continuously seek ways to reduce the production costs, while still offering attractive products. This in effect has shifted the manufacturing focus from mass production to mass customization; a trend no longer limited to high value products. This phenomenon is demonstrated by the fact that from 1973 to 1989, there has been a 70% increase in the number of car models produced in the US with a commensurate drop in the volume of production per model (McDuffie *et al.*, 1996).

An added modern complication of mass customization is the need to manage the high variety of products in an environment where demands are stochastic and the lead times to fulfill those demands are short.

In response to this need various strategies have been developed including, the concept of delayed differentiation (Lee, 1996; Lee and Tang, 1998; Swaminathan and Tayur, 1998), exploiting commonality at the product design stage (Ulrich and Pearson, 1993), use of lean manufacturing concepts (Womack *et al.*, 1990), and product platform strategy (Meyer and Lehnerd, 1997).

As the focus of this chapter, the platform approach has gained acceptance by many corporations as the means to increase product count without increasing the cost per part. Black and Decker applied this idea to its power tool products (Meyer and Lehnerd, 1997). Volkswagen used a platform architecture strategy and reduced development and production costs (Wilhelm, 1997). Sony applied this approach to its product development process (Sanderson and Uzumeri, 1995). Also, AeroAstro Inc. used a platform architecture with their multipurpose radio platform, and solved many of the communication problems faced by spacecraft system designers (Caffrey *et al.*, 2002). These are just a few examples of industrial applications of the platform concept.

In this chapter, we demonstrate a platform based approach for the production of a product family under demand uncertainty. Using this approach, every product variant in the family may either be assembled directly from its components, or from any platform whose component set resembles those required by the product. The methodology seeks to find an optimal platform that will minimize the overall production costs of the products, which include the costs of production, holding cost of unused platform inventory and shortage cost of lost demands of products, while considering the stochastic demand of each product type.

The chapter presents three solution approaches to the problem – an exact solution and two heuristic approaches. The results obtained from the exact method are used to validate the model formulation and measure the significance of using a stochastic program for modeling the problem. Only small instances of the problem could be solved by the exact approach. Therefore, the two heuristic methods that can provide good solutions to large instances of the problem more quickly are developed. The first heuristic method combines a genetic search process and integer programming to provide a near optimal solution. Using this approach the solution time increases exponentially with the increase in number of possible demand scenarios. To deal with a large number of demand scenarios a pure probability based genetic search process is proposed. This approach is faster even when a large number of demand scenarios is considered, with slightly inferior solution quality. Both of these methods are illustrated with an example of a product family of cordless drills.

The rest of the chapter is organized as follows. Section 12.2 provides a review of the work related to the platform formation problem. Section 12.3 defines the problem, and gives the description and formulation of the problem. In Section 12.4, the heuristic solution methodologies are explained. Section 12.5 provides an example to illustrate the methodologies, and presents computational results and insights into the platform based production approach. Finally, Section 12.6 provides conclusions and future recommendations for the research.

12.2 Background

Ulrich and Eppinger (2003) define a *platform* as a collection of assets, including component designs, shared by multiple products. It can also be defined as a set of shared functionality, components, subsystems and manufacturing processes across the product family (Robertson and Ulrich 1998).

Figure 12.1 A product platform and a product family.

As an example, Figure 12.1 shows a product family which shares several components with a possibility for a platform based strategy.

Various streams of research in the area of product platforms were greatly influenced by contributions from Pine (1993) in the area of mass customization, Meyer and Lehnerd (1997) in the area of platform concepts, Sanderson and Uzmeri (1995) in the area of managing product families (Allada *et al.*, 2006). Krishnan and Ulrich (2001), Simpson (2004), Jose and Tollenaere (2005), and more recently Allada *et al.* (2006) have provided a review of various aspects of product platform development methodologies. Simpson *et al.* (2006) provides an overview of the platform concept, application areas, and ongoing research and expanding views on platforms in academia and in industry.

Research on qualitative approaches to the platform problem include works of Maier and Fadel (2001), Dahmus *et al.* (2001), Shil and Allada (2005), Wilson and Norton (1989), Martin and Ishii (1997), Martin and Ishii (2002), Kota *et al.* (2000), and Park and Simpson (2005).

Various quantitative solution methods to the platform optimization problem were implemented including (but not limited to) Branch and Bound algorithm (Fujita and Yoshida, 2001), Dynamic programming (Allada and Jiang, 2002), agent based techniques (Rai and Allada, 2003), Simulated Annealing (Fujita *et al.*, 1999), genetic algorithms (Fujita and Yoshida, 2001; Li and Azaram, 2002;

Simpson and D'Souza, 2002; Simpson and D'Souza, 2004; Jiao and Zhang, 2005).

Industry has used platform scaling to develop the product family to satisfy various market niches. For example, Rolls Royce (Rothwell and Gardiner, 1990), Boeing (sabbagh, 1996), and Honda (Naughton *et al.*, 1997) have successfully applied this strategy to produce a product family on the basis of platforms to satisfy target market niches. Additional research works in this area include Simpson *et al.* (2001), Messac *et al.* (2002a), Messac *et al.* (2002b), and Hernandez *et al.* (2003).

12.3 Problem Description

In this chapter, a platform is considered to be a set of shared components among multiple products. A specific product belonging to a product family is produced by adding or removing some of the components that are assembled using the platform. Figure 12.2 illustrates a hypothetical product family with four products (P_1, P_2, P_3, and P_4), each consisting of a different collection of components from the set $\{A, B, C, \ldots, H\}$. A possible platform for this set of products is as shown in Figure 12.3. In this example, P_1 would be created by using the platform and removing G and adding C.

A platform is only justified if the assembly of the components to the platform can be done efficiently using mass production methods. Thus, adding and removing components from a platform to fit a particular product typically costs more than if the component is included in the platform.

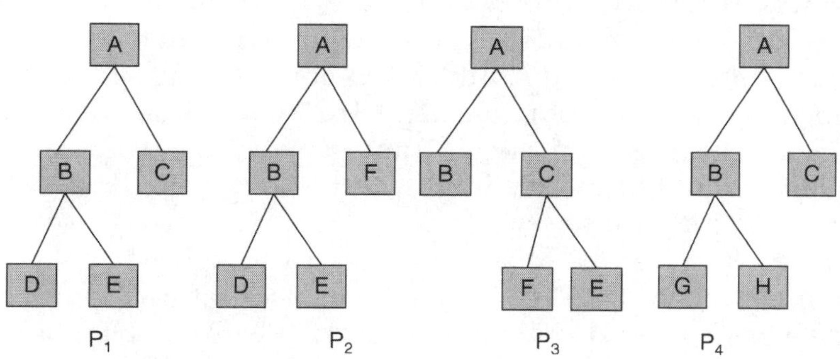

Figure 12.2 Example of a product family.

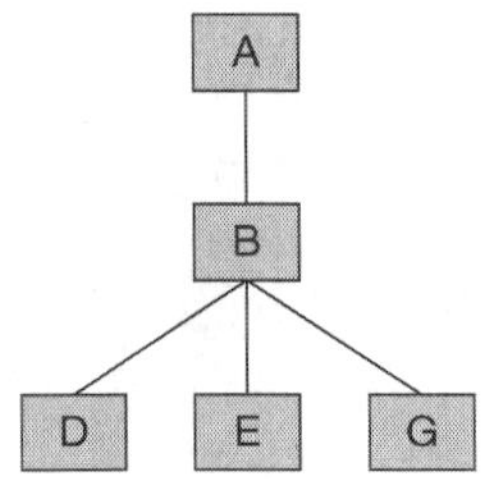

Figure 12.3 A platform for the product family.

The objective of the analysis presented in this chapter is to determine the optimal configuration of the platform (its component set), the optimal inventory level of the platform, and the optimal number of each of the products that should be produced in each demand scenario, while minimizing the overall production costs.

One essential factor is maintaining the product assembly structure while deciding on the configuration of the platform. The assembly relationships are represented using a Part Assembly Relationship matrix (PAR). An element of PAR, f_{jl}, represents the fact that component j precedes component l i.e. component j is needed to be present in the platform for l to be included in the platform. As an example, a PAR for the product family shown in Figure 12.2 is presented in Table 12.1.

12.3.1 Problem Statement

Assume that the platform supports N types of products. The facility mass-produces w units of a single type of platform. The manufacturer experiences stochastic demand for each of the products. If the actual total demand of all the product types is higher than the inventory level of the mass produced platforms there would be some demand losses and shortage cost would be incurred. On the other hand if the actual total demand of the product is less than the inventory level of the platforms, the demands would be satisfied but with the added holding cost for the unused platforms.

The problem can be formulated as a two stage stochastic programming model with recourse. The demand of each product is modeled as a set of demand scenarios each with some probability of occurrence.

Table 12.1 The PAR for Product family.

	A	B	C	D	E	F	G	H
A		1	1			1		
B				1	1		1	1
C					1	1		
D								
E								
F								
G								
H								

First stage decision variables are:

1. The configuration (components set of a platform)
2. The quantity of the platforms (inventory level) to be produced.

Second stage decision variables would be to:

1. The additional components that would be added manually to the platform to make a particular product type
2. The components that would be manually removed from the platform to make a particular product type, and
3. The quantity of each product type to be produced for each scenario.

The objective is to minimize the total production cost that includes the cost of production of platforms, cost of producing the products using the platforms, holding cost of unused platforms and stock-out cost of lost demands, in addition to the cost of manually adding and removing components.

12.3.2 Formulation of the Model

Let

- $k = 1, 2, \ldots, N$ index of products
- $j, l = 1, 2, \ldots, M$ index of components

- $s = 1, 2, \ldots, S$ index of demand scenarios
- C_j = cost of the component j (purchasing price)
- CP_j = cost of assembling a component j to form the platform (mass assembly)
- CA_j = cost of manually adding a component to the platform to form a product $(CA_j > CP_j)$
- CR_j = cost of removing a component from the platform to form a product
- h = per unit holding cost for the platforms
- q_k = per unit stock-out cost for product k
- ξ_s = Vector of demands $(\xi_{1s}, \xi_{1s}, \ldots, \xi_{NS})$ in scenario s
- p_s = probability of occurrence of scenario s

- $$\nu_{jk} = \begin{cases} 1 & \text{if Product } K \text{ requires component } j \\ 0 & \text{otherwise} \end{cases}$$

- $$f_{jl} = \begin{cases} 1 & \text{if component j precedes component l according to part assembly relationship matrix} \\ 0 & \text{otherwise} \end{cases}$$

12.3.2.1 Decision Variables

- $$x_j = \begin{cases} 1 & \text{if component j participates in the platform} \\ 0 & \text{otherwise} \end{cases}$$

- w = quantity of the platforms to be made
- y_{ks} = amount of product k to be produced using platforms in scenario s

- $$a_{jk} = \begin{cases} 1 & \text{if component } j \text{ is added manually t o the platform to make product } k \\ 0 & \text{otherwise} \end{cases}$$

- $$r_{jk} = \begin{cases} 1 & \text{if component } j \text{ is removed manually from the platform to make product } k \\ 0 & \text{otherwise} \end{cases}$$

- u_{ks}^{-} = Lost demand of product k in scenario s
- v_s^{+} = Unused inventory of platforms in scenario s

The following model provides an optimal solution:

12.3.2.2 Model 1

$$Min\ w \times \sum_{j=1}^{M} \left(CP_j + C_j\right) \cdot x_j +$$

$$\sum_{s=1}^{S} p_s \left(\sum_{k=1}^{N} \sum_{j=1}^{M} \left(CA_j + C_j\right) \cdot a_{jk} \cdot y_{ks} + \sum_{k=1}^{N} \sum_{j=1}^{M} \left(CR_j - C_j\right) \cdot r_{jk} \cdot y_{ks} \right) +$$

$$\sum_{s=1}^{S} p_s \sum_{k=1}^{N} q_k \times u_{ks}^{-} +$$

$$\sum_{s=1}^{S} p_s \times h \times v_s^{+} \tag{12.1}$$

S.t.

$$a_{jk} + x_j \le 1 \quad \forall j \in \{1,2,...,M\}, k \in \{1,2,...,N\} \tag{12.2}$$

$$a_{jk} + x_j \ge v_{jk} \quad \forall j \in \{1,2,...,M\}, k \in \{1,2,...,N\} \tag{12.3}$$

$$\nu_{jk} \ge a_{jk} \quad \forall j \in \{1,2,...,M\}, k \in \{1,2,...,N\} \tag{12.4}$$

$$x_j \ge r_{jk} \quad \forall j \in \{1,2,...,M\}, k \in \{1,2,...,N\} \tag{12.5}$$

$$r_{jk} + x_j + v_{jk} \le 2 \quad \forall j \in \{1,2,...,M\}, k \in \{1,2,...,N\} \tag{12.6}$$

$$w - v_s^{+} = \sum_{k=1}^{N} y_{ks} \quad \forall s \in \{1,2,...,S\} \tag{12.7}$$

$$y_{ks} + u_{ks}^{-} = \xi_{ks} \quad \forall s \in \{1,2,...,S\}, k \in \{1,2,...,N\} \tag{12.8}$$

$$1 + x_j \ge f_{jl} + x_l \quad \forall j,l \in \{1,2,...,M\} \tag{12.9}$$

$$x_j \in \{0,1\};\ a_{jk} \in \{0,1\};\ r_{jk} \in \{0,1\};\ y_{ks} \ge 0;\ w \ge 0;\ u_{ks}^{-} \ge 0;\ v_s^{+} \ge 0 \tag{12.10}$$

The objective function (equation 12.1) represents the total production cost that includes cost of making the platforms, cost of assembling the products using the platforms, total stock-out costs, and total holding cost under all possible scenarios. Constraints 12.2–12.4 state that component j must be added to the platform to make product k if j is not in the platform and is required in product k. Constraints

12.5–12.6 state that component j may be removed from the platform to make product k if that component is in the platform and is not required in product k. Constraint 12.7 expresses that for any scenario s, the total number of products produced cannot exceed the platform inventory level. Constraint 12.8 limits the total quantity of product k produced to the random demand value of product k for any scenario s plus the lost demand. Constraint 12.9 checks the assembly feasibility of the platform while deciding the configuration of platform. These constraints state that if component l is in the platform and according to the part assembly relationship matrix j precedes $l(f_{jl} = 1)$ then j must also be present in the platform. Constraint 12.10 ensures the binary and non-negativity nature of the decision variables.

12.4 Evolutionary Solution Approaches

While small-dimensioned problems can be solved optimally using standard solvers, larger and practical sized problems necessitate looking for heuristic approaches. In this section we present two evolutionary based algorithms.

Genetic Algorithms have been studied in Goldberg (1989), Mitchell (1998), and a review of GA applications is provided in Aytug *et al.* (2003).

The solution methodologies presented herein exploit the principles of evolution and mutation and the concept of fitness. In general, the genetic search procedure starts with the generation of a random population of solutions (chromosomes) represented as strings. The number of strings, i.e., the population size remains constant throughout this genetic search process. The cost function, which should be minimized is converted to a *fitness value* which decides which solutions will contribute to the next generation of solutions. The population then evolves through successive generations using the genetic operators of reproduction, cross-over and mutation. The algorithm continues until a pre-specified condition is met (here the maximum number of iterations).

It is assumed that the reader is sufficiently familiar with the mechanism of genetic algorithms.

In this chapter, we propose two types of Genetic algorithms. The first uses a simplified Integer Programming model (GAIP) to provide a near optimal solution. The heuristic can solve large instances of the problem reasonably fast; however the solution time increases exponentially with the number of possible demand scenarios. The second algorithm uses sampling to choose a population of good solutions representing each scenario. This approach can solve large instances of the problem with a large number of demand scenarios, supporting the extension of the solution to independent demand distribution for each product as shown later.

12.4.1. Solution Encoding

For both heuristic approaches, the chromosome is represented as a string of length M (number of distinct components). A value of '1'at any position j in the string represents that component j is in the platform.

Each chromosome (or configuration of a platform) in the population is feasible if it follows the part assembly relationship for the product family under consideration (such as in Table 12.1). Initially a population of chromosomes is randomly generated; therefore some of the solutions may not be feasible. Also, applying the crossover, and mutation operators may create new temporary solutions that may not be feasible. An algorithm to check for and correct the solution's feasibility is presented in (Ben-Arieh *et al.* 2007).

12.4.2 Genetic Algorithm with Integer Programming (GAIP)

In this method a genetic search algorithm, presented in Figure 12.4, is used to explore the search space and an integer program (Model 2), and is solved for each chromosome to calculate its fitness value.

A chromosome string, x_j, is generated probabilistically (randomly or using genetic operators) and the value of x_j is fed to the Model 1 and the model reduces to a linear integer program (Model 2). Solving this integer program provides the total cost of production (objective function value of Model 2) and values of other decision variables such as the number of platforms to be mass-produced, the components that should be added or removed to produce each particular product, and the number of each product produced using the mass-produced platforms. Now the genetic

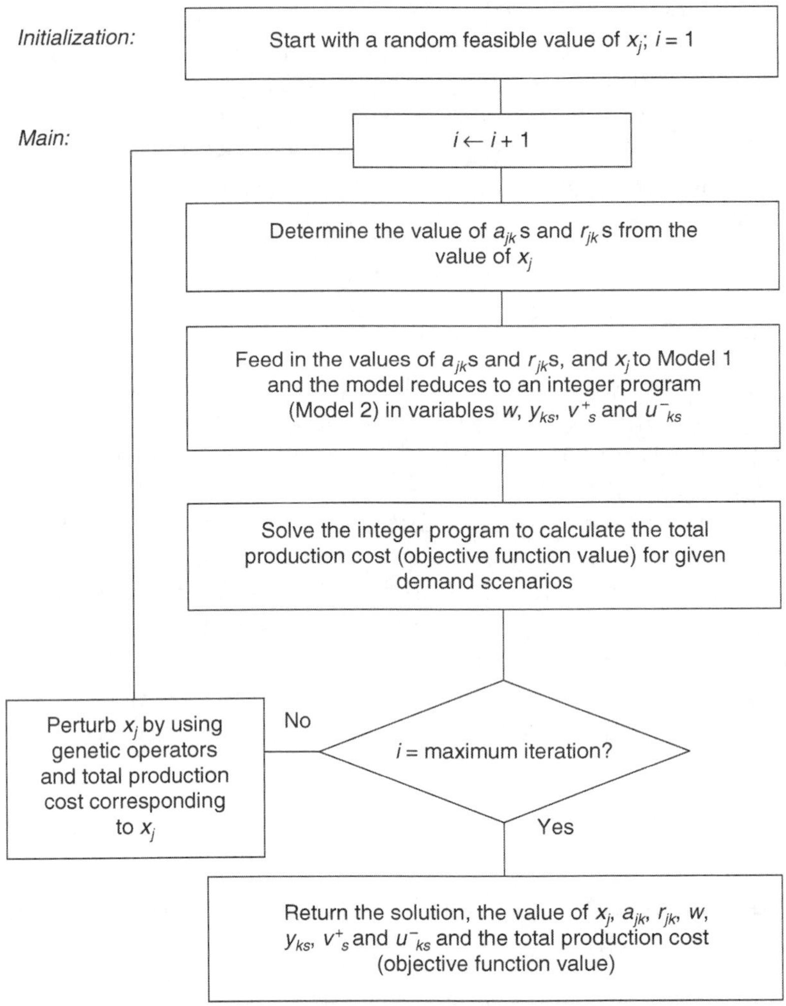

Figure 12.4 Flow chart of the solution strategy.

operators are applied to alter the chromosome information (x_j) and the process is repeated until the maximum number of iterations is reached.

Once the genetic algorithm generates the platform structure (vector x_j) Model 1 is reduced to Model 2 by using in the values of x_j, a_{jk} and r_{jk}. The reduced model 2 is shown below.

Model 2 (Underlined variables are known)

$$
\begin{aligned}
Min \quad & w \times \sum_{j=1}^{M} \left(CP_j + C_j \right) \cdot x_j \\
& + \sum_{s=1}^{S} p_s \left(\sum_{k=1}^{N} \sum_{j=1}^{M} \left(CA_j + C_j \right) \cdot \underline{a_{jk}} \cdot y_{ks} \right. \\
& \left. + \sum_{k=1}^{N} \sum_{j=1}^{M} \left(CR_j - C_j \right) \cdot \underline{r_{jk}} \cdot y_{ks} \right) \\
& + \sum_{s=1}^{S} p_s \sum_{k=1}^{N} q_k \times u_{ks}^{-} + \sum_{s=1}^{S} p_s \times h \times v_s^{+}
\end{aligned} \tag{12.11}
$$

S.t.

$$w - v_s^{+} = \sum_{k=1}^{N} y_{ks} \forall s \in \{1,2,\ldots,S\} \tag{12.12}$$

$$y_{ks} + u_{ks}^{-} = \xi_{ks} \quad \forall s \in \{1,2,\ldots,S\}, k \in \{1,2,\ldots,N\} \tag{12.13}$$

$$y_{ks} \geq 0; \; w \geq 0 \; ; v_s^{+} \geq 0; u_{ks}^{-} \geq 0 \tag{12.14}$$

This integer program now can be solved as a linear program yielding integer solutions. The solution provides the fitness value of that chromosome.

12.4.3 Pure Probability Based Heuristic Approach

The time taken by the GAIP approach to solve the problem increases exponentially with the number of demand scenarios. To alleviate this problem we propose one more heuristic which is a pure probability based genetic search process. The new algorithm is based on the following concept:

1. The number of solutions (chromosomes) selected as a solution to each scenario is proportional to the probability of that scenario.
2. These selected "good" chromosomes are mixed together and undergo crossover and mutation in anticipation that children chromosomes would demonstrate overall good performance for all the scenarios together. These solutions are then ranked based on overall cost.

The algorithm for the approach is presented next.

Step 1: Using x_j, determine the values of a_{jk} and r_{jk}.
Step 2: Assign these values in the following equation to get the cost value denoted Obj_s.

$$Obj_s = \sum_{k=1}^{N}\sum_{j=1}^{M}\left(C_p + C_j\right)\cdot x_j \cdot D_k + \sum_{k=1}^{N}\sum_{j=1}^{M}\left(\left(C_a + C_j\right)\cdot a_{jk} + \left(C_r - C_j\right)\cdot r_{jk}\right)\cdot D_k$$

Where, D_k is the demand of product k in scenario s.
Step 3: The total cost of a solution is calculated using the expected demand. The cost calculation follows the same steps presented for the calculation of Obj_s except that in place of D_k, the expected value of D_k, $E(D_k)$, is used, where. $E(D_k) = \sum_{s=1}^{S}(p_s \times D_k)$.

12.4.4 Extension to Independent Demand for Each Product

This extension allows the demand to be represented as an independent demand for each product rather than a scenario as presented before. However, this case can be converted back to a scenario based solution in the following way:

Given N products, the demand for each product can be discretized into (2m + 1) values as shown for example in (Delft and Vial, 2001). Once this is done, the combinations of all possible demands for each product result in $(2m + 1)^N$ scenarios. This large number of scenarios can be solved using the Pure Probability Based Heuristic.

12.5 Example Problem – Results and Discussions

12.5.1 Example

To validate and illustrate the solution approaches we use an example of a family of cordless drills. These drills have varied construction and household applications. There are various types of drills available; however, for our purpose we consider five types (products): heavy duty, high performance, high value, standard, and multipurpose power drills as shown in Figure 12.5 (Sudjiato and Otto, 2001). The objective in this example is to demonstrate the use

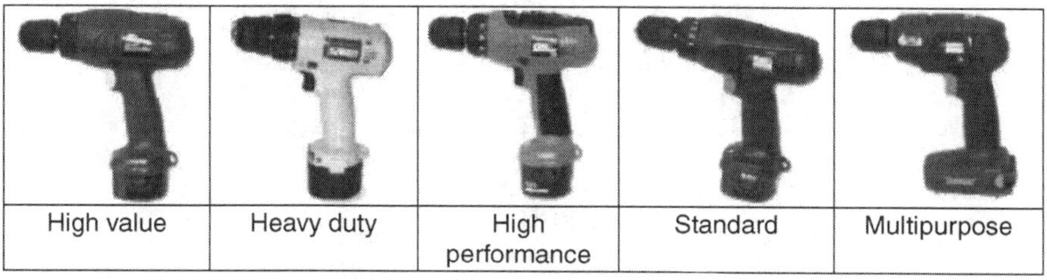

Figure 12.5 The product family of the cordless drills.

of the heuristics to determine the optimal platform for the product family. The information about the products and the components is provided in Table 12.2 and Table 12.3. Table 12.4 provides the information regarding the part assembly relationship of the whole family.

The maximum numbers of components for this case is 23.

12.5.2 Results and Discussions

Initially, the genetic search heuristics proposed in the last section were run several times to make the appropriate choice of population size, number of generations, and mutation rate (tune-up phase).

12.5.3 Results and Analysis Using GAIP

Table 12.5 provides the data for the example in which CP_j = \$2, CA_j = \$4, and CR_j = \$2 (for all j). Table 12.6 presents the solution obtained using the GAIP.

Figure 12.6 shows the convergence plot of GAIP for the above case.

Figure 12.7 shows the solution quality as a function of the number of generations. It is evident that the algorithm converges efficiently. The heuristic is run 20 times, and the mean and 'mean + standard deviation' of the best solutions for all the runs at each iteration is plotted *vs.* number of iterations. From this Figure it is clear that the standard deviation kept decreasing with the number of iterations and finally it becomes negligible which means that in almost all the runs the heuristic hit the same solution at the end which in turn supports the global convergence and repeatability of the heuristic.

Table 12.2 Binary bill of material for the products.

Component # →	**1**	**2**	**3**	**4**	**5**	**6**	**7**	**8**	**9**	**10**	**11**	**12**	**13**	**14**	**15**	**16**	**17**	**18**	**19**	**20**	**21**	**22**	**23**
Component index → Products	A	B	C	D	E	F	G	H	I	J	K	L	M	N	O	P	Q	R	S	T	U	V	W
Heavy duty	1	1	0	0	1	0	1	0	1	1	1	1	0	1	0	0	0	1	0	1	0	1	0
High performance	1	0	1	0	0	1	0	1	1	1	1	0	1	0	1	0	0	0	1	0	1	1	0
High value	1	0	0	1	0	1	0	1	1	1	1	0	0	0	0	1	0	0	0	0	1	1	0
Standard	1	0	0	1	0	1	0	1	1	1	1	0	0	0	0	1	0	0	0	0	1	0	1
Multi-purpose	1	0	0	1	0	1	0	1	1	1	1	0	0	0	0	0	1	0	0	0	1	0	1

Table 12.3 The possible components for the product family with their cost.

Comp #	Comp. Index	Comp. Name	Comp. Cost ($)
1	A	Encasing	2
2	B	Rough palm to permit positioning	1
3	C	Padded palm to permit positioning	1.5
4	D	Diamond palm to permit positioning	2
5	E	Bevel 2 point to lock/unlock battery	3
6	F	Straight 2 point to lock/unlock r battery	2
7	G	Square, 9.6 V, 2 pt. Electricity transmission	4
8	H	Open, 9.6 V, 2 pt. Electricity transmission	4
9	I	Chuck to secure/ unsecure bit	2
10	J	Chuck teeth to register/ unregister bit	1
11	K	Bit to act on object	2
12	L	Thin button to input speed	1
13	M	Wide button to input speed	1
14	N	16 slip clutch to transmit power	5
15	O	22 slip clutch to transmit power	6
16	P	6 slip clutch to transmit power	5
17	Q	Solid shaft to transmit power	3
18	R	Fine ring gear to switch speed	5
19	S	Ring gear to switch speed	4
20	T	Black oval button to unlock switch	1
21	U	Black button to unlock switch	1
22	V	Variable speed to switch power	6
23	W	2 speed to switch power	4

12.5.4 The Solution Quality of PHA and Comparison with the GAIP Approach

Figure 12.8 shows a convergence plot of PHA for any instance of the example. This approach requires more iterations than the GAIP, but it converges efficiently too.

Table 12.4 The part assembly relationship for the example.

	A	B	C	D	E	F	G	H	I	J	K	L	M	N	O	P	Q	R	S	T	U	V	W
A		1	1	1	1	1			1			1	1	1	1	1	1						
B																							
C																							
D																							
E							1																
F								1															
G																							
H																							
I										1													
J											1												

(Continued)

Table 12.4 The part assembly relationship for the example. (Continued)

	A	B	C	D	E	F	G	H	I	J	K	L	M	N	O	P	Q	R	S	T	U	V	W
K																							
L																							
M																							
N																		1		1			
O																			1		1		
P																					1		
Q																					1		
R																							
S																							
T																						1	
U																						1	1
V																							
W																							

Table 12.5 Data for the example.

Scenarios	Pr.	Demand				
		P1	**P2**	**P3**	**P4**	**P5**
S1	0.2	100	100	100	100	100
S2	0.8	200	200	200	200	200
Holding cost (Platforms)		Shortage cost				
		P1	P2	P3	P4	P5
$10		100	100	100	100	100

Table 12.6 Solution obtained for the example using GAIP.

Objective value = $ 54710
Number of platforms made = 600

Scenarios	Number of products made					Shortages of products					Leftover platforms
	P1	**P2**	**P3**	**P4**	**P5**	**P1**	**P2**	**P3**	**P4**	**P5**	
S1	100	100	100	100	100	0	0	0	0	0	100
S2	200	200	0	0	200	0	0	200	200	0	0

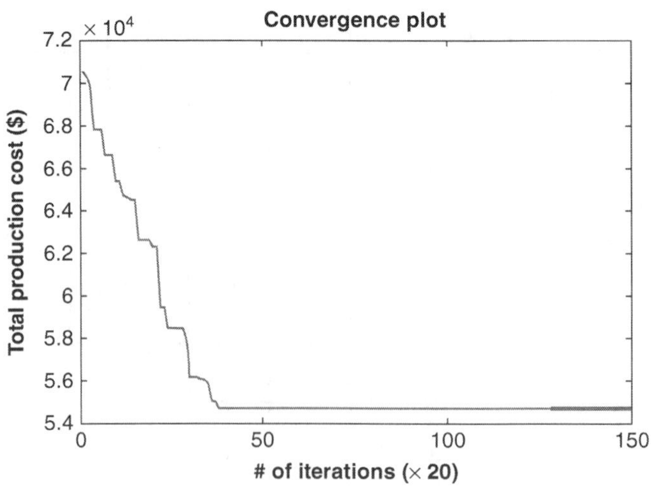

Figure 12.6 Convergence plot of GAIP.

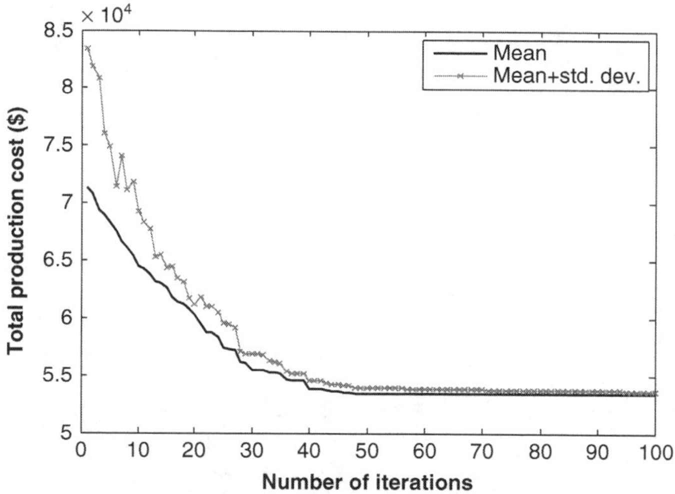

Figure 12.7 The plot of mean and standard deviation using GAIP approach.

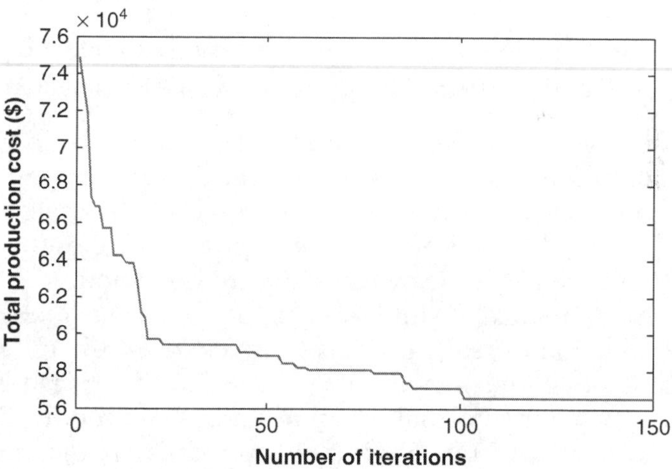

Figure 12.8 A convergence plot for the PHA.

The graph shown in Figure 12.7 for the GAIP approach is also plotted for the PHA approach and is presented in Figure 12.9. This Figure supports the global convergence and repeatability of the PHA approach.

Table 12.7 shows the comparison of results obtained from both heuristics. The results show that GAIP performs slightly better than PHA.

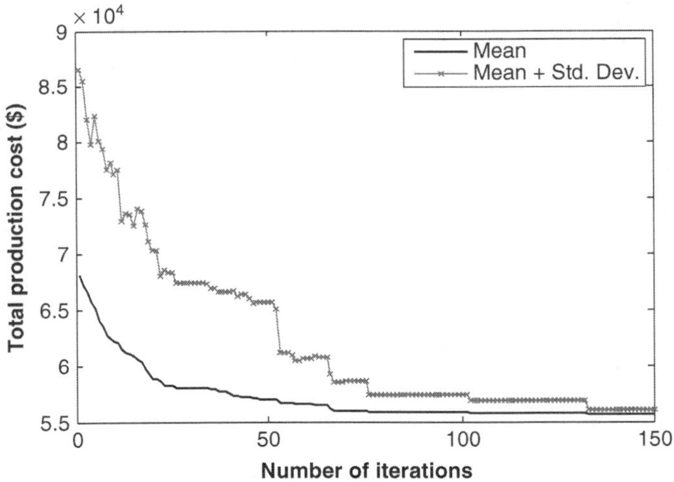

Figure 12.9 The plot of mean and standard deviation convergence for the PHA approach.

12.5.5 Results When Demand of Each Product is Represented as a Probability Distribution

So far we have considered the cases of scenario based demand realization. In this section we consider the demand distribution of each product instead of demand scenarios. For the cases here we assume that the demand of each of the products follows a normal distribution with some mean and standard deviation.

As the mathematical formulation of the model considers the scenario based demand realization, we convert the probability distribution based demand information into a scenario based demand realization. Since the normal distribution is a continuous distribution we divide it into several discrete ranges of demand and calculate the probability values of those demand ranges. If the number of discrete demand points of a product is m, after converting this information into a scenario based demand realization the total number of scenarios would be equal to $S = m^{(\text{number of products})}$.

A small example with a family of three products is used to demonstrate the methodology. The demand for each of the products follows a normal probability distribution with a known mean and standard deviation.

Table 12.8 presents the results obtained by both heuristic approaches for the cases where the mean of the demand of each

Table 12.7 Comparison of results obtained from both heuristics.

Case # 1		q_1	q_2	q_3	q_4	q_5	**GAIP**					**PHA**				
		$ 200	**$ 100**	**$ 200**	**$ 100**	**$ 100**	**W=500; Cost value=26895**					**W=500; Cost value=27985**				
h= $ 5 Scenarios	Pr.	D1	D2	D3	D3	D5	Y1	Y2	Y3	Y4	Y5	Y1	Y2	Y3	Y4	Y5
S1	0.8	100	100	100	100	100	100	100	100	100	100	100	100	100	100	100
S2	0.2	50	50	50	50	50	50	50	50	50	50	50	50	50	50	50
Case # 2		q_1	q_2	q_3	q_4	q_5	W=250; Cost value = 19720					W=250; Cost value = 20220				
		$ 200	$ 100	$ 200	$ 100	$ 100										
h= $ 5 Scenarios	Pr.	D1	D2	D3	D3	D5	Y1	Y2	Y3	Y4	Y5	Y1	Y2	Y3	Y4	Y5
S1	0.8	50	50	50	50	50	100	100	100	100	100	100	100	100	100	100
S2	0.2	100	100	100	100	100	100	0	100	50	0	100	0	100	50	0
Case # 3		q_1	q_2	q_3	q_4	q_5	W=250; Cost value= 21183					W=250; Cost value= 21758				
		$ 100	$ 100	$ 100	$ 100	$ 100										
h= $ 50 Scenarios	Pr.	D1	D2	D3	D3	D5	Y1	Y2	Y3	Y4	Y5	Y1	Y2	Y3	Y4	Y5
S1	**0.1**	200	200	200	200	200	0	0	0	50	200	0	0	0	50	200
S2	**0.9**	50	50	50	50	50	50	50	50	50	50	50	50	50	50	50

(Continued)

Table 12.7 Comparison of results obtained from both heuristics. (Continued)

Case # 4		q_1	q_2	q_3	q_4	q_5	W=250; Cost value=50525					W=250; Cost value=52025				
		$ 100	**$ 100**	**$ 100**	**$ 100**	**$ 100**										
h=$ 50/80/100 Scenarios	Pr.	D1	D2	D3	D3	D5	Y1	Y2	Y3	Y4	Y5	Y1	Y2	Y3	Y4	Y5
S1	**0.5**	200	200	200	200	200	0	200	50	0	0	0	0	0	50	200
S2	**0.5**	50	50	50	50	50	50	50	50	50	50	50	50	50	50	50
Case # 5		q_1	q_2	q_3	q_4	q_5	W=1000; Cost value=57518					W=1000; Cost value=57368				
		$ 100	**$ 100**	**$ 100**	**$ 100**	**$ 100**										
h= $ 10 Scenarios	Pr.	D1	D2	D3	D3	D5	Y1	Y2	Y3	Y4	Y5	Y1	Y2	Y3	Y4	Y5
S1	**0.9**	200	200	200	200	200	200	200	200	200	200	200	200	200	200	200
S2	**0.1**	50	50	50	50	50	50	50	50	50	50	50	50	50	50	50

Table 12.8 Results obtained by both the approaches.

Case #	Normal Demand Vector {(mean, std. dev.), (mean, std. dev.), (mean, std. dev.)}	GAIP Cost Value ($)	PHA Cost Value ($)
1	{(100, 5), (100, 5), (100, 5)}	19726	21428
2	{(100, 10), (100, 10), (100, 10)}	20708	24820
3	{(100, 20), (100, 20), (100, 20)}	23324	28318
4	{(100, 30), (100, 30), (100, 30)}	26895	30425

of the products is fixed but with an increasing standard deviation. Also provided are Shortage cost = [$200, $100, $200] and Holding cost = $10, and the number of discrete demand points on the probability distribution of each product = 5.

From Table 12.8 it is obvious that with an increase in standard deviation of the demand, with all other things fixed, the cost of production increases, i.e. with an increase in variability in demand the stochastic model provides a higher cost. Also, from the table it is clear that GAIP provides a slightly better solution than PHA. However, we can see from Table 12.9 that the time taken by GAIP to solve the problem increases exponentially with an increase in the number of demand points considered on the probability distribution, i.e. with an increase in number of scenarios. PHA is a very fast approach and more suitable where the number of scenarios considered is large.

Figure 12.10 shows the effect of the variance in demand and the number of scenarios considered in the stochastic model. This Figure depicts a plot of total production cost *vs.* number of demand points considered on the demand distribution of each of the product *vs.* the standard deviation in normal demand. The mean demand for each product is kept fixed at 100 units. We can see from the figure that with an increase in the number of demand points (more scenarios) the stochastic cost decreases; and with an increase in the normal standard deviation, the cost increases.

12.6. Conclusion and Recommendations for Future Research

The chapter presents the platform based optimization approach for the economic production of a product family with demand

Table 12.9* Comparison of computational time for both the approaches.

Case #	# of Demand Points on the Probability Distribution	GAIP Time in Min. (Approx.)	PHA Time in Min. (Approx.)
1	3	61	12
2	5	438	31
3	7	1181	65

*Demand vector for all the three cases = {(100, 10), (100, 10), (100, 10)}.

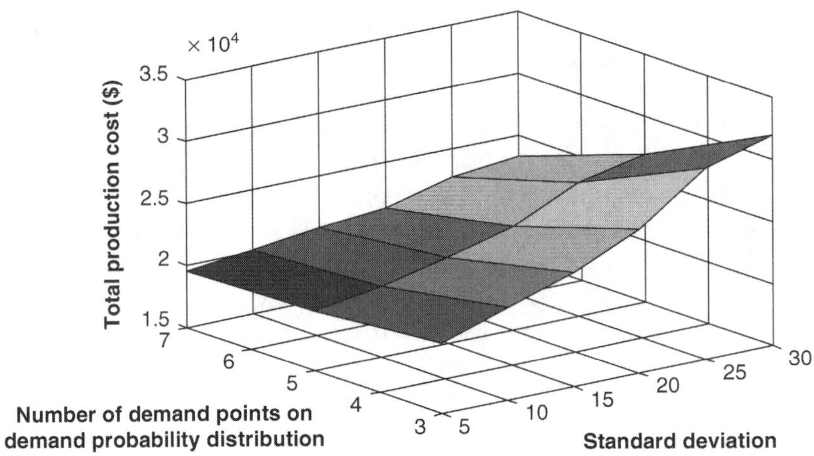

Figure 12.10 The effect of demand variance and number scenarios on total cost.

uncertainty. The problem is formulated as a two-stage stochastic integer program with recourse.

Only a very small instance of the problem could be solved by an exact approach using OPL 3.5. Therefore, two heuristic methods, that can provide good solutions to large instances of the problem more quickly, are developed. The first heuristic method that combines a genetic search process and integer programming provides a near optimal solution and solves a large instance of the problem in a reasonable time, yet this approach takes a long time to solve problems with a large number of demand scenarios. The second method, a pure probability based genetic search heuristic, solves the problems with a large number of demand scenarios very quickly but with slightly inferior solution quality than the first heuristic approach.

The chapter establishes the use and importance of the stochastic model for the platform based production approach especially when the variance in demand is significant. The platform based production approach is explained and illustrated with an example of a family of cordless drills. The chapter provides insight into the effects of demand variance, and various cost components on the optimal platform strategy, and discusses the effectiveness of the heuristics tailored for that application.

Future work in this area includes consideration of more complex cost structures, and multi-period demand settings with some inventory management policy such as base stock policy. The correlation in demands of the products can be used to capture *cannibalization* effects or to make the problem more tractable for optimization by reducing the number of independent demand scenarios.

References

Allada, V. and Jiang, L., "New modules launch planning for evolving modular product families", *ASME Design Engineering Technology Conf.*, 2002, Paper No. DETC20020DFM-34190.

Allada, V., Choudhury, A., Pakala, P.K., Simpson, T.W., Scott, M.J. and Valliyappan, S., "Product Platform Problem Taxonomy: Classification and Identification of Benchmark Problems", *ASME Design Engineering Technical Conferences - Design Automation Conference*, Philadelphia, PA, 2006, September 10–13, Paper No. DETC2006/DAC-99569.

Aytug, H., Khouja, M. and Vergara, F.E., "Use of genetic algorithms to solve production and operations management problems: a review", *International Journal of Production Research*, 2003, 41 (17), pp. 3955–4009.

Ben Arieh, D., Easton, T., Choubey A., Solving the Multiple Platform Configuration Problem, *Int. J. Prod. Res.*, 1–21, iFirst, 2007.

Caffrey, R., Mitchell, G., Wahl, Z. and Zenick, R., "Product Platform Concepts Applied to Small Satellites: A New Multipurpose Radio Concept by AeroAstro Inc.", *16th Annual AIAA/USU Conf. on Small Satellites*, 2002.

Dahmus, J.B., Gonzalez–Zugasti, J.P., and Otto, K.N., "Modular product architecture", *Design Studies*, 2001, 22(5), pp. 409–424.

Delft CH. Van and Vial J.-Ph, Quantitative analysis of multi-periodic supply chain contracts with options via stochastic programming, Working Paper, Department of Management Studies, University of Geneva, 2001.

Fujita, K., Sakaguchi, H. and Akagi, S., "Product variety deployment and its optimization under modular architecture and module communalization," *ASME Design Engineering Technical Conferences*, 1999, Paper No. DETC1999/DFM-8923.

Fujita, K. and Yoshida, H., "Product Variety optimization: simultaneous optimization of module combination and module attributes," *ASME Design Engineering Technical Conferences*, 2001, Paper No. DETC2001/DAC-21058.

Goldberg, E. D., "*Genetic Approach in Search Optimization and Machine Learning*", Addison Wesley Publishing Company Inc., New York, 1989.

Hernandez, G., Allen, J. K., Mistree, F., "Platform design for customizable products as a problem of access in a geometric space," *Journal of Engineering Optimization*, 2003, 35 (3), pp. 229–254.

Hollins, B. and Pugh, S., "*Successful Product Design*", Butterworths, Boston, MA, 1990.

Jiao, J. and Zhang, Y., "Product portfolio planning with customer-engineering interaction", *IIE Transactions*, 2005, 37 (9), pp. 801–814.

Jose and Tollenaere, "Modular and platform methods for product family design: literature analysis", *Journal of Intelligent Manufacturing*, 2005, 16 (3), pp. 371–390.

Kota, S., Sethuraman, K. and Miller, R., "A Metric for Evaluating Design Commonality in Product Families", *ASME Journal of Mechanical Design*, 2000, 122 (4), pp. 403–410.

Krishnan, V. and Ulrich, K.T., "Product Development Decisions: A review of the literature," *Management Science*, 2001, 47 (1), pp. 1–21.

Lee, H. L., "Effective inventory and service management through product and process redesign", *J. Oper. Res.*, 1996, 44 (1), pp. 151–159.

Lee, H. L., and C. S. Tang, "Variability reduction through operations reversal" *Management Sci.*, 1998, 44 (2), pp. 162–172.

Li, H. and Azarm, S., "An approach for product line design selection under uncertainty and competition", *ASME Journal of Mechanical Design*, 2002, 124(3), pp. 385–392.

Maier J. R.A, and Fadel G. M., "Strategic Decisions in the early stages of product family design," *ASME Design Engineering Technical Conference*, 2001, September 9–12, Pittsburgh, PA, DETC2001/DFM-21200.

Martin, M. V. and Ishii, K., "Design for Variety: Development of Complexity Indices and Design Charts", *Advances in Design Automation, Sacramento, CA*, 1997, Paper No. DETC97/DFM-4359.

Martin, M.V., and Ishii, K., "Design for variety: Developing standardized and modularized product platform architectures", *Research in Engineering Design*, 2002, 13(4), pp. 213–235.

McDuffie, J. P., Sethuraman K., and Fisher M.L., "Product Variety and Manufacturing Performance: Evidence from the International Automotive Assembly Plant Study", *Management Science*, 1996, 42 (3), pp. 350–369.

Messac, A., Martinez, M. P., and Simpson, T. W., "Effective product family design using physical programming," *Engineering Optimization*, 2002a, 34 (3), pp. 245–261.

Messac, A., Martinez, M.P. and Simpson, T.W., "A penalty function for product family design using physical programming", *ASME Journal of Mechanical Design*, 2002b, 124(2), pp. 164–172.

Meyer, M. H. and Lehnerd, A. P., "*The Power of Product Platforms: Building Value and Cost Leadership*", The Free Press, New York, NY, 1997.

Mitchell, M., "An Introduction to Genetic Algorithms (Complex Adaptive Systems)", The MIT Press, Boston, 1998.

Naughton, K., Thornton, E., Kerwin, K. and Dawley, H., "Can Honda build a world car"? *Business Week*, 1997, 100(7).

Park, J. and Simpson, T. W., "Development of a Production Cost Estimation Framework to Support Product Family Design," *International Journal of Product Research*, 2005, 43 (4), pp. 731–772.

Pine, B. J., "*Mass Customization: The New Frontier in Business Competition*", Harvard Business School Press, Boston, MA, 1993.

Rai, R, and Allada, V., "Modular product families design: agent-based Pareto-optimization and quality loss function-based post-optimal analysis," *International Journal Production Research*, 2003, 41 (17), pp. 4075–4098.

Robertson, D., and Ulrich, K., "Planning product platforms", *Sloan Management Review*, 1998, 39(4), pp. 19–31.

Rothwell, R., and Gardiner, P., "Robustness and product design families", *Design Management: A Handbook of Issues and Methods*, 1990, Cambridge, MA: Basil Blackwell Inc, pp. 279–292.

Sabbagh, K., "*Twenty-First Century Jet: The Making and Marketing of the Boeing 777.*" New York: Scribner, 1996.

Sanderson, S. and Uzumeri, M., "Managing product families: The case of the Sony Walkman", *Research Policy*, 1995, 24 (5), pp. 761–782.

Shil, P., and Allada, V., "Evaluating new product platform development projects: A Game theoretic real option approach," *Industrial Engineering Research Conference*, Atlanta, GA, 2005.

Simpson, T. W. and D'Souza, B., "A Variable Length Genetic Algorithm for Product Platform Design", *9th AIAA/ISSMO Symposium on Multidisciplinary Analysis and Optimization*, 2002, Atlanta, GA, September 4–6, AIAA, AIAA-2002–5427.

Simpson, T. W. and D'Souza, B., "Assessing Variable Levels of Platform Commonality Within a Product Family Using a Multiobjective Genetic Algorithm", *Concurrent Engineering*, 2004, Vol. 12, No. 2, pp. 119–129.

Simpson, T. W., "Product Platform Design and Customization: Status and Promise", *Artificial Intelligence for Engineering Design, Analysis and Manufacturing*, 2004, 18 (1), pp. 3–20.

Simpson, T. W., Marion, T. J., de Weck, O., Holtta-Otto, K., Kokkolaras, M. and Shooter, S. B., "Platform-Based Design and Development: Current Trends and Needs in Industry", *ASME Design Engineering Technical Conferences - Design Automation Conference*, Philadelphia, PA, 2006, September 10–13, Paper No. DETC2006/DAC-99229.

Simpson, T. W., Seepersad, C. C. and Mistree, F., "Balancing Commonality and Performance within the Concurrent Design of Multiple Products in a Product Family", *Concurrent Engineering: Research and Applications*, 2001, 9 (3), pp. 177–190.

Sudjitanto, A. and Otto, K. N., "Modularization to Support Multiple Brand Platforms", *ASME Design Engineering Technical Conferences - Design Theory & Methodology Conference, Pittsburgh, PA*, 2001, Paper No. DETC2001/DTM-21695.

Swaminathan and Tayur, "Managing Broader Product Lines through Delayed Differentiation Using Vanilla Boxes", *Management Science*, 1998, 44 (12.2), pp. 161–172.

Ulrich, T.K. and Eppinger, S.D., *Product Design and Development*, 2nd edition, McGraw-Hill Inc., New York, 2003.

Ulrich, K., S. A. Pearson, "Does product design really determine 80% of manufacturing cost?", Working Paper. Sloan School, MIT, MA, 1993.

Wilhelm, B., "Platform and Modular Concepts at Volkswagen – Their Effect on the Assembly Process," *Transforming Automobile Assembly: Experience in Automation and Work Organization*, Springer, New York, 1997, pp. 146–156.

Wilson, L. O., and Norton, J. A., "Optimal entry timing for a product line extension," *Marketing Science*, 1989, 8(1), pp. 89–112.

Womak, J. P. , Jones, D. T., and Roos, D., T h e Machine That Changed the World, Rawson Associates, New York, 1990.

13

A Hybrid Particle Swarm and Ant Colony Optimizer for Multi-attribute Partnership Selection in Virtual Enterprises

S.H. Niu, S.K. Ong and A.Y.C. Nee

Mechanical Engineering Department, National University of Singapore, Singapore

Abstract
With an increasing level of global competition, enterprises would need to focus on their core strengths for survival. Virtual enterprise (VE) is an important form of enterprise alliance to cope with global competition and mass customization. In this form of alliance, the selection of the right partners is crucial. In this research, a mathematical model based on Particle Swarm Optimization (PSO) and Ant Colony Optimization (ACO) is presented, and five factors, namely, cost, time, quality, risk, and reputation are considered in this model. A hybrid PSO-ACO algorithm is proposed to obtain the optimal solution for partner selection for a VE. The fuzzy analytical hierarchy process is adopted to resolve the subjective and imprecise information, as well as the vague preferences in the partner selection process.

Keywords: Multi-attribute partnership selection, virtual enterprises, particle swarm optimization, ant colony optimization, fuzzy analytical hierarchy process

13.1 Introduction

In today's production environment with high product mix, fluctuating production batch size and emerging new technologies, responsiveness

Manoj Tiwari and Jenny A. Harding (eds.) Evolutionary Computing in Advanced Manufacturing, (289–326) © Scrivener Publishing LLC

to dynamic market changes in a cost-effective manner is becoming a key success factor for any manufacturing enterprise. The concept of virtual enterprises (VE) has been introduced to react quickly and effectively to competitive market demands. A VE is a temporary alliance of enterprises that come together sharing costs, resources and skills to enhance their competence and competitiveness in the fast changing global economy (Camarinha-Matos and Afsarmanesh 2003, Ip *et al.* 2003). Unlike traditional enterprises, a VE is a special organization of manufacturing units and enterprises that can make full use of its member enterprise's collective strengths, thus achieving product design and manufacturing efficiency. A VE is characterized by more frequent reconfiguration, and this distinguishes it from other long-term collaborations, such as supply chains or extended enterprises. This new concept of manufacturing entity is also becoming one of the most promising paradigms for the Small and Medium-sized Enterprises (SME) to better respond to business opportunities. It is important for SMEs to form VEs for the following reasons.

1. Faster response to job opportunities
   When a market opportunity arrives, the SMEs may not always have the capability to complete the tasks required. However, they have to respond swiftly in order to secure the jobs. If they try to expand their capacity by acquiring new factories and equipment, it will often be too late for them to close the deal. Hence, the most feasible approach is to find partners and leverage the partner's capabilities and capacities to respond to the new opportunities.
2. Meet the requirements of the increasing trend of mass customization
   Mass customization is a new production trend where products are made in smaller batches and with larger varieties. As a result, the enterprises, especially the SMEs, which usually do not have the capabilities to meet the demand for mass customization, would need to cooperate with suitable partners to achieve this. With increasing market pressure, more enterprises are collaborating with other factories to form a VE in order to manufacture products with greater varieties at a lower cost and in a shorter time.

The life cycle of a VE involves four phases, namely, creation, operation, evolution and dissolution (Liang *et al.* 2008). When a dominant enterprise cannot complete a task by itself, it will need to search for potential partners or coordinators and negotiate with them to form a VE. After contracts between them are signed, a VE is created. The newly formed VE will coordinate with its member enterprises to manufacture the products. Evolution may happen during the operation of a VE; hence, normally the operation phase and the evolution phase will be integrated as a dynamic operation phase. When the current task is completed and a new market opportunity is created, the VE can be reconfigured to meet the evolving resource requirements. When the mission of the VE is fulfilled, it will be dissolved eventually.

Before contracts between enterprises are signed, a key factor for the success of a VE is the partner selection issue for the business opportunity. In this context, the selection of the right partners is crucial. Partner selection is a multi-criteria decision making problem that involves assessing trade-offs between conflicting tangible and intangible criteria, and stating preferences based on incomplete or non-available information (Crispim and de Sousa 2010). In this selection process, the candidates should be evaluated based on qualitative, quantitative and stochastic aspects, the technical specifications, the reputation and credibility of the service provider, the cost of production and delivery, the time constraints, the availability and cost of logistics, service providers, precedence requirements, due date constraints, the nature of the discrete decisions, the failure risk of the coalition, etc. These factors should be combined to determine the partners to be selected. The criteria can be tuned according to the different aspects and the emphasis of the enterprises. Thus, the evaluation results of the same set of candidate enterprises can be quite different. However, the key factors addressed mainly in the literature are cost and time. Partner selection decisions should not only be based on least cost and due date or delivery time, other factors should be incorporated in the decision-making process. Among these factors, the qualitative factors and the stochastic events are not easily considered in a mathematical model and optimization method unless quantitative values for these factors are known.

In this research, five factors involved in partner selection are categorized and considered in the selection process. A hybrid PSO (Particle Swarm Optimization) and ACO (Ant Colony Optimization) algorithm is used for partner selection.

The rest of the paper is organized as follows. Section 13.2 presents a literature review on the latest research on partner selection issues in a VE. Section 13.3 presents the partner selection problem formulation and the factors involved in this process. Section 13.4 describes the hybrid PSO-ACO algorithm. Section 13.5 presents a case study using the hybrid PSO-ACO algorithm. Section 13.6 concludes the chapter.

13.2 Literature Review

In recent years, there are two partner selection approaches to form a VE. The first approach chooses the appropriate partners to form the VE, while the second approach embeds the partner selection issue in project scheduling (Brucker *et al.* 1999). The methodologies on partner selection can be grouped into four categories, namely, (1) exact algorithms which can solve the problem exactly, (2) heuristic and meta-heuristic algorithms, e.g., Genetic Algorithms (GA), Simulated Annealing (SA), ACO, PSO, Tabu Search (TA), etc., (3) fuzzy decision making algorithms, e.g., fuzzy analytical hierarchy process, multi-attribute utility theory, fuzzy set theory, etc., and (4) mathematical modeling and programming, e.g., goal programming, integer programming, and their enhanced types, etc. Although many researchers are continually developing new methods, the partner selection problem is still a challenge to VEs and this is an active area of academic research (Zhao *et al.* 2008).

Zhao *et al.* (2006a) proposed a Rule-based Genetic Algorithm, (R-GA) method to solve the partner selection problem considering the risk of failure, due date and the precedence of the sub-projects. Ip *et al.* (2004) modeled the partner selection problem using a 0–1 integer programming approach with non-analytical objective, and used the branch and bound (B&B) algorithm to find the solution. The results showed that the customized B&B algorithm can solve medium-sized problems optimally. However, their method cannot solve large-sized problems efficiently. Chen *et al.* (2007) presented qualitative search algorithms to select partners, aiming to reduce the redundant number of candidates in the early stages.

Among the aforementioned approaches to solve the partner selection problem, PSO and ACO are two newly emerged evolutionary algorithms that adopt the Artificial Intelligence (AI) related techniques to select the best partners automatically. In addition, these two algorithms can provide satisfactory solutions to large-sized problems.

PSO has been used across a wide range of applications. Poli *et al.* (2007) divided the applications into 26 different categories, among them, scheduling and combinational optimization problems are some of the important application areas. Partner selection is a typical combinational optimization problem. Wang *et al.* (2009) analyzed the different corporation types of partners, and considered the cost for collaboration in the optimization goal. A discrete PSO is proposed to solve the partner selection problem in a VE. This approach deals with discrete variables and each component of the particle takes an integer number (Bu *et al.* 2008). A multi-dimensional PSO was adopted by (Kiranyaz *et al.* 2008). In their approach, the particles can navigate through a range of dimensions instead of operating at a fixed dimension. Thus, the swarm particles can seek both positional and dimensional optima.

Several researchers have combined the use of the PSO evolutionary computation with other methods to enhance the PSO algorithm. SA has been employed to create a hybrid system with PSO since SA employs certain probability tools to avoid becoming trapped in a local optimal and the search process can be controlled by the SA cooling schedule. This new approach is used for the task allocation problem in a holonic manufacturing system (Yang *et al.* 2006, Zhao *et al.* 2006b), and the partner selection problem in a VE (Zhao *et al.* 2005, Yao *et al.* 2006). PSO combined with GA is proposed to model the scheduling holons in a holonic manufacturing system (Zhao *et al.* 2006c). PSO has been combined with K-means clustering for partner selection in supply chain management (Qi *et al.* 2008).

Zhao *et al.* (2008) presented the partner selection problem with precedence requirements and due date constraints. They analyzed the problem theoretically, transformed partner selection into the knapsack problem, and used the particle swarm algorithm (PSA) to find the solution. Their results demonstrated that the PSA is able to solve a large-sized partner selection problem. However, there are some common problems with the population-based algorithm. The PSA is prone to local optimal solutions, hence it is unable to perform rigorous search to obtain the global optimal solution when it is trapped in a local optimum. Two approaches can be adopted to solve this problem. One is to use an initial population with higher quality or larger variety. To maintain the initial diversity, a randomly generated initial population is usually adopted. However, using high quality initial solutions is a better approach compared with the randomly generated initial particles. Thus, solutions from other heuristics with higher quality can be used as the initial population. The other way is to use specially designed algorithms to guide the search process for values outside the local optimum.

ACO has a slow search speed in the earlier search stages due to the lack of the initial pheromone. With the availability of the pheromone information, the ACO search process can be sped up (Yao *et al.* 2008). Thus, in this research, a hybrid PSO-ACO optimizer is proposed, taking advantage of both algorithms to improve the search efficiency and precision.

13.3 Partner Selection Problem Formation

13.3.1 Fundamental Variables Discussion

The factors considered in the partner selection process in this research, namely, cost, time, quality, risk and reputation, are analyzed and evaluated according to their different characteristics, as shown in Figure 13.1.

Among the five categories of factors considered, cost, time and quality are the conflicting quantitative factors. When selecting a candidate, cost is largely based on the individual components, such as operational cost, delivery cost, material cost and logistics overhead. The time constraints involved include project duration, time sequence, transportation/delivery time, due date, processing time, etc. Quality can be reflected by the candidates' technological level, performance and service level, etc. Quantitative information can be expressed in numbers, percentages, performance indices or numerical scales. It can be modeled easily in the partner selection optimization process. However, risk and reputation are typical qualitative aspects which cannot be presented easily by unambiguous numbers or percentages. Risk is highly related to political stability, economy status of the region, financial health, market fluctuations, and competency of a company. Reputation is also a typical qualitative term; it is related to the candidate's past and present performance, financial status, the image of the enterprise and the cooperation history of the candidates and the dominant enterprise. The modeling and analysis of the qualitative terms is still a challenging research

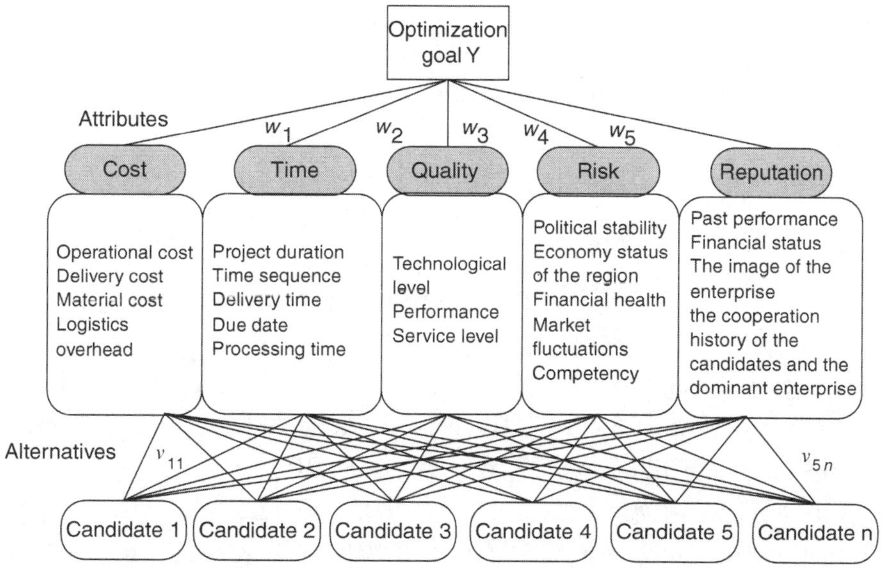

Figure 13.1 Hierarchy for the partner selection problem.

issue. These terms can be described in linguistic terms to facilitate the expression of the decision maker's preferences. After that, fuzzy numbers can be used to deal with the vagueness of the preferences. Through this process, the qualitative terms can be incorporated in the mathematical model of the partner selection process.

13.3.2 Partner Selection Problem Description

When a dominant company does not have the capability to complete a large project by itself, it would need to invite bidders to cooperate in order to complete the project. The entire project *P* can be divided into *n* sub-projects, $P=\{p_1, p_2, \ldots p_n\}$. For each sub-project p_i, there are Ca_i candidates bidding for the jobs, and the candidate set can be presented as $CA=\{ca_1, ca_2, \ldots . ca_n\}$.

In this research, five factors, namely, cost, quality, delivery time, reputation and reliability are considered to evaluate the performance of the candidates in the partner selection process. The core issue is to select the best set of *n* candidates for the sub-project p_i $(i=1,2,3\ldots n)$ minimizing the cost, risk and delivery time while maximizing the quality and reputation of the candidates. For the dominant enterprise, the selected best set of candidates should have the lowest cost and risk and the minimum delivery time while having the best quality and reputation. This is an NP-hard multi-objective optimization problem, with the goals:

$$\min\ CT = f(C_O, C_D, C_M, C_L), \min\ TE = f(T_{du}, T_{sq}, T_{de}, T_{ED}, T_{pt}),$$

$$\max\ QT = f(TL, PF, SV), \min\ RK = f(\widetilde{PS}, \widetilde{ES}, \widetilde{FH}, \widetilde{MF}, \widetilde{CP}),$$

$$\max\ RP = f(\widetilde{PP}, \widetilde{FS}, \widetilde{EI}, \widetilde{CH}).$$

Since $\max\ QT = f(TL, PF, SV) = \min\ (-QT) = -\min\ QT$, and $\max\ RP = f(\widetilde{PP}, \widetilde{FS}, \widetilde{EI}, \widetilde{CH}) = \min\ (-RP) = -\min\ RP$, thus the various objectives

$\{\min\ CT, \min\ TE, \max\ QT \min\ RK, \max\ RP\}$ can be rewritten as

{min *CT*, min *TE*, –min *QT*,min *RK*, –min *RP*}, i.e.,

min {CT, TE, –QT, RK, –RP}.

In this research, min {*CT, TE, –QT, RK,–RP*} can be synthesized by adding weights to each objective to simplify the problem, and the synthetic goal (the fitness function to be minimized later) can be derived as:

$$\begin{aligned}\min Y &= \omega_1 CT + \omega_2 TE - \omega_3 QT + \omega_4 RK - \omega_5 RP \\ &= \omega_1 f(C_O, C_D, C_M, C_L) \\ &\quad + \omega_2 f(T_{du}, T_{sq}, T_{de}, T_{ED}, T_{pt}) - \omega_3 f(TL, PF, SV) \\ &\quad + \omega_4 f(\widetilde{PS}, \widetilde{ES}, \widetilde{FH}, \widetilde{MF}, \widetilde{CP}) - \omega_5 f(\widetilde{PP}, \widetilde{FS}, \widetilde{EI}, \widetilde{CH})\end{aligned} \tag{13.1}$$

CT, TE, QT, RK and *RP* are formed based on the components and factors as described in Section 13.1. All these factors are discussed and evaluated in Section 13.5.

$$CT = \sum_{i=1}^{I}\sum_{j=1}^{J} ct_{ij} H_{ij} \tag{13.2}$$

$$TE = \sum_{i=1}^{I}\sum_{j=1}^{J} te_{ij} H_{ij} \tag{13.3}$$

$$QT = \sum_{i=1}^{I}\sum_{j=1}^{J} qt_{ij} H_{ij} \tag{13.4}$$

$$RK = \sum_{i=1}^{I}\sum_{j=1}^{J} rk_{ij} H_{ij} \tag{13.5}$$

$$RP = \sum_{i=1}^{I} \sum_{j=1}^{J} rp_{ij} H_{ij} \tag{13.6}$$

$$\sum_{j=1}^{J} H_{ij} = 1, \text{where} \quad H_{ij} = \begin{cases} 1 \text{ when the candidate is chosen} \\ 0 \text{ when the candidate is not chosen} \end{cases} \tag{13.7}$$

A constraint of the problem is $\omega_1 + \omega_2 + \omega_3 + \omega_4 + \omega_5 = 1$. The weight of each main criterion is obtained using the fuzzy comparison matrix, which will be described in Section 13.4.4.

For the time sequence and the duration of the sub-projects, the following constraints are implemented.

1. The actual start time st_i of each sub-project p_i should not be earlier than the planned start time ST of the entire project: $st_i \geq ST, \forall i = 1, 2, \ldots, p_n$.
2. The end time ed_i of each sub-project p_i should not be later than the planned end time ED of the entire project: $ed_i \leq ED, \forall i = 1, 2, \ldots, p_n$.
3. The relation between the end time ed_i, the start time st_i and the processing time pt_i is as follows: $ed_i = st_i + pt_i$.
4. The time sequence of the sub-projects should be satisfied: $ed_i \leq st_{i+1}$.

If $H_{ij} = 1$ and $st_i \geq CST_i, \forall i = 1, 2, \ldots, P_n$, $ed_i \leq CED_i, \forall i = 1, 2, \ldots, p_n$, the actual start time st_i and end time ed_i of each sub-project p_i should fall in the same time slot $[CST_i, CED_i]$ provided by the candidates.

13.4 Solution Methodology

13.4.1 Particle Swarm Optimization

PSO was introduced as a population-based stochastic search and optimization model in 1995. It is inspired by artificial intelligence and the natural behavior of animals, such as bird flocking, fish schooling, and the swarm theory in particular. It is also related to evolutionary computing, and has ties to both GAs and evolutionary programming (Kennedy and Eberhart 1995).

PSO conducts a search using a population of particles. Each particle represents a candidate solution of the problem. In PSO, a population of the particles is randomly generated and placed in the search space of the target problem, and each evaluates the objective function in its current location. Each particle is characterized by its position and velocity, and each particle determines its movement through the search space using the knowledge of its own current best locations (called local best *pbest*) and the best experience or position ever found by all the particles (called global best *gbest*). The next iteration takes place after all the particles have flown along their continuously adjusted trajectories. Eventually, the swarm as a whole, like a flock of birds foraging collectively for food, is likely to move closer to an optimum of the fitness function (Poli *et al.* 2007).

Each individual in the particle swarm is composed of three *D*-dimension vectors, where *D* is the dimensionality of the search space, and the three vectors are the current position $\vec{x}_i$, the previous best position, and the velocity $\vec{v}_i$. The *D*-dimension position for the ith particle in the tth iteration can be denoted as $x_i^t = \{x_{i1}^t, x_{i2}^t, \ldots, x_{iD}^t\}$. It can be considered as a set of coordinates describing a point in space. The current position of the particle is evaluated as a problem solution after the current iteration. If that position is better than any positions obtained in the search history, the coordinates will be stored in $\overrightarrow{pbest_i}$ (for "previous best"). The $\overrightarrow{pbest_i}$ of each particle is compared to obtain the global best position, and the best value of the entire population in the tth iteration is called $gbest^t$ (for "global best" during the tth generation). The objective is to continue to find better positions and update $pbest_i$ and $gbest_i$. New points of the particles are determined by adding $\vec{v}_i$ coordinates to, $\vec{x}_i$ i.e., $\overrightarrow{x_i^{t+1}} \longleftarrow \overrightarrow{x_i^t} + \overrightarrow{v_i^{t+1}}$. The algorithm operates by adjusting $\vec{v}_i$, where the velocity $\vec{v}_i$ is also a *D*-dimensional vector. The ith particle in the tth iteration can be denoted as $v_i^t = \{v_{i1}^t, v_{i2}^t, \ldots v_{iD}^t,\}$. In this research, the method proposed by Shi and Eberhart (1998) is used to update $\vec{v}_i$:

$$\overrightarrow{v_i^{t+1}} \longleftarrow \omega\overrightarrow{v_i^t} + c_1\vec{U}(0,\phi_1) \otimes (\overrightarrow{pbest_i^t} - \overrightarrow{x_i^t}) + c_2\vec{U}(0,\phi_2) \otimes (\overrightarrow{gbest^t} - \overrightarrow{x_i^t}) \tag{13.8}$$

In the initial stage, the velocities and positions of the particles are randomly generated in the range [0,1], and are updated during each iteration. The *D*-dimensional vector $x_i^t = \{x_{i1}^t, x_{i2}^t, \ldots, x_{iD}^t\}$ represents one set of partners selected for the VE in the tth iteration. The dimensionality number is equal to the number of sub-projects, and the elements $x_{i1}^t, x_{i2}^t, \ldots, x_{iD}^t \in [0,1]$ represent the choices for the projects $p_1, p_2, \ldots p_n$ respectively. The decoding procedures are as follows. For a project p_k, there are ca_k candidates. In the tth iteration, the corresponding element is x_{ik}^t. If $x_{ik}^t \in [0, 1/ca_k]$, the first candidate is selected for project p_k. The second candidate is selected if $x_{ik}^t \in (1/ca_k, 2/ca_k)$, and the Ca_k candidate is chosen when $x_{ik}^t \in (ca_k - 1/ca_k, 1)$.

13.4.2 Ant Colony Optimization

ACO is a search-based algorithm for solving NP-hard combinatorial optimization problems (Dorigo *et al.* 1996, Dorigo *et al.* 1999). The first form of ACO, namely the Ant System (AS) (Colorni *et al.* 1992) is based on the foraging behavior observed in a real ant colony. The cooperation of ants and how they find the shortest routes efficiently have been formulated into an algorithm used to solve combinatorial optimization problems. The ants in the colony move towards the local optimum and global optimum through the feedback of simulated pheromones and pheromone intensity. Different from PSO, where the position of the particle is evaluated as a problem solution, each path in ACO that is followed by an ant is associated with a candidate solution for a given problem. Each ant deposits its pheromone on the path it follows; the amount of pheromone deposited is proportional to the quality of the corresponding candidate solution for the problem. The ants are prone to choose the path(s) with a larger amount of pheromone. Thus, the ants will converge to a shorter path, which is expected to be the optimum or near-optimum solution for the target problem after several iterations.

During the search process, each ant finds a route according to the probability:

$$p_{ij}^t(k) = \frac{\left[\tau_{ij}^t\right]^a \left[\frac{1}{d_{ij}}\right]^\beta}{\sum\limits_{j \in allowed\_nodes} \left[\tau_{ij}^t\right]^a \left[\frac{1}{d_{ij}}\right]^\beta} \tag{13.9}$$

where $\tau_{ij}^{t+1} = \rho\tau_{ij}^t + \Delta\tau_{ij}^t$, $\Delta\tau_{ij}^t = \sum_{k=1}^{NO_p} \Delta\tau_{ij}^t(k)$, and

$$\Delta \tau_{ij}^{t}(k)=\begin{cases}\dfrac{Q}{Y^{t}} & \textit{if the } k^{th} \textit{ ant goes from } i \textit{ to } j\ d_{ij}^{t}=Y^{t}. \\ 0, \textit{otherwise} \end{cases}$$

$p_{ij}^{t}(k)$ is the probability for an ant to travel from node i to node j at t^{th} iteration; d_{ij} is the heuristic distance between nodes i and j; ρ is the evaporation coefficient, which is a real number between 0 and 1.0; $\tau_{ij}^{\ t}$ is the quantity of pheromone on the edge connecting nodes i and j at t^{th} iteration; $\Delta\tau_{ij}^{\ t}$ is the quantity of increased pheromone on the edge connecting nodes i and j at t^{th} iteration; $\Delta\tau_{ij}^{t}(k)$ is the quantity of increased pheromone on the edge connecting nodes i and j at t^{th} iteration by the ant k; Q is a constant representing the total quality of pheromone on a route; Y^t is the fitness function of the partner selection; α is the relative importance of the trial, $\alpha \geq 0$; β is the relative importance of the visibility, $\beta \geq 0$; ρ is the trail persistence, $0 \leq \rho \prec 1$.

13.4.3 Hybrid PSO-ACO

Both ACO and PSO optimization stem from AI techniques, and they share some common characteristics. PSO is problem-independent and more robust than many other search algorithms. However, PSO lacks global search ability at the end of a search. The convergence speed of ACO is slow when there is a lack of information in the initial searching stage, while PSO is prone to convergence in the early stage and can become trapped in the local optimum. With the availability of the pheromone information, the ACO will have a faster convergence rate (Yao *et al.* 2008). To speed up the search and improve the searching quality, ACO and PSO are hybridized in this research. The proposed hybrid algorithm adopts PSO to generate the initial available solutions, and these initial solutions provide the initial pheromone for the ACO. Thus, the PSO is employed to search for the optimal result before the search reaches early convergence and slows down, after which the ACO will take over the search using the results from the first stage as the initial solutions.

The initialization of the ants is important in the second stage of this hybrid approach. The number of ants NO_{ant} is equal to the number of particles NO_p in PSO. Each ant is located in a small area AR

around the global best position g_{best}. The area AR is determined by the Gaussian distribution with mean g_{best} and standard deviation σ. The initial position of each of the ants consists of two parts:

1. If the current personal best positions of the particles p_{best}^{t} fall within this small neighborhood area, namely, $p_{best}^{t} \in AR$, the position of the ant is set to be located at the personal best position p_{best}^{t} of these particles, and the ant chooses its next position PN_i^t according to the probability in Eq. (13.9).
2. Otherwise, when $p_{best}^{t} \notin AR$, the corresponding positions of the remaining ants will be determined by the vector $PN_i^t = N(p_{best}, \sigma)$.

After the initialization of the ants, the ACO will take over the search until the termination criterion is met. The relationships of the three types of positions are illustrated in Figure 13.2.

The hybrid approach can be viewed as a p^{best}-oriented ACO or PSO-reinforced ACO optimization. The PSO-ACO algorithm is described as follows and illustrated in Figure 13.3.

1. Initialize the swarm population, i.e., initialize the particles with random positions and velocities in the search space, and evaluate each particle's fitness;

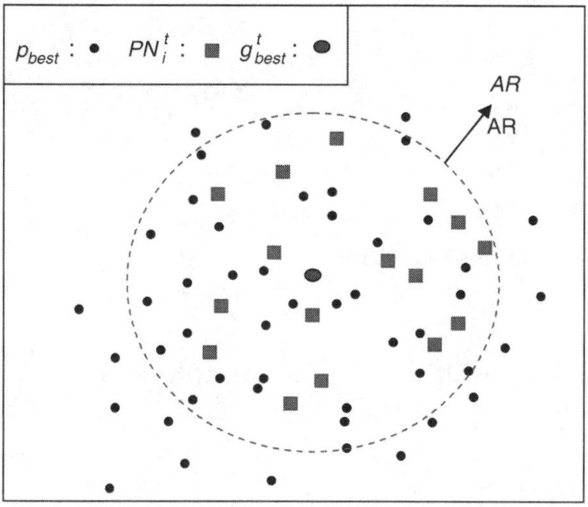

Figure 13.2 Initialization of ant position.

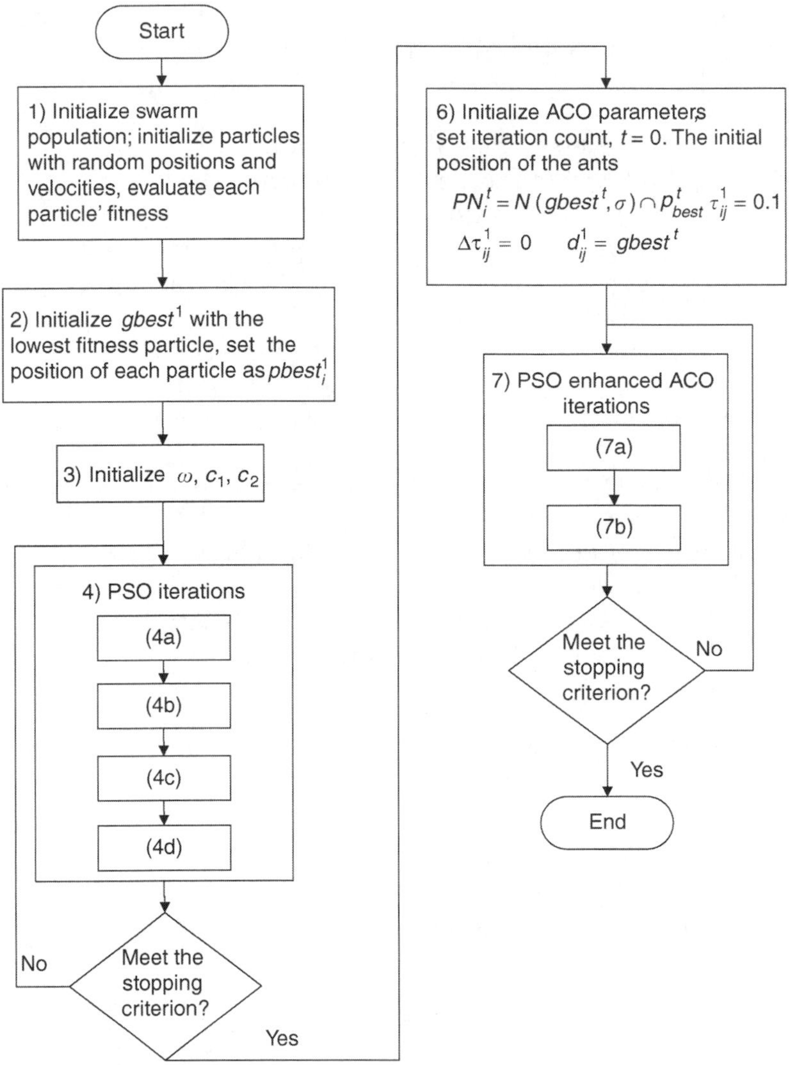

Figure 13.3 The PSO-ACO algorithm.

2. Initialize $gbest^1$ with the lowest fitness particle in the population, and set the position of each particle as $pbest_i^1$;
3. Initialize ω, c_1, c_2;
4. Loop1, while (the stopping criterion is not met)

   4a. For each particle, evaluate its fitness using Eq. (13.1).

4b. Compare the fitness of the particle with its $pbest_i^t$. If the current value is better than $pbest_i^t$, set the current value as $pbest_i^t$ and $\overrightarrow{p_i}$ equal to the current location $\overrightarrow{x_i}$ in the D-dimensional space.

4c. Identify the particle in the neighborhood with the best fitness so far, and assign its index to the variable g, i.e., set the best $pbest_i^t$ to $gbest^t$.

4d. Change the velocity $\overrightarrow{v_i^{t+1}}$ and position $\overrightarrow{x_i^{t+1}}$ of the particle according to Eq. (13.8),

$$\text{if } \overrightarrow{v_i^{t+1}} \succ V_{max}, \text{ then } \overrightarrow{v_i^{t+1}} = V_{max};$$
$$\text{if } \overrightarrow{v_i^{t+1}} \prec V_{min}, \text{ then } \overrightarrow{v_i^{t+1}} = V_{min};$$
$$\overrightarrow{x_i^{t+1}} \longleftarrow \overrightarrow{x_i^t} + \overrightarrow{v_i^{t+1}}$$

5. If the criterion (the search speed slows down or the particles are trapped in local optima) is met, exit Loop 1.
6. Initialize ACO parameters, set iteration count $t=0$, the initial position of the ants

$$PN_i^t = N(gbest^t, \sigma) \cap p_{best}^t, \ \tau_{ij}^1 = 0.1 \ \ \Delta\tau_{ij}^1 = 0, \ d_{ij}^1 = gbest^t$$

7. Loop 2:

   7a. Each ant finds a route according to Eq. (13.9).

   7b. Find the minimum d_{ij}^t and update.

8. If $(Y_i - Y_{i-1})/Y_i \le e^{-10}$, exit loop.
9. End Loop 2.

13.4.4 Weights of the Criteria and the Qualitative Variables

In this research, different criteria are considered and analyzed through studying their corresponding sub-criteria. Two issues are addressed. The first issue is the relative importance of the criteria and the sub-criteria, and the second is to describe the candidate enterprises numerically in the mathematical model by considering the qualitative aspects. The assessment of the relative importance of the criteria and the numerical description of the qualitative aspects are typically subjective and imprecise. In practice, normally a group of experts is formed by the dominant enterprise to handle those issues using vague linguistic representations. Fuzzy set theory is used to transform the linguistic representations into numeric values. Fuzzy set theory resembles human reasoning in using approximate information and uncertainty to generate decisions, and it can represent uncertainty and vagueness mathematically to provide a formalized tool for dealing with imprecision intrinsic to many problems (Chan *et al.* 2008).

13.4.4.1 *Representation of the Linguistic Terms Using Positive Triangular Fuzzy Numbers*

The triangular fuzzy number technique is used in this research to deal with the linguistic terms, the vagueness in the criteria comparison and the evaluation of the qualitative assessments of the enterprises The triangular fuzzy number technique is the most widely adopted technique to represent fuzzy numbers as it is easier to understand and use, and can be easily mapped to the real situations to deal with the ambiguity in a system.

A linguistic term can be expressed by the positive triangular fuzzy number M_{gi}^{j}, where $M=(l, m, u)$. l, m and u. are the parameters and $l \leq m \leq u$. When $l=m=u$, it is a non-fuzzy number. l, m and u are the least possible value, the most possible value and the largest possible value respectively (Buyukozkan 2004, Chan *et al.* 2008, Percin 2008). The membership function of M can be expressed as:

$$\mu_M(x) = \begin{cases} (x-l)/(m-l) & l \leq x \leq m \\ (x-u)/(m-u) & m \leq x \leq u \\ 0 & x \prec l \ or \ x \succ u \end{cases} \quad \text{where } \mu_M(x): \longrightarrow [0,1]$$

The triangular fuzzy number is used to evaluate the qualitative objectives, namely, risk and reputation in this research. In addition, it is also used to obtain the weights of the main criteria and the weights of the sub-criteria for risk and reputation. To evaluate the relative importance of the criteria with respect to each other, the fuzzy number in Table 13.1 (Chan *et al.* 2008) is adopted to represent numerically the linguistics 'absolutely important', 'very important', 'fairly important', 'equal', 'almost equal', 'fairly unimportant', 'very important' and 'absolutely unimportant'. The criteria are compared with each other to get the relative importance, the triangular representation are shown in Figure 13.4.

Table 13.1 Importance/preference of one criterion over another (Chan *et al.* 2008).

Absolutely Important	Very Important	Fairly Important	Equal	Almost Equal	Fairly Unim-portant	Very Unim-portant	Absolutely Unimportant
(3.5, 4.0, 4.5)	(2.5, 3.0, 3.5)	(1.5, 2.0, 2.5)	(1.0, 1.0, 1.0)	(0.67, 1.0, 1.5)	(0.4, 0.5, 0.67)	(0.28, 0.33, 0.4)	(0.22, 0.25, 0.28)

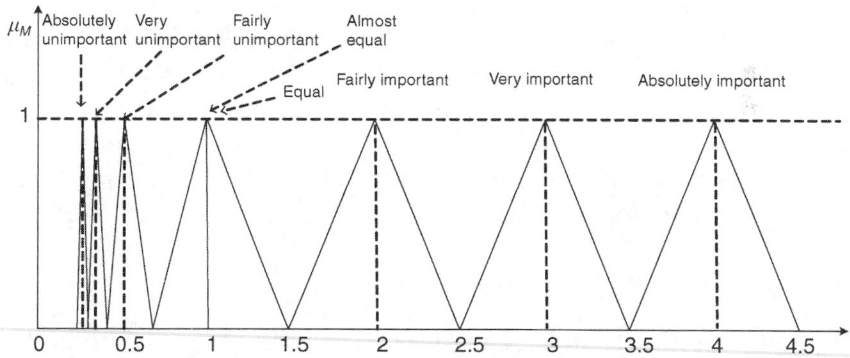

Figure 13.4 Triangular fuzzy number representation of the importance of one criterion over another.

Risk and reputation are the two important objectives for the evaluation of the candidates. However, they are difficult to represent quantitatively. The risk of choosing one candidate can be analyzed from five aspects, namely, the political stability $\widetilde{PS}$, the economy status of the region $\widetilde{ES}$, the financial health $\widetilde{FH}$, the market fluctuations $\widetilde{MF}$ and the competency $\widetilde{MF}$. These aspects form the five sub-criteria to assess the risk factor of a candidate enterprise. Similarly, the reputation of a candidate enterprise can be evaluated from four aspects, namely, the past performance $\widetilde{PP}$, the financial status $\widetilde{FS}$, the image $\widetilde{EI}$ and the cooperation history with the dominant enterprise $\widetilde{CH}$.

The various aspects to evaluate the risk of cooperation and the reputation of an enterprise can be described using linguistic variables, which can be interpreted using fuzzy sets. In linguistics, fundamental atomic terms are often modified with adjectives (nouns) or adverbs (verbs) like *very, low, slight, more-or-less, fairly, slightly, almost, barely, mostly, roughly,* etc. These modifiers can be called 'linguistic hedges', and the singular meaning of an atomic term can be modified or hedged from its original interpretation. Thus, the operations of the modifiers to represent different meanings are defined as follow.

$\Theta = (6.5,7,7.5)$, [Weakly very Θ] = $\Theta + 1$, [very Θ]= $\Theta + 2$, [Absolutely very Θ]= $\Theta + 3$, [Weakly Θ]= $\Theta - 1$, [Weakly not Θ] = $\Phi + 1 = \Theta - 2$, [*not* Θ]= Φ, [Weakly not very Θ] = $\Phi - 1$, [Very *not* Θ]= $\Phi - 2$, [Absolutely *not* very Θ]= $\Phi - 3$.

For example, $\Theta = good\ x, \Phi = bad\ x$. $\Theta(x)$ can represent good political stability, good economy status of the region, good financial health, low market fluctuations and good competency of the candidate enterprise. It can also represent good past performance, good financial status, good enterprise image and satisfactory cooperation history of the candidates and the dominant enterprise. $\Phi(x)$ represents the bad situations of all the listed terms. Thus, the triangular fuzzy conversion scale in Table 13.2 and Figure 13.5 can be obtained.

13.4.4.2 *Synthetic Evaluation and Defuzzification*

To determine the relative importance (weights) of the criteria, the various criteria have to be compared. Since many qualitative aspects can affect the outcome of the weights, designing a crisp mathematical model to include all these aspects can be complicated. One consideration is that it is easier to conduct subjective pairwise comparison of these criteria rather than all at the same time. The matrix containing all the pairwise relative values is the C matrix. The C matrix (a fuzzy matrix in this research) is used to represent the relative importance (weights) of the criteria. The C matrix can be evaluated synthetically. The term synthetic is used to connote the process

Table 13.2 Triangular fuzzy conversion scale.

Linguistic Scale	**Absolutely Worse**	**Worse**	**Weakly Worse**	**Bad**	**Weakly Bad**
PTFN scale	(0.5, 1, 1.5)	(1.5, 2, 2.5)	(2.5, 3, 3.5)	(3.5, 4, 4.5)	(4.5, 5, 5.5)
Linguistic scale	Weakly good	Good	Weakly better	Better	Absolutely better
PTFN scale	(5.5, 6, 6.5)	(6.5, 7, 7.5)	(7.5, 8, 8.5)	(8.5, 9, 9.5)	(9.5, 10, 10)

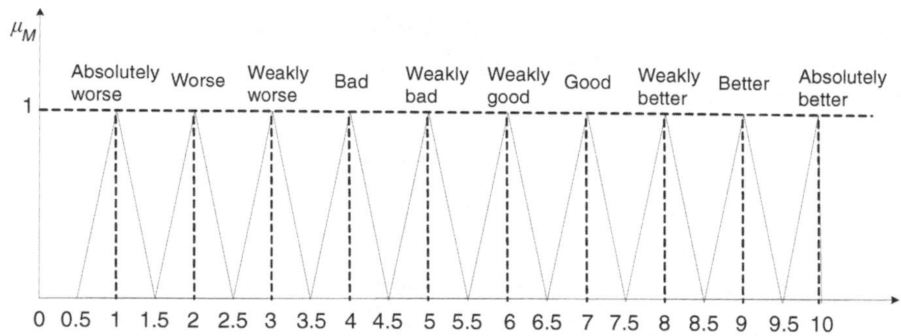

Figure 13.5 The linguistic scale of the triangular numbers.

of evaluation whereby several individual elements and components of an evaluation are synthesized into an aggregate form, i.e., the whole is a synthesis of the parts (Timothy 2010).

Fuzzy transformation laws for the two positive triangular fuzzy numbers M_1 and M_2 are defined as follows:

$$M_1 + M_2 = (l_1 + l_2, m_1 + m_2, u_1 + u_2)\, M_1 \otimes M_2 \approx (l_1 l_2, m_1 m_2, u_1 u_2),$$
$$\lambda M_1 = (\lambda l_1, \lambda m_1, \lambda u_1),\ M_1^{-1} \approx (1/u_1, 1/m_1, 1/l_1)$$

In this research, the m extent analysis values for each criterion or sub-criterion are denoted as: $N^{i,1}, N^{i,2}, \ldots N^{i,m}$, where $i = 1, 2, \ldots n$. For the ith criterion or sub-criterion, the value of the fuzzy synthetic extent can be derived based on the method proposed by Chan *et al.* (2008), given in Eq. (13.10). $F^i = (f_1^i, f_2^i, f_3^i)$ is a fuzzy number. $N^{i,j}$ represents the value of the evaluation of the ith object for the jth goal.

$$F^i = \sum_{j=1}^{m} N^{i,j} \otimes \left[\sum_{i=1}^{n} \sum_{j=1}^{m} N^{i,j} \right]^{-1}$$
$$= \left(\sum_{j=1}^{m} n_1^{i,j}, \sum_{j=1}^{m} n_2^{i,j}, \sum_{j=1}^{m} n_3^{i,j} \right) \otimes \left(\frac{1}{\sum_{i=1}^{n}\sum_{j=1}^{m} n_3^{i,j}}, \frac{1}{\sum_{i=1}^{n}\sum_{j=1}^{m} n_2^{i,j}}, \frac{1}{\sum_{i=1}^{n}\sum_{j=1}^{m} n_1^{i,j}} \right) \quad (13.10)$$

The fuzzy value F^i obtained can be transformed into a non-fuzzy value using the centroid method (Yager 1980). The centroid method (also called center of area, center of gravity) is the most prevalent and physically appealing of all the defuzzification methods. It is represented as

$$x^{*} = \frac{\int \mu_M(x) \bullet x dx}{\int \mu_M(x) dx}$$

(Timothy 2010). For the fuzzy number F^i, the numeric value after defuzzification is given as:

$$\overline{F^i} = (f_1^i + f_2^i + f_3^i)/3 \tag{13.11}$$

The non-fuzzy weights of the criteria are $W = (\overline{F^1}, \overline{F^2}, \overline{F^3} ... \overline{F^m})^T$. To normalize them,

$$NW = (\overline{F^1}, \overline{F^2}, \overline{F^3} ... \overline{F^m})^T \Big/ \sum_{j=1}^{m} \overline{F^j} \quad j = 1, 2, 3, ...m \tag{13.12}$$

After normalization, the elements in NW represent the weights of the criteria. In this research, the weights of the main criteria cost, time, quality, risk and reputation are determined using this method. The weights of the sub-criteria in risk and reputation are also determined using this method. This method is not applied to determine the weights for the quantitative criteria.

13.5 Experimental Analysis

Experiments have been conducted to demonstrate the efficiency and performance of the hybrid PSO-ACO algorithm. In these experiments, a dominant enterprise has one large project p consisting of seven sub-projects $\{p_1, p_2, p_3, p_4, p_5, p_6, p_7\} \in P$, and it invites bids for these seven sub-projects. The sub-projects $p_1, p_2, p_3, p_4, p_5, p_6,$ p_7 have time and sequence constraints and different operation times (durations), which are shown in Figure 13.6. The start-time of each sub-project should meet the time constraints. The planned start-time and end-time of the sub-projects are: p_1 [20, 40], p_2 [40, 55], p_3 [55, 75], p_4 [75, 100], p_5 [100, 120], p_6 [120, 135] and p_7 [135, 145].

The quantitative information, such as the cost, quality and time, are expressed as numbers, percentages, performance indices or numerical scales, and are modeled in the partner selection optimization

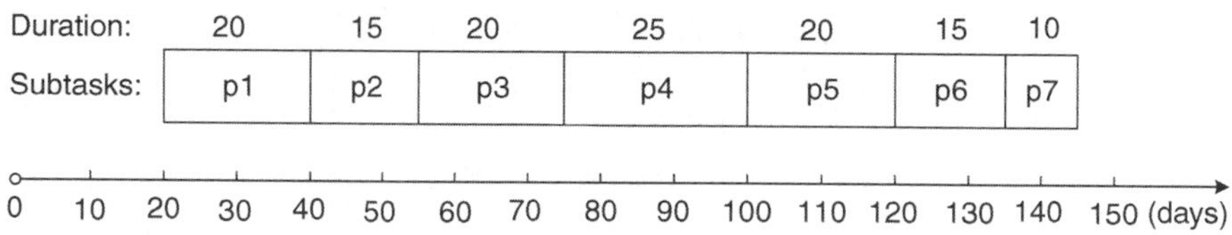

Figure 13.6 The time sequence and duration of each sub-project.

process. The qualitative aspects are described in linguistic terms as discussed in the earlier sections. After that, the linguistic terms are transformed into fuzzy numbers which can finally be transformed to crisp numbers using the centroid method.

13.5.1 Determine the Weights of the Main Criteria and Sub-Criteria

This section describes the method to obtain the weights of the main criteria, and the weights of the sub-criteria of two qualitative criteria, namely, risk and reputation.

1. Weights of the main criteria

To determine the relative importance (weights) of the five main criteria, these criteria, namely, cost, time, quality, risk and reputation, are compared in pairs to form a C matrix. Table 13.2 gives the conversion scale for the importance/preference of one criterion over another using the triangular fuzzy numbers. The fuzzy comparison matrix in Table 13.3 for the comparison of the criteria is created based on Table 13.2. The different values of the fuzzy synthetic extent with respect to the five criteria are denoted by F^1, F^2, F^3, F^4 and F^5, and they are determined using the approach described in Section 13.4.4.

According to Eq. (13.10),

$$F^1 = (10,12,14) \otimes (40.33,33.74,27.9)^{-1} = (0.25,0.36,0.50)$$

$$F^2 = (7.78,9.33,10.9) \otimes (40.33,33.74,27.9)^{-1} = (0.19,0.28,0.39)$$

$$F^3 = (3.97,5.0,6.34) \otimes (40.33,33.74,27.9)^{-1} = (0.10,0.15,0.23)$$

$$F^4 = (3.85,4.83,6.07) \otimes (40.33,33.74,27.9)^{-1} = (0.10,0.14,0.22)$$

Table 13.3 The fuzzy comparison matrix of the main criteria with respect to each other.

Objective	O_1 Cost	O_2 Time	O_3 Quality	O_4 Risk	O_5 Reputation	Weight
O_1	(1.0, 1.0, 1.0)	(2.5, 3.0, 3.5)	(1.5, 2.0, 2.5)	(2.5, 3.0, 3.5)	(2.5, 3.0,3.5)	0.35
O_2	(0.28, 0.33, 0.4)	(1.0, 1.0, 1.0)	(1.5, 2.0, 2.5)	(1.5, 2.0, 2.5)	(3.5, 4.0, 4.5)	0.28
O_3	(0.4, 0.5, 0.67)	(0.4, 0.5, 0.67)	(1.0, 1.0, 1.0)	(0.67, 1.0, 1.5)	(1.5, 2.0, 2.5)	0.15
O_4	(0.28, 0.33, 0.4)	(0.4, 0.5, 0.67)	(0.67, 1, 1.5)	(1.0, 1.0, 1.0)	(1.5, 2.0, 2.5)	0.14
O_5	(0.28 ,0.33, 0.4)	(0.22, 0.25, 0.28)	(0.4, 0.5, 0.67)	(0.4, 0.5, 0.67)	(1.0, 1.0, 1.0)	0.08

$$F^5 = (2.3, 2.58, 3.02) \otimes (40.33, 33.74, 27.9)^{-1} = (0.06, 0.08, 0.11)$$

Using Eq. (13.11), the weights are $\overline{F^1} = 0.37$, $\overline{F^2} = 0.29$, $\overline{F^3} = 0.16$, $\overline{F^4} = 0.15$, $\overline{F^5} = 0.08$. After normalization using Eq. (13.12), the final weight vector used in the hybrid PSO-ACO is

$$NW = \frac{(0.37,\ 0.29,\ 0.16, 0.15, 0.08)^T}{(0.37 + 0.29 + 0.16 + 0.15 + 0.08)} = (0.35, 0.28, 0.15, 0.14, 0.08)^T,$$

where $w_1 = 0.35, w_2 = 0.28, w_3 = 0.15, w_4 = 0.14, w_5 = 0.08$.

2. Weights of the sub-criteria of risk

The risk criterion consists of several sub-criteria. The C matrix of the sub-criteria with respect to each other is shown in Table 13.4.

According to Eq. (13.10),

$$F_{RK}^1 = (5.56, 6.66, 7.80) \otimes (40.33, 33.74, 27.9)^{-1} = (0.14, 0.20, 0.28)$$

$$F_{RK}^2 = (8.90, 10.5, 12.17) \otimes (40.33, 33.74, 27.9)^{-1} = (0.22, 0.31, 0.44)$$

$$F_{RK}^3 = (5.07, 6.50, 8.17) \otimes (40.33, 33.74, 27.9)^{-1} = (0.13, 0.19, 0.29)$$

$$F_{RK}^4 = (2.75, 3.33, 4.24) \otimes (40.33, 33.74, 27.9)^{-1} = (0.07, 0.10, 0.15)$$

$$F_{RK}^5 = (5.62, 6.75, 7.95) \otimes (40.33, 33.74, 27.9)^{-1} = (0.14, 0.20, 0.28)$$

Using Eq. (13.11), the weights are $\overline{F_{RK}^1} = 0.21$, $\overline{F_{RK}^2} = 0.32$, $\overline{F_{RK}^3} = 0.20$, $\overline{F_{RK}^4} = 0.11$, $\overline{F_{RK}^5} = 0.21$. After normalization using Eq. (13.12), the final weight vector to evaluate the risk of selecting an enterprise is

$$NW = \frac{(0.21, 0.32, 0.20, 0.11, 0.21)^T}{(0.21 + 0.32 + 0.20 + 0.11 + 0.21)} = (0.20, 0.30, 0.19, 0.11, 0.20)^T,$$

where $w_1^{RK} = 0.20, w_2^{RK} = 0.30, w_3^{RK} = 0.19, w_4^{RK} = 0.11, w_5^{RK} = 0.20$

Table 13.4 The C matrix of the sub-criteria of risk with respect to each other.

Objective	$\widetilde{PS}$	$\widetilde{ES}$	$\widetilde{FH}$	$\widetilde{MF}$	$\widetilde{CP}$	Weight
$\widetilde{PS}$	(1.0, 1.0, 1.0)	(0.28, 0.33, 0.4)	(1.5, 2.0, 2.5)	(2.5, 3.0, 3.5)	(0.28, 0.33, 0.4)	0.20
$\widetilde{ES}$	(2.5, 3.0, 3.5)	(1.0, 1.0, 1.0)	(0.4, 0.5, 0.67)	(1.5, 2.0, 2.5)	(3.5, 4.0, 4.5)	0.30
$\widetilde{FH}$	(0.4 ,0.5, 0.67)	(1.5, 2.0, 2.5)	(1.0, 1.0, 1.0)	(0.67, 1.0, 1.5)	(1.5, 2.0, 2.5)	0.19
$\widetilde{MF}$	(0.28, 0.33, 0.4)	(0.4, 0.5, 0.67)	(0.67, 1, 1.5)	(1.0, 1.0, 1.0)	(0.4, 0.5, 0.67)	0.11
$\widetilde{CP}$	(2.5, 3.0, 3.5)	(0.22, 0.25, 0.28)	(0.4, 0.5, 0.67)	(1.5, 2.0, 2.5)	(1.0, 1.0, 1.0)	0.20

3. Weights of the sub-criteria of reputation

The reputation of one enterprise can be evaluated based on four sub-criteria. The C matrix of the sub-criteria with respect to each other is shown in Table 13.5.

According to Eq. (13.10),

$$F_{RP}^{1} = (7.5,9,10.5) \otimes (26.82,23.24,19.86)^{-1} = (0.28,0.39,0.53)$$

$$F_{RP}^{2} = (5.28,6.33,7.4) \otimes (26.82,23.24,19.86)^{-1} = (0.20,0.27,0.37)$$

$$F_{RP}^{3} = (5.3,6.0,6.84) \otimes (26.82,23.24,19.86)^{-1} = (0.20,0.26,0.34)$$

$$F_{RP}^{4} = (1.78,1.91,2.08) \otimes (26.82,23.24,19.86)^{-1} = (0.07,0.08,0.10)$$

According to Eq. (13.11), the weights are given as $\overline{F_{RP}^{1}} = 0.40$, $F_{RP}^{2} = 0.28$, $F_{RP}^{3} = 0.27$, $F_{RP}^{4} = 0.08$. After normalization, the final weight vector used to evaluate the reputation of one candidate enterprise is,

$$NW = \frac{(0.40, 0.28, 0.27, 0.08)^{T}}{(0.40 + 0.28 + 0.27 + 0.08)} = (0.39, 0.27, 0.26, 0.08)^{T}$$

Table 13.5 The C matrix of the sub-criteria of reputation with respect to each other.

Objective	$\widetilde{PP}$	$\widetilde{FS}$	$\widetilde{EI}$	$\widetilde{CH}$	Weight
$\widetilde{PP}$	(1.0, 1.0, 1.0)	(2.5, 3.0, 3.5)	(1.5, 2.0, 2.5)	(2.5, 3.0, 3.5)	0.39
$\widetilde{FS}$	(0.28, 0.33, 0.4)	(1.0, 1.0, 1.0)	(1.5, 2.0, 2.5)	(2.5, 3.0, 3.5)	0.27
$\widetilde{EI}$	(0.4, 0.5, 0.67)	(0.4, 0.5, 0.67)	(1.0, 1.0, 1.0)	(3.5, 4.0, 4.5)	0.26
$\widetilde{CH}$	(0.28, 0.33, 0.4)	(0.28, 0.33, 0.4)	(0.22, 0.25, 0.28)	(1.0, 1.0, 1.0)	0.08

where $w_1^{RP} = 0.39, w_2^{RP} = 0.27, w_3^{RP} = 0.26, w_4^{RP} = 0.08.$

13.5.2 Evaluation of Qualitative Attributes

Risk and reputation cannot be easily represented as crisp values. However, they can be described easily using linguistic terms, which can be expressed using fuzzy numbers. The risk of choosing one enterprise and its reputation can be evaluated based on the 'rules' in Table 13.2 and Figure 13.5. Table 13.6 shows the evaluation results of risk and reputation of each enterprise with respect to the sub-criteria.

From Section 13.5.1, the weights for risk and reputation sub-criteria are:

$$w_1^{RK} = 0.20, w_2^{RK} = 0.30, w_3^{RK} = 0.19, w_4^{RK} = 0.11, w_5^{RK} = 0.20$$

$$w_1^{RP} = 0.39, w_2^{RP} = 0.27, w_3^{RP} = 0.26, w_4^{RP} = 0.08$$

The weighted risk and reputation are determined as

$$\begin{aligned}\widetilde{RK} &= w_1^{RK} * \widetilde{PS} + w_2^{RK}\widetilde{ES} + w_3^{RK}\widetilde{FH} + w_4^{RK}\widetilde{MF} + w_5^{RK}\widetilde{CP} \\ &= (RK_l, RK_m, RK_u)\end{aligned}$$

$$\widetilde{RP} = w_1^{RP} * \widetilde{PP} + w_2^{RP}\widetilde{FS} + w_3^{RP}\widetilde{EI} + w_4^{RP}\widetilde{CH} = (RP_l, RP_m, RP_u)$$

Next, $\widetilde{RK}$ and $\widetilde{RP}$ canbedefuzzifiedas $RK = (RK_l + RK_m + RK_u)/3$ and $RP = (RP_l + RP_m + RP_u)/3$.

Table 13.6 The fuzzy evaluation of $\widetilde{RK}$ and $\widetilde{RP}$ for the candidates.

Candidates	$\widetilde{PS}$	$\widetilde{ES}$	$\widetilde{FH}$	$\widetilde{MF}$	$\widetilde{CP}$	$\widetilde{PP}$	$\widetilde{FS}$	$\widetilde{EI}$	$\widetilde{CH}$
P11	(2.5,3,3.5)	(1.5,2,2.5)	(0.5,1,1.5)	(1.5,2,2.5)	(4.5,5,5.5)	(4.5,5,5,5)	(2.5,3,3.5)	(3.5,4,4.5)	(4.5,5,5.5)
P12	(0.5,1,1.5)	(1.5,2,2.5)	(3.5,4,4.5)	(2.5,3,3.5)	(1.5,2,2.5)	(7.5,8,8.5)	(6.5,7,7.5)	(6.5,7,7.5)	(7.5,8,8.5)
P13	(2.5,3,3.5)	(3.5,4,4.5)	(6.5,7,7.5)	(4.5,5,5.5)	(7.5,8,8.5)	(7.5,8,8.5)	(8.5,9,9.5)	(7.5,8,8.5)	(2.5,3,3.5)
P21	(0.5,1,1.5)	(7.5,8,8.5)	(6.5,7,7.5)	(3.5,4,4.5)	(5.5,6,6.5)	(8.5,9,9.5)	(9.5,10,10)	(8.5,9,9.5)	(7.5,8,8.5)
P22	(5.5,6,6.5)	(7.5,8,8.5)	(9.5,10,10)	(8.5,9,9.5)	(8.5,9,9.5)	(7.5,8,8.5)	(8.5,9,9.5)	(9.5,10,10)	(2.5,3,3.5)
P23	(2.5,3,3.5)	(1.5,2,2.5)	(3.5,4,4.5)	(4.5,5,5.5)	(1.5,2,2.5)	(0.5,1,1.5)	(3.5,4,4.5)	(5.5,6,6.5)	(7.5,8,8.5)
P31	(2.5,3,3.5)	(1.5,2,2.5)	(1.5,2,2.5)	(4.5,5,5.5)	(1.5,2,2.5)	(4.5,5,5.5)	(7.5,8,8.5)	(5.5,6,6.5)	(6.5,7,7.5)
P32	(0.5,1,1.5)	(3.5,4,4.5)	(2.5,3,3.5)	(1.5,2,2.5)	(0.5,1,1.5)	(6.5,7,7.5)	(7.5,8,8.5)	(7.5,8,8.5)	(3.5,4,4.5)
P33	(3.5,4,4.5)	(7.5,8,8.5)	(2.5,3,3.5)	(4.5,5,5.5)	(5.5,6,6.5)	(0.5,1,1.5)	(1.5,2,2.5)	(2.5,3,3.5)	(1.5,2,2.5)
P34	(0.5,1,1.5)	(1.5,2,2.5)	(2.5,3,3.5)	(1.5,2,2.5)	(2.5,3,3.5)	(8.5,9,9.5)	(9.5,10,10)	(8.5,9,9.5)	(5.5,6,6.5)
P35	(5.5,6,6.5)	(6.5,7,7.5)	(7.5,8,8.5)	(9.5,10,10)	(3.5,4,4.5)	(4.5,5,5.5)	(5.5,6,6.5)	(6.5,7,7.5)	(3.5,4,4.5)
P41	(2.5,3,3.5)	(3.5,4,4.5)	(1.5,2,2.5)	(4.5,5,5.5)	(1.5,2,2.5)	(6.5,7,7.5)	(7.5,8,8.5)	(2.5,3,3.5)	(7.5,8,8.5)
P42	(7.5,8,8.5)	(4.5,5,5.5)	(5.5,6,6.5)	(1.5,2,2.5)	(0.5,1,1.5)	(1.5,2,2.5)	(2.5,3,3.5)	(1.5,2,2.5)	(1.5,2,2.5)

(Continued)

Table 13.6 The fuzzy evaluation of $\widetilde{RK}$ and $\widetilde{RP}$ for the candidates.(Continued)

Candidates	$\widetilde{PS}$	$\widetilde{ES}$	$\widetilde{FH}$	$\widetilde{MF}$	$\widetilde{CP}$	$\widetilde{PP}$	$\widetilde{FS}$	$\widetilde{EI}$	$\widetilde{CH}$
P43	(6.5,7,7.5)	(6.5,7,7.5)	(2.5,3,3.5)	(5.5,6,6.5)	(1.5,2,2.5)	(3.5,4,4.5)	(5.5,6,6.5)	(7.5,8,8.5)	(2.5,3,3.5)
P51	(8.5,9,9.5)	(5.5,6,6.5)	(3.5,4,4.5)	(6.5,7,7.5)	(8.5,9,9.5)	(2.5,3,3.5)	(3.5,4,4.5)	(5.5,6,6.5)	(0.5,1,1.5)
P52	(0.5,1,1.5)	(2.5,3,3.5)	(0.5,1,1.5)	(1.5,2,2.5)	(1.5,2,2.5)	(0.5,1,1.5)	(4.5,5,5.5)	(5.5,6,6.5)	(0.5,1,1.5)
P53	(0.5,1,1.5)	(1.5,2,2.5)	(3.5,4,4.5)	(6.5,7,7.5)	(1.5,2,2.5)	(7.5,8,8.5)	(4.5,5,5.5)	(5.5,6,6.5)	(6.5,7,7.5)
P54	(9.5,10,10)	(8.5,9,9.5)	(9.5,10,10)	(7.5,8,8.5)	(8.5,9,9.5)	(9.5,10,10)	(8.5,9,9.5)	(8.5,9,9.5)	(7.5,8,8.5)
P61	(0.5,1,1.5)	(1.5,2,2.5)	(2.5,3,3.5)	(4.5,5,5.5)	(1.5,2,2.5)	(7.5,8,8.5)	(4.5,5,5.5)	(5.5,6,6.5)	(2.5,3,3.5)
P62	(2.5,3,3.5)	(5.5,6,6.5)	(7.5,8,8.5)	(6.5,7,7.5)	(7.5,8,8.5)	(8.5,9,9.5)	(6.5,7,7.5)	(3.5,4,4.5)	(6.5,7,7.5)
P63	(2.5,3,3.5)	(5.5,6,6.5)	(4.5,5,5.5)	(3.5,4,4.5)	(1.5,2,2.5)	(1.5,2,2.5)	(4.5,5,5.5)	(5.5,6,6.5)	(3.5,4,4.5)
P64	(0.5,1,1.5)	(0.5,1,1.5)	(0.5,1,1.5)	(3.5,4,4.5)	(0.5,1,1.5)	(3.5,4,4.5)	(1.5,2,2.5)	(5.5,6,6.5)	(4.5,5,5.5)
P65	(2.5,3,3.5)	(0.5,1,1.5)	(1.5,2,2.5)	(2.5,3,3.5)	(0.5,1,1.5)	(7.5,8,8.5)	(6.5,7,7.5)	(2.5,3,3.5)	(9.5,10,10)
P71	(0.5,1,1.5)	(2.5,3,3.5)	(4.5,5,5.5)	(6.5,7,7.5)	(5.5,6,6.5)	(6.5,7,7.5)	(5.5,6,6.5)	(4.5,5,5.5)	(2.5,3,3.5)
P72	(7.5,8,8.5)	(4.5,5,5.5)	(5.5,6,6.5)	(9.5,10,10)	(7.5,8,8.5)	(2.5,3,3.5)	(1.5,2,2.5)	(5.5,6,6.5)	(4.5,5,5.5)
P73	(1.5,2,2.5)	(2.5,3,3.5)	(1.5,2,2.5)	(3.5,4,4.5)	(5.5,6,6.5)	(6.5,7,7.5)	(2.5,3,3.5)	(1.5,2,2.5)	(2.5,3,3.5)
P74	(8.5,9,9.5)	(9.5,10,10)	(7.5,8,8.5)	(9.5,10,10)	(8.5,9,9.5)	(6.5,7,7.5)	(9.5,10,10)	(8.5,9,9.5)	(7.5,8,8.5)
P75	(3.5,4,4.5)	(8.5,9,9.5)	(8.5,9,9.5)	(6.5,7,7.5)	(7.5,8,8.5)	(7.5,8,8.5)	(6.5,7,7.5)	(1.5,2,2.5)	(1.5,2,2.5)

13.5.3 Evaluation of the Quantitative Aspects of the Enterprise

When a candidate enterprise bids for the sub-projects, the bids should include information of the cost, time and quality for processing the jobs. These quantitative aspects can be expressed with crisp numbers. Cost, time and quality consist of many components; the data used in the PSO-ACO are the synthetic values, which are the weighted sum of each sub-criterion. For example, the cost for candidate P11 consists of four parts, namely, the operational cost, the delivery cost, the material cost and the logistic overhead, as shown in Table 13.7.

Since cost, time and quality are quantitative variables that can be obtained from the bids submitted by the candidates, the data for cost, time and quality used in this research are randomly generated. The data falls in the range of [0, 10] with a uniform distribution as shown in Table 13.8.

13.5.4 Results

The hybrid PSO-ACO algorithm is used to select the best set of partners. The parameters and data used in the experiments are given in Table 13.6, Table 13.8 and Table 13.9. The convergence speed of the fitness function with respect to the number of iterations is shown in Figure 13.7. PSO-ACO is compared with PSO and ACO individually using the same set of input data and parameters settings. The experiments showed that PSO-ACO has a faster convergence speed than ACO and PSO individually. The hybrid PSO-ACO algorithm decreases the fitness by 7.86% and 4.19% as compared to the PSO and ACO algorithms. The improvements are mainly because the hybrid algorithm searches more thoroughly during the optimization process to avoid being trapped in the local optima, and the advantages of each algorithm are fused together.

Table 13.7 The bidding price of cost for candidate P11.

Sub-criteria	C_o	C_D	C_M	C_L
Weights	0.30	0.20	0.35	0.15
P11	7.3	5.4	5.8	6.0

Table 13.8 Performance data and available time slots for the candidates.

Subproject	P1			P2			P3					P4		
Candidate	1	2	3	1	2	3	1	2	3	4	5	1	2	3
Cost	6.2	6.7	8.7	6.2	4.5	8.1	6.9	3.9	7.2	5.7	6.1	5.7	4.5	9
Time	9.4	8.2	7.2	8.3	5.6	6.6	5.8	6.4	3.4	6.7	3.3	6.2	4.8	5.1
Quality	5.2	7.6	8.1	5.3	6.5	5.5	5.4	6.4	5.8	8.4	7.1	6.2	7.2	8.2
Avail time Slot	[0, 50]	[0, 30]	[10, 50]	[30, 60]	[20, 55]	[20, 60]	[60, 80]	50, 80	[55, 75]	[50, 80]	[60, 80]	[70, 100]	[75, 110]	[70, 100]
Subproject	P5				P6					P7				
Candidate	1	2	3	4	1	2	3	4	5	1	2	3	4	5
Cost	6.2	6.7	8.7	5.8	7.2	4.5	8.1	9.3	4.4	9.4	6.9	8	4.7	6.3
Time	9.4	8.2	7.2	7.7	8.3	5.6	6.5	3.3	8.2	7	6.4	3.4	6.7	4.9
Quality	5.2	7.6	8.1	5.9	6.3	6	5.5	7.7	4.5	5.8	7.4	3.4	8.8	4.7
Available time slot	[90, 130]	[100, 115]	[100, 120]	[100, 130]	[120, 140]	[110, 130]	[110, 130]	[120, 140]	[120, 140]	[130, 150]	[130, 140]	[135, 150]	[100, 180]	[130, 150]

Table 13.9 The parameters in the hybrid PSO-ACO algorithm.

Parameter	D	C_1	C_2	V_{max}	V_{Min}	ϕ	ω	α	β	ρ	No_{ant}	No_p
Value	7	2	2	2	–2	1	[0.5, 0.9]	1	1	0.5	28	28

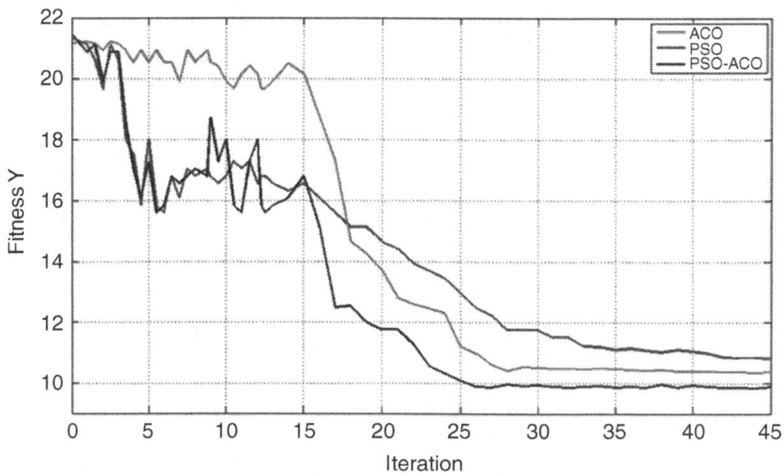

Figure 13.7 Fitness comparisons of PSO-ACO, ACO and PSO.

To test the robustness and sensitivity of the hybrid PSO-ACO algorithm, experiments were conducted to evaluate the influence of each factor using the same set of data in Table 13.6 and Table 13.8. The influence of the time constraints and duration is investigated. Figure 13.8 and Figure 13.9 show the results with and without the consideration of these two factors. The results show that the hybrid PSO-ACO algorithm can converge quickly under both conditions. Figure 13.10 shows the influence of cost, time, quality, risk, reputation and time sequence and duration. Figure 13.10(a) shows the results with the five criteria considered *CT*, *TE*, *QT*, *RK*, *RP*, Figure 13.10(b) with *CT* excluded, Figure 13.10(c) with *TE* excluded, Figure 13.10(d) with *QT* excluded, Figure 13.10(e) with *RK* excluded, and Figure 13.10(f) with *RP* excluded. Figure 13.10 demonstrates that the hybrid PSO-ACO algorithm can achieve good performance under different conditions, and the time sequence and duration can largely affect the results. The selected partner sets for different conditions are given in Table 13.10.

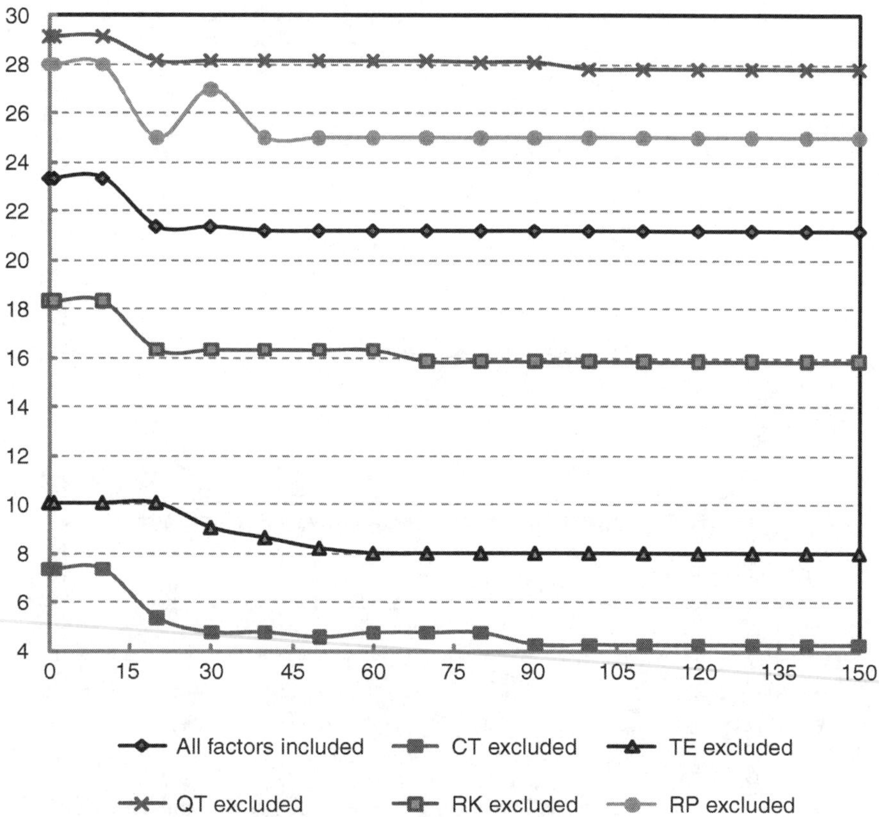

Figure 13.8 Fitness comparison considering time constraint and sequence.

13.6 Conclusion

In the global manufacturing environment, with increasing competition and market changes, companies would need to produce customized products in smaller batches with higher quality, greater varieties, lower cost, and faster delivery. Customers' needs also change over time. This new situation compels enterprises to organize themselves to form VEs in order to maintain their competitiveness. Partner selection is a critical issue related to the success of a VE.

The paper presents a mathematical model that can obtain the optimal set of candidates efficiently. In this research, a mathematical model considering both qualitative and quantitative factors is presented, and a hybrid PSO-ACO algorithm is applied. In addition,

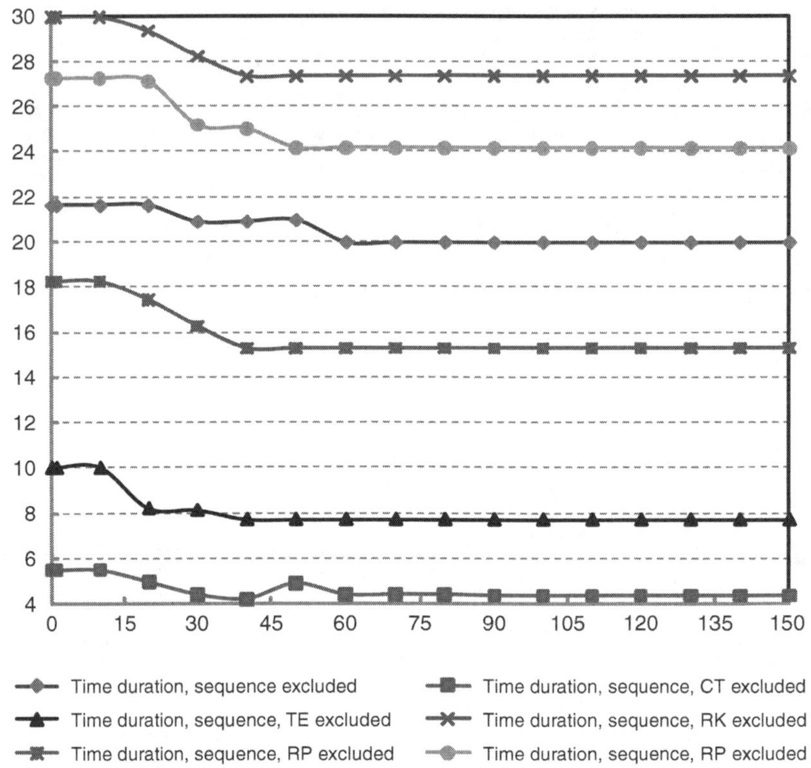

Figure 13.9 Fitness comparison with time duration and sequence excluded.

fuzzy set theory is adopted to handle subjective, vague and imprecise information in the partner selection process. The case study shows that the proposed hybrid PSO-ACO approach can shorten the search time, improve the convergence speed and the accuracy of the results with a decrease in the fitness of 7.86% and 4.19% as compared to the PSO and ACO algorithms.

Nomenclature

t	is the index of iteration number;
NO_p	is the number of particles;
D	is the dimensionality of the search space of the particles;
v_i^t	is the velocity of particle i of the tth iteration;
x_i^t	is the current position of particle i of the tth iteration;

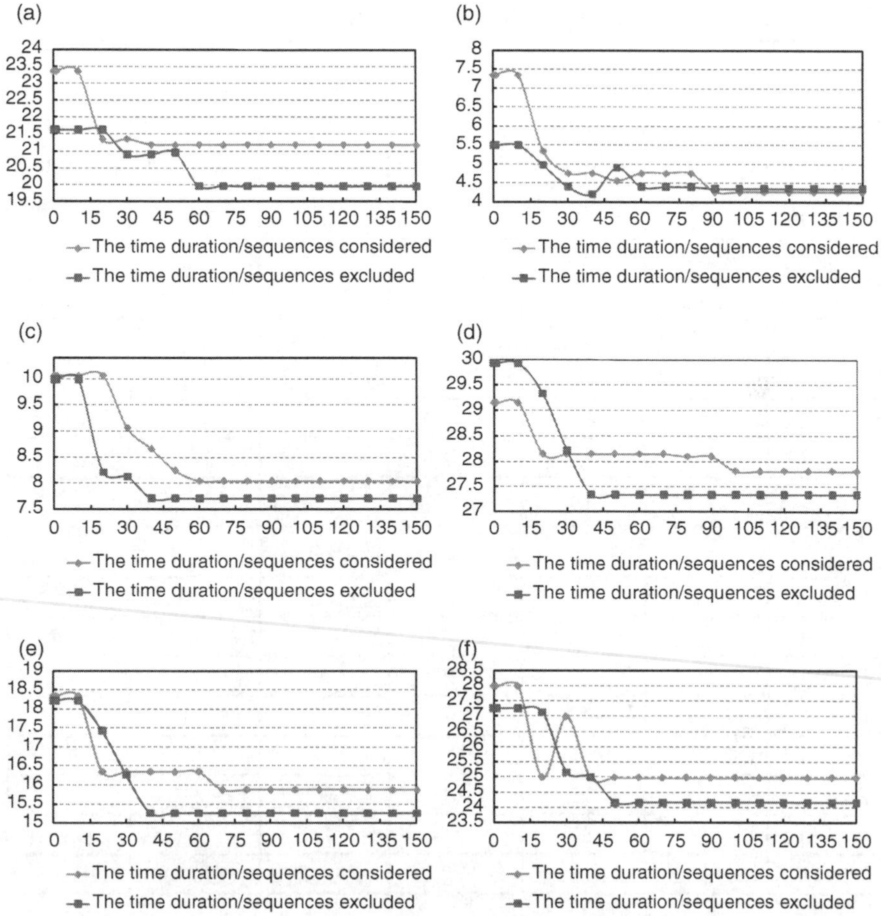

Figure 13.10 Fitness comparison under different conditions.

$v_i^t = \{v_{i1}^t, v_{i2}^t, \ldots v_{iD}^t,\}$:	the D component of the ith particle in the tth iteration;
$x_i^t = \{x_{i1}^t, x_{i2}^t, \ldots, x_{iD}^t\}$:	the *D*-dimension position for the ith particle in the tth iteration;
V_{min}	is the minimum velocity of each particle;
V_{max}	is the maximum velocity of each particle;
ω	is the inertia weight;

$c_1$and c_2 are the acceleration constraints;

$\vec{U}(0, \phi_i)$	represents a vector of random numbers uniformly distributed in $[0,\phi_i]$ which is randomly generated at each iteration and for each particle;

Table 13.10 Selected partners for each project under different conditions.

Time sequence/duration considered						
	All criteria included	***CT* excluded**	***TE* excluded**	***QT* excluded**	***RK* excluded**	***RP* excluded**
Project 1	3	3	1	1	3	1
Project 2	5	6	6	5	5	5
Project 3	8	8	10	10	8	10
Project 4	13	14	13	13	13	13
Project 5	18	17	17	15	15	17
Project 6	19	22	19	19	19	22
Project 7	27	27	27	27	24	27
MinY	20.687	4.166	8.04	27.804	15.862	24.985
Time sequence/duration excluded						
Project 1	2	2	2	2	1	2
Project 2	5	6	5	5	6	5
Project 3	9	9	9	9	10	8
Project 4	13	12	13	13	13	13
Project 5	18	17	16	15	16	16
Project 6	19	22	20	20	23	20
Project 7	27	27	28	14	27	25
MinY	19.954	4.133	7.709	27.337	14.622	24.155

$pbest_i^t$ is the personal best position of particle i in the tth iteration;

$gbest^t$ is the global best position of the tth generation;

P_{ij} is the probability for an ant to travel from node i to node j at t^{th} iteration;

d_{ij} is the heuristic distance between nodes i and j;

ρ is the evaporation coefficient, which can be a real number between 0 and 1.0;

τ_{ij}^t is the quantity of pheromone on the edge connecting nodes i and j at t^{th} iteration;

$\Delta\tau_{ij}^t$ is the quantity of increased pheromone on the edge connecting nodes i and j at t^{th} iteration;

$\Delta\tau_{ij}^{t}(k)$	is the quantity of increased pheromone on the edge connecting nodes i and j at t^{th} iteration by the ant k;
Q	is a constant representing the total quality of pheromone on a route;
Y^t	is the fitness function of the partner selection;
α	is the relative importance of the trial, $\alpha \geq 0$;
β	is the relative importance of the visibility, $\beta \geq 0$;
ρ	is the trail persistence, $0 \leq \rho \prec 1$;
NO_{ant}	is the number of the ant;
Q	is a constant related to the quantity of trail laid by ants;
CT	is the cost of each candidate enterprise;
C_0	is the operational cost of candidate enterprise;
C_D	is the delivery cost of candidate enterprise;
C_M	is the material cost of the candidate enterprise;
C_L	is the logistic overhead of the candidate enterprise;
TE	is the time related criteria of the candidate enterprise;
T_{du}	is the project duration of the candidate enterprise;
T_{sq}	is the time sequence of the project;
T_{de}	is the delivery time of the candidate enterprise;
T_{ED}	is the due date of one subproject;
T_{pt}	is the processing time of one subproject in a candidate enterprise;
QT	is the quality criterion to optimize;
TL	is the technological level of the candidate enterprise;
PF	is the performance of the candidate enterprise;
SV	is the service level of the candidate enterprise;
RK	is the risk criterion to optimize;
$\widetilde{PS}$	is a fuzzy number representing the political stability of candidate enterprise;
$\widetilde{ES}$	is a fuzzy number representing the economy status of the region of candidate enterprise;

$\widetilde{FH}$	is a fuzzy number representing the financial health of candidate enterprise;
$\widetilde{MF}$	is a fuzzy number representing the market fluctuations of candidate enterprise;
$\widetilde{CP}$	is a fuzzy number representing the competency of candidate enterprise;
RP	is the reputation criterion to maximize;
$\widetilde{PP}$	is a fuzzy number representing the past performance of candidate enterprise;
$\widetilde{FS}$	is a fuzzy number representing the financial status of candidate enterprise;
$\widetilde{EI}$	is a fuzzy number representing the image of the enterprise of candidate enterprise;
$\widetilde{CH}$	is a fuzzy number representing the cooperation history of the candidates and the dominant enterprise

References

Brucker, P., Drexl, A., Mohring, R., Neumann, K. & Pesch, E., 1999. Resource-constrained project scheduling: Notation, classification, models, and methods. *European Journal of Operational Research,* 112 (1), 3–41.

Bu, Y., Zhou, W. & Yu, J., 2008. A discrete pso algorithm for partner selection of virtual enterprise. *The 2nd International Symposium on Intelligent Information Technology Application,* Shanghai, China: IEEE, 814–817.

Buyukozkan, G., 2004. Multi-criteria decision making for e-marketplace selection. *Internet Research: Electronic Networking Applications and Policy,* 14 (2), 139–154.

Camarinha-Matos, L.M. & Afsarmanesh, H., 2003. Elements of a base VE infra-structure. *Computers in Industry,* 51 (2), 139–163.

Chan, F.T.S., Kumar, N., Tiwari, M.K., Lau, H.C.W. & Choy, K.L., 2008. Global supplier selection: A fuzzy-AHP approach. *International Journal of Production Research,* 46 (14), 3825–3857.

Chen, Q.X., Chen, X. & Lee, W.B., 2007. Qualitative search algorithms for partner selection and task allocation in the formulation of virtual enterprise. *International Journal of Computer Integrated Manufacturing,* 20 (2–3), 115–126.

Colorni, A., Dorigo, M. & Maniezzo, V., 1992. Distributed optimization by ant colonies. *Proceedings of the 1st European conference on artificial life.* Paris, France: MIT Press, 134–142.

Crispim, J.A. & De Sousa, J.P., 2010. Partner selection in virtual enterprises. *International Journal of Production Research,* 48 (3), 683–707.

Dorigo, M., Di Caro, G. & Gambardella, L.M., 1999. Ant algorithms for discrete optimization. *Artificial Life,* 5 (2), 137–172.

Dorigo, M., Maniezzo, V. & Colorni, A., 1996. Ant system: Optimization by a colony of cooperating agents. *IEEE Transactions on Systems, Man and Cybernetics, Part B (Cybernetics)*, 26 (1), 29–41.

Ip, W.H., Huang, M., Yung, K.L. & Wang, D., 2003. Genetic algorithm solution for a risk-based partner selection problem in a virtual enterprise. *Computers and Operations Research*, 30 (2), 213–231.

Ip, W.H., Yung, K.L. & Dingwei, W., 2004. A branch and bound algorithm for sub-contractor selection in agile manufacturing environment. *International Journal of Production Economics*, 87 (2), 195–205.

Kennedy, J. & Eberhart, R., 1995. Particle swarm optimization. *International Conference on Neural Networks*, Perth, Australia: IEEE, 1942–1948.

Kiranyaz, S., Pulkkinen, J. & Gabbouj, M., 2008. Multi-dimensional particle swarm optimization for dynamic environments. *The 5th International Conference on Innovations in Information Technology*, Al Ain, U.A.E.: IEEE, 34–38.

Liang, F., Fung, R.Y.K., Jiang, Z. & Wong, T.N., 2008. A hybrid control architecture and coordination mechanism in virtual manufacturing enterprise. *International Journal of Production Research*, 46 (13), 3641–3663.

Percin, S., 2008. Use of fuzzy AHP for evaluating the benefits of information-sharing decisions in a supply chain. *Journal of Enterprise Information Management*, 21 (3), 263–284.

Poli, R., Kennedy, J. & Blackwell, T., 2007. Particle swarm optimization. *Swarm Intelligence*, 1 (1), 33–57.

Qi, E.-S., Chen, J.-Y. & Liu, L., 2008. Supply chain partners selection based on rvpk algorithm. *The 4th International Conference on Wireless Communications, Networking and Mobile Computing (WiCOM)*. Dalian, China: Piscataway, NJ, USA, 1–4.

Timothy, J.R., 2010. *Fuzzy logic with engineering applications*: New Jersey: John Wiley and Sons.

Wang, Z.J., Xu, X.F. & Zhan, D.C., 2009. Genetic algorithm for collaboration cost optimization-oriented partner selection in virtual enterprises. *International Journal of Production Research*, 47 (4), 859–881.

Yager, R.R., 1980. On a general class of fuzzy connectives. *Fuzzy Sets and Systems*, 4 (3), 235–242.

Yang, Y., Zhao, F., Yao, Y. & Zhu, A., 2006. A PSO and simulated annealing hybrid algorithm to task allocation problem for holonic manufacturing system. *The 6th World Congress on Intelligent Control and Automation*, Dalian, China: IEEE, 6767–6771.

Yao, Y., Chen, H. & Li, Y., 2006. A hybrid algorithm with particle swarm optimization for partner selection problem in virtual enterprise. Eds., Xinjiang, China: Inst. of Elec. and Elec. Eng. Computer Society, 386–390.

Yao, Z., Liu, J. & Wang, Y.G., 2008. Fusing genetic algorithm and ant colony algorithm to optimize virtual enterprise partner selection problem. *IEEE congress on evolutionary computation (CEC 2008)*, Hong Kong, China: IEEE, 3614–3620.

Zhao, F., Hong, Y. & Yu, D., 2006a. A multi-objective optimization model of the partner selection problem in a virtual enterprise and its solution with genetic algorithms. *International Journal of Advanced Manufacturing Technology*, 28 (11–12), 1246–1253.

Zhao, F., Zhang, Q. & Yang, Y., 2006b. A novel task allocation problem solution with pso algorithm for holonic manufacturing system. *10th International Conference on Computer Supported Cooperative Work in Design (CSCWD 2006).* Nanjing, China: Inst. of Elec. and Elec. Eng. Computer Society, Piscataway, NJ, USA, 1308–1313.

Zhao, F., Zhang, Q. & Yang, Y., 2006c. A scheduling holon modeling method with petri net and its optimization with a novel PSO-GA algorithm. *10th International Conference on Computer Supported Cooperative Work in Design (CSCWD 2006).* Nanjing, China: Inst. of Elec. and Elec. Eng. Computer Society, Piscataway, NJ, USA, 1302–1307.

Zhao, F., Zhang, Q., Yu, D., Chen, X. & Yang, Y., 2005. A hybrid algorithm based on PSO and simulated annealing and its applications for partner selection in virtual enterprise. *International Conference on Intelligent Computing (ICIC 2005).* Hefei, China: Springer Verlag, 380–389.

Zhao, Q., Zhang, X. & Xiao, R., 2008. Particle swarm optimization algorithm for partner selection in virtual enterprise. *Progress in Natural Science,* 18 (11), 1445–1452.

Index

Also of Interest

Check out these published and forthcoming related titles from Scrivener Publishing

Grid Manufacturing and Resource Service Management
By Fei Tao
Forthcoming Fall 2011. ISBN 978-1-118-12231-0

RFID-Enabled Real-Time Manufacturing
By George Huang, Qy Dai, and Andrew Y.C. Nee
Forthcoming Fall 2011. ISBN 978-0-470-90165-6

Cost Reduction and Optimization for Manufacturing and Industrial Companies
By Joseph Berk
Published 2010. ISBN 978-0-470-60957-6

Construction Management for Industrial Projects
By Mohamed El-Reedy
Published 2011. ISBN: 978-0-470-87816-3

The Basics of Troubleshooting in Plastics Processing
By Muralisrinivasan Subramanian
Published 2011. ISBN 978-0-470-62606-1

Roll-to-Roll Vacuum Deposition of Barrier Coatings
By Charles A. Bishop
Published 2010. ISBN 978-0-470-60956-9

Introduction to Industrial Polyethylene
Properties, Catalysts, Processes
By Dennis B. Malpass
Published 2010. ISBN 978-0-470-62598-9